T0329701

PRINCIPLES OF ELECTROMAGNETIC COMPATIBILITY

PRINCIPLES OF ELECTROMAGNETIC COMPATIBILITY

Laboratory Exercises and Lectures

Bogdan Adamczyk
Grand Valley State University
Michigan, USA

WILEY

IEEE PRESS

Registered Offices
John Wiley & Sons, Inc., 111 River Street, Hoboken, NJ 07030, USA
John Wiley & Sons Ltd, The Atrium, Southern Gate, Chichester,West Sussex, PO19 8SQ, UK

For details of our global editorial offices, customer services, and more information about Wiley products visit us at www.wiley.com.

Library of Congress Cataloging-in-Publication Data Applied for:

Hardback ISBN: 9781119718710

Cover Design: Wiley
Cover Image: © sakkmesterke/Shutterstock

Set in 10/12pt HelveticaLTStd by Straive, Chennai, India
Printed and bound by CPI Group (UK) Ltd, Croydon, CR0 4YY

C9781119718710_161023

Contents

Preface

Seven years ago, I started writing a monthly column "EMC Concepts Explained" for an EMC trade magazine "In Compliance". The content of the tutorial articles in that column has been based on the educational research performed at the Grand Valley State University (GVSU) EMC Center in collaboration with its industrial partner, E3 Compliance, LLC. This collaboration not only led to this book but also led to the "Principles of Electromagnetic Compatibility" certificate course for industry that I regularly teach at GVSU.

This textbook is intended for both a university/college course in electromagnetic compatibility and practicing professionals. There is currently no textbook on EMC that consists of each tutorial chapter followed by laboratory exercises at the end of it. A great amount of effort has been taken by the author to use off-the-shelf readily available equipment and components so that one can easily replicate the laboratory exercises. Several such exercises require a custom PCB, for which Altium files with the supporting documentation are provided. These files can be customized by the user to account for the topology or component changes, if necessary.

The author owes a great deal of gratitude to many colleagues at E3 Compliance for their time and effort spent on joint research, publications, and the design of custom PCBs. Of primary mention are Scott Mee, Jim Teune, Krzysztof Russa, Nick Koeller, and Mat French. Many thanks go to the former graduate students, Dimitri Haring and Ryan Aldridge. Finally, the author would like to acknowledge the support of GVSU, its Engineering Dean Paul Plotkowski, and the Engineering School Director Wael Mokhtar.

Grand Valley State University   *Bogdan Adamczyk*
Michigan, USA
September 2023

About the Companion Website

This book is accompanied by a companion website:

www.wiley.com/go/principlesofelectromagneticcompatibility

This website includes:

- Case Studies

Chapter 1: Frequency Spectra of Digital Signals

1.1 EMC Units

1.1.1 Logarithm and Decibel Definition

In mathematics, base 10 logarithm of a positive number A is defined as $\log_{10}A$. The resulting number is negative, zero, or positive according to

$$0 < A < 1 \Rightarrow \log_{10}A < 0$$
$$A = 1 \Rightarrow \log_{10}A = 0 \qquad (1.1)$$
$$A > 1 \Rightarrow \log_{10}A > 0$$

The only requirement in mathematics for this operation to be defined is that the number A is a positive number. There is no other implicit assumption about this number.

In engineering, especially in EMC, we use a unit of decibel (dB) which is related to the base 10 logarithm and is defined as

$$A_{\mathrm{dB}} = 10\log_{10}A \qquad (1.2)$$

when A represents power ratio (gain), or

$$A_{\mathrm{dB}} = 20\log_{10}A \qquad (1.3)$$

when A represents voltage or current ratios (gains). In Section 1.1.2, we will discuss the difference between these two definitions.

Note the very important aspect of a number A when expressed in dB: it is always a ratio of two quantities, not just a positive number. What if this ratio is negative? In this case we treat the negative quantity as a complex quantity with a positive magnitude and a phase of $180°$.

$$-A = -1 \times A = (1\angle180°)(A\angle0°) = A\angle180° \qquad (1.4)$$

and simply consider the magnitude only.

1.1.2 Power and Voltage (Current) Gain in dB

Consider the circuit shown in Figure 1.1. A sinusoidal source consisting of an open-circuit rms voltage $\hat{V}_S$ and source resistance R_S delivers a signal to an amplifier whose load is represented by R_L.

Principles of Electromagnetic Compatibility: Laboratory Exercises and Lectures, First Edition. Bogdan Adamczyk.
© 2024 John Wiley & Sons Ltd. Published 2024 by John Wiley & Sons Ltd.
Companion website: www.wiley.com/go/principlesofelectromagneticcompatibility

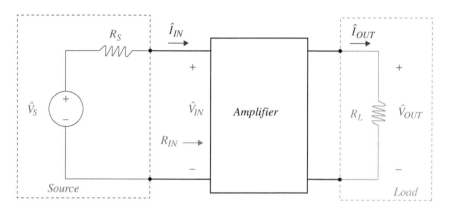

Figure 1.1 Circuit used to define decibel (dB).

Comment on the notation: In sinusoidal steady state we represent voltages and currents by complex quantities, or phasors [Adamczyk, 2022]. To differentiate between a complex quantity and its magnitude we use the "hat" notation. Thus, a complex voltage $\hat{V}$ has a magnitude V and a corresponding angle θ,

$$\hat{V} = V e^{j\theta} \tag{1.5}$$

Returning to Figure 1.1, the (average) input power to the amplifier is

$$P_{IN} = \frac{V_{IN}^2}{R_{IN}} \tag{1.6}$$

while power delivered to the load is

$$P_{OUT} = \frac{V_{OUT}^2}{R_L} \tag{1.7}$$

The power gain in dB, PG_{dB}, is defined as

$$PG_{\mathrm{dB}} \triangleq 10\log_{10}\frac{P_{OUT}}{P_{IN}} \tag{1.8}$$

Using Eqs. (1.6) and (1.7) this gain can also be expressed as

$$PG_{\mathrm{dB}} \triangleq 10\log_{10}\frac{\frac{V_{OUT}^2}{R_L}}{\frac{V_{IN}^2}{R_{IN}}} = 10\log_{10}\left(\frac{V_{OUT}}{V_{IN}}\right)^2\frac{R_{IN}}{R_L} \tag{1.9}$$

The voltage and current gains in dB, VG_{dB}, and IG_{dB}, respectively, are defined as

$$VG_{\mathrm{dB}} \triangleq 20\log_{10}\frac{V_{OUT}}{V_{IN}} \tag{1.10}$$

$$IG_{\mathrm{dB}} \triangleq 20\log_{10}\frac{I_{OUT}}{I_{IN}} \tag{1.11}$$

Note that if *the input resistance to the amplifier equals the load resistance, $R_L = R_{IN}$* (which is often the case in practice) then the power gain in dB is the same as the voltage (or current) gain in dB.

$$PG_{\mathrm{dB}} = 10 \log_{10} \left(\frac{V_{OUT}}{V_{IN}} \right)^2 = 20 \log_{10} \frac{V_{OUT}}{V_{IN}}, \quad R_L = R_{IN} \tag{1.12}$$

1.1.3 EMC dB Units

In EMC dB units we don't simply express the ratio of two powers, voltages, etc., we express the ratio of a given nominal value to a base quantity, as shown in Figures 1.2 and 1.3 [Adamczyk and Teune, 2019].

One of the most common measurement units expressed in decibels is dBm. Many spectrum analyzers display voltage amplitudes both in dBμV and in dBm, even though dBm is a unit of power. The two units are related by the impedance of the measuring equipment, namely 50 Ω.

Let's derive the relationship between these two units. We start with the power–voltage relationship, related by the 50 Ω impedance.

$$P = \frac{V^2}{50} \tag{1.13}$$

Taking the logarithm of both sides gives

$$10 \log_{10} P = 10 \log_{10} \frac{V^2}{50} \tag{1.14}$$

or

$$10 \log_{10} P = 10 \log_{10} V^2 - 10 \log_{10} 50 \tag{1.15}$$

Voltage units	Current units	Power units
dBV = $20 \log_{10} \left(\frac{\mathrm{volts}}{1\ \mathrm{V}} \right)$	dBA = $20 \log_{10} \left(\frac{\mathrm{amperes}}{1\ \mathrm{A}} \right)$	dBW = $10 \log_{10} \left(\frac{\mathrm{watts}}{1\ \mathrm{W}} \right)$
dBmV = $20 \log_{10} \left(\frac{\mathrm{volts}}{1\ \mathrm{mV}} \right)$	dBmA = $20 \log_{10} \left(\frac{\mathrm{amperes}}{1\ \mathrm{mA}} \right)$	dBm = $10 \log_{10} \left(\frac{\mathrm{watts}}{1\ \mathrm{mW}} \right)$
dBμV = $20 \log_{10} \left(\frac{\mathrm{volts}}{1\ \mathrm{\mu V}} \right)$	dBμA = $20 \log_{10} \left(\frac{\mathrm{amperes}}{1\ \mathrm{\mu A}} \right)$	dBμW = $10 \log_{10} \left(\frac{\mathrm{watts}}{1\ \mathrm{\mu W}} \right)$

Figure 1.2 Voltage, current, and power EMC units.

Electric field intensity units	Magnetic field intensity units	Impedance units
dBμV/m = $20 \log_{10} \left(\frac{\mathrm{V/m}}{1\ \mathrm{\mu V/m}} \right)$	dBμA/m = $20 \log_{10} \left(\frac{\mathrm{A/m}}{1\ \mathrm{\mu A/m}} \right)$	dBΩ = $20 \log_{10} \left(\frac{\mathrm{ohms}}{1\ \Omega} \right)$

Figure 1.3 Field intensities and impedance EMC units.

leading to

$$10 \log_{10} P = 20 \log_{10} V - 17 \tag{1.16}$$

Equivalently,

$$10 \log_{10} \left(\frac{P}{10^{-3}} \times 10^{-3} \right) = 20 \log_{10} \left(\frac{V}{10^{-6}} \times 10^{-6} \right) - 17 \tag{1.17}$$

or

$$10 \log_{10} \left(\frac{P}{10^{-3}} \right) + 10 \log_{10}(10^{-3}) = 20 \log_{10} \left(\frac{V}{10^{-6}} \right) + 20 \log_{10}(10^{-6}) - 17 \tag{1.18}$$

and thus,

$$P_{\text{dBm}} - 30 = V_{\text{dB}\mu V} - 120 - 17 \tag{1.19}$$

resulting in the relationship between the two units as

$$V_{\text{dB}\mu V} = P_{\text{dBm}} + 107$$
$$P_{\text{dBm}} = V_{\text{dB}\mu V} - 107 \tag{1.20}$$

Figure 1.4 shows a spectrum analyzer measurement result in dBμV, while Figure 1.5 shows the same measurement in dBm.

Note that the corresponding measurements are offset by 107 dB.

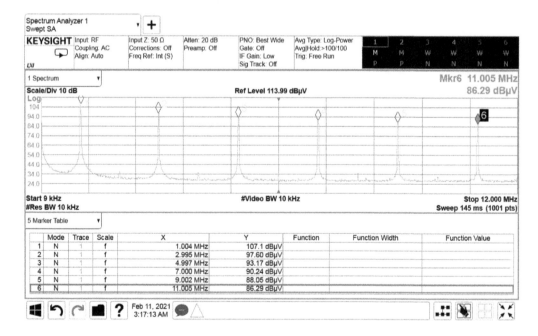

Figure 1.4   Spectrum analyzer measurement in dBμV.

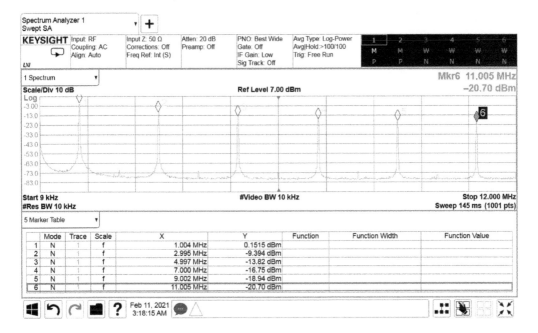

Figure 1.5 Spectrum analyzer measurement in dBm.

Often, we need to convert a unit given in dB to its absolute (normal) value. To do this we use the definition of the base m logarithm of a number A:

$$\log_m A = n \tag{1.21}$$

Equivalently,

$$m^n = A \tag{1.22}$$

Let's show the conversion from $108\,\mathrm{dB\mu V}$ to the absolute value:

$$108\,\mathrm{dB\mu V} = 20\log_{10}\left(\frac{V}{1\times 10^{-6}}\right) \tag{1.23}$$

or

$$\frac{108\,\mathrm{dB\mu V}}{20} = \log_{10}\left(\frac{V}{1\times 10^{-6}}\right) \tag{1.24}$$

thus

$$10^{\frac{108\,\mathrm{dB\mu V}}{20}} = \frac{V}{10^{-6}} \tag{1.25}$$

Thus, the absolute value of V is

$$V = 10^{\frac{108}{20}}\times 10^{-6} = 0.2512\,\mathrm{V} \tag{1.26}$$

The common conversions are

$$Volts = 10^{\frac{dB\mu V}{20}} \times 10^{-6} \tag{1.27}$$

$$Volts = 10^{\frac{dBmV}{20}} \times 10^{-3} \tag{1.28}$$

$$Watts = 10^{\frac{dB\mu W}{10}} \times 10^{-6} \tag{1.29}$$

$$Watts = 10^{\frac{dBm}{10}} \times 10^{-3} \tag{1.30}$$

$$Ohms = 10^{\frac{dB\Omega}{20}} \times 10^{-6} \tag{1.31}$$

1.2 Fourier Series Representation of Periodic Signals

Any periodic function can be represented as an infinite sum of sinusoidal components as

$$x(t) = a_0 + \sum_{n=1}^{\infty} (a_n \cos 2\pi n f_0 t + b_n \sin 2\pi n f_0 t) \tag{1.32}$$

or

$$x(t) = a_0 + \sum_{n=1}^{\infty} (a_n \cos n\omega_0 t + b_n \sin n\omega_0 t) \tag{1.33}$$

An expansion of this type is known as a Fourier Series Expansion. Note that each sinusoidal component has a frequency that is a multiple of the fundamental frequency

$$f_0 = \frac{1}{T} \tag{1.34}$$

or the radian fundamental frequency

$$\omega_0 = 2\pi f_0 \tag{1.35}$$

The multiples of the fundamental frequency, $n f_0$, or $n\omega_0$ are called *harmonics* of that fundamental frequency. The coefficients a_0, a_n, b_n are called the Fourier coefficients and are given by

$$a_0 = \frac{1}{T} \int_T x(t) dt \tag{1.36}$$

$$a_n = \frac{2}{T} \int_T x(t) \cos n\omega_0 t \, dt \tag{1.37}$$

$$b_n = \frac{2}{T} \int_T x(t) \sin n\omega_0 t \, dt \tag{1.38}$$

The Fourier series as expressed in Eq. (1.33) is called the *trigonometric* Fourier Series. The trigonometric Fourier Series can be put in much simpler and more convenient *complex* form as [Adamczyk, 2017a]

$$x(t) = \sum_{-\infty}^{\infty} \hat{c}_n e^{jn\omega_0 t} \, dt \tag{1.39}$$

where

$$\hat{c}_n = \frac{1}{T} \int_T x(t) e^{-jn\omega_0 t} \, dt \tag{1.40}$$

$\hat{c}_n$ is a complex Fourier coefficient with its magnitude and angle denoted by

$$\hat{c}_n = c_n \angle \theta_n \tag{1.41}$$

Using the notation in Eq. (1.41) the two-sided Fourier series, or spectrum, in Eq. (1.39) can be written in an equivalent form as a one-sided spectrum given by [Adamczyk, 2017a]

$$x(t) = c_0 + \sum_{n=1}^{\infty} 2c_n \cos(n\omega_0 t + \theta_n) \tag{1.42}$$

where

$$c_0 = \frac{1}{T} \int_T x(t) dt \tag{1.43}$$

and $\hat{c}_n$ given by Eq. (1.40).

1.3 Spectrum of a Clock Signal

Clock signals can be represented as periodic trains of trapezoid-shaped pulses, as shown in Figure 1.6.

Each clock pulse is described by an amplitude A, a pulse rise time t_r, a pulse fall time t_f, and the pulse width τ. If the pulse rise time equals the pulse fall time, $t_r = t_f$, the Fourier coefficients are given by [Paul, 2006]

$$\hat{c}_n = \frac{A\tau}{T} \left(\frac{\sin \frac{1}{2} n\omega_0 \tau}{\frac{1}{2} n\omega_0 \tau} \right) \left(\frac{\sin \frac{1}{2} n\omega_0 t_r}{\frac{1}{2} n\omega_0 t_r} \right) \tag{1.44}$$

The magnitudes of these coefficients are given by

$$c_0 = 2A \frac{\tau}{T} \tag{1.45}$$

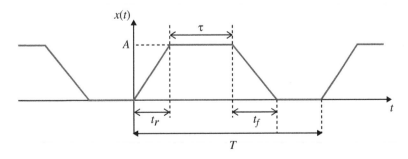

Figure 1.6   Trapezoidal clock signal.

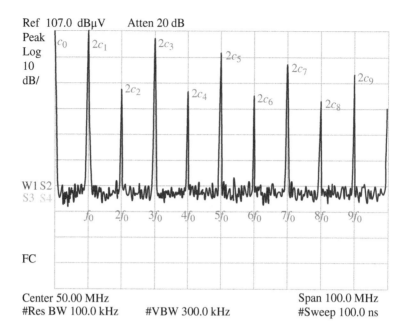

Ref 107.0 dBμV   Atten 20 dB

Peak
Log
10
dB/

W1 S2
S3 S4

f_0 $2f_0$ $3f_0$ $4f_0$ $5f_0$ $6f_0$ $7f_0$ $8f_0$ $9f_0$

FC

Center 50.00 MHz   Span 100.0 MHz
#Res BW 100.0 kHz   #VBW 300.0 kHz   #Sweep 100.0 ns

Figure 1.7   Magnitudes of the Fourier coefficients.

and for $n \neq 0$

$$c_n = 2A\frac{\tau}{T}\left|\frac{\sin\frac{n\pi\tau}{T}}{\frac{n\pi\tau}{T}}\right|\left|\frac{\sin\frac{n\pi t_r}{T}}{\frac{n\pi t_r}{T}}\right| \tag{1.46}$$

The angle of Fourier coefficients is

$$\theta_n = -\frac{n\omega_0(\tau + t_r)}{2} \pm \pi \tag{1.47}$$

These coefficients (spectral components) exist only at the discrete frequencies $f = n/T$, as shown in Figure 1.7 for a 1 V, 10 MHz clock signal with a rise time of 5 ns, and a duty cycle of 49%.

When the duty cycle $D = \tau/T = 50\%$ the even coefficients are zero, as shown in Figure 1.8.

Figures 1.8 and 1.9 reveal another important fact.

When the duty cycle changed not only the even harmonics appeared but the magnitudes of the odd harmonics also changed. Why? Because the time domain signal changed, so its Fourier coefficients also changed.

We can obtain the continuous envelope of the spectral components by replacing $n/T = f$ in Eq. (1.46)

$$Envelope = 2A\frac{\tau}{T}\left|\frac{\sin\pi\tau f}{\pi\tau f}\right|\left|\frac{\sin\pi t_r f}{\pi t_r f}\right| \tag{1.48}$$

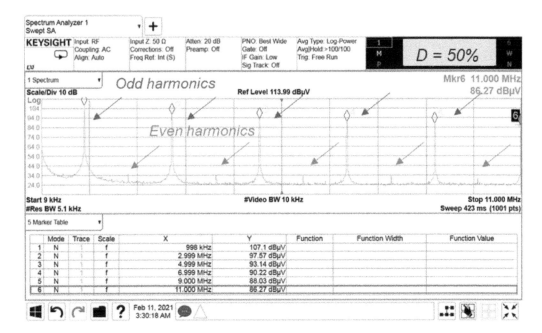

Figure 1.8　Fourier spectrum for D = 50%.

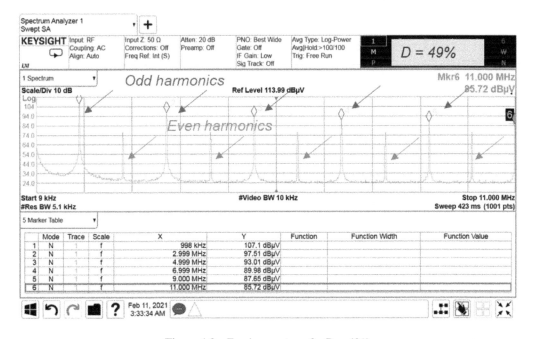

Figure 1.9　Fourier spectrum for D = 49%.

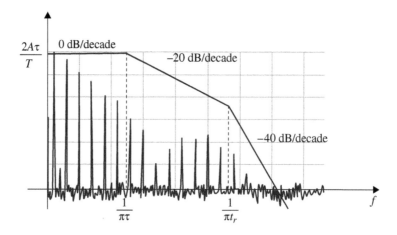

Figure 1.10   Bounds of the magnitude spectrum of a clock signal.

or in dB

$$(Envelope)_{\text{dB}} = \left(2A\frac{\tau}{T}\right)_{\text{dB}} + \left|\frac{\sin \pi\tau f}{\pi\tau f}\right|_{\text{dB}} + \left|\frac{\sin \pi t_r f}{\pi t_r f}\right|_{\text{dB}} \tag{1.49}$$

These bounds are shown in Figure 1.10.

There is one extremely important observation we can make from the plots in Figure 1.10. Note that above the frequency $f = 1/\pi t_r$ the amplitudes of spectral components are attenuated at a rate of 40 dB/decade.

It seems reasonable, therefore, to postulate that somewhere beyond this frequency these amplitudes are negligible (as compared to the magnitudes of the components at lower frequencies) and can be neglected in the Fourier series expansion

$$x(t) = c_0 + \sum_{n=1}^{\infty} 2c_n \cos(n\omega_0 t + \theta_n) \tag{1.50}$$

A reasonable choice for this frequency is [Paul, 2006]

$$f_{max} = 3\frac{1}{\pi t_r} \simeq \pi\frac{1}{\pi t_r} = \frac{1}{t_r} \tag{1.51}$$

This is the origin of one of the EMC rules which states that the bandwidth (BW) of a trapezoidal signal (the highest significant frequency) is

$$BW = \frac{1}{t_r} \tag{1.52}$$

Figure 1.11 shows the frequency spectrum of a 10 MHz, 50% duty cycle, 1 V trapezoidal waveform with a rise time of 2.5 ns.

Superimposed on the spectrum are two straight lines connecting the peaks of the spectrum. Note that the slopes of these lines conform to the envelope slopes shown

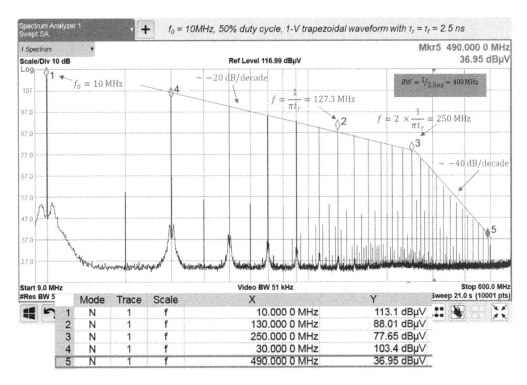

Figure 1.11 Measured frequency spectrum of a clock signal with $t_r = 2.5\,\text{ns}$.

in Figure 1.10. The break point where the peak slope changes from 20 to 40 dB is $2\frac{1}{\pi t_r}$ confirming that past the frequency of $\frac{1}{\pi t_r}$ the amplitudes of the frequency spectrum are attenuated at 40 dB/decade.

With this choice of the highest significant frequency present in the signal given by Eq. (1.52), infinite summation in Eq. (1.50) can be replaced by the upper limit of N, where N corresponds to the highest harmonic up to the bandwidth in Eq. (1.52).

$$x(t) = c_0 + \sum_{n=1}^{N} 2c_n \cos(n\omega_0 t + \theta_n) \qquad (1.53)$$

Let's verify this criterion for a 1 V, 50% duty cycle, 10 MHz, trapezoidal signal with a rise time of 2.5 ns. Figures 1.12 and 1.13 show the result of the summation in Eq. (1.53) with $N = 20$ (harmonics up to the half of the highest significant frequency) [Adamczyk and Gilbert, 2020].

With $N = 40$ (up to the highest significant harmonics) the results are shown in Figures 1.14 and 1.15.

Finally, with $N = 80$ (twice the highest significant harmonics) the results are shown in Figures 1.16 and 1.17.

Comparing the results in Figures 1.15 and 1.17 we observe that little is gained by adding the summation terms past the highest significant frequency defined by Eq. (1.52).

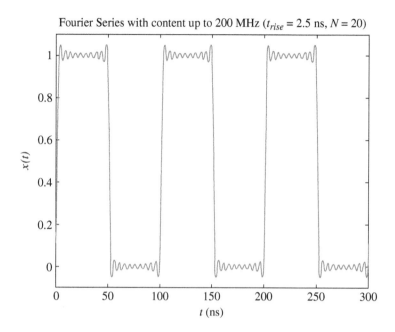

Figure 1.12   Signal reconstruction with $N = 20$.

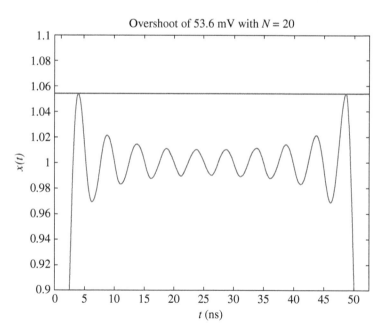

Figure 1.13   Peak overshoot with $N = 20$.

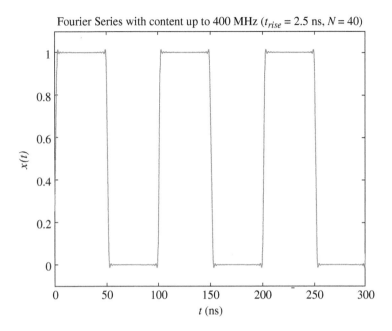

Figure 1.14   Signal reconstruction with $N = 40$.

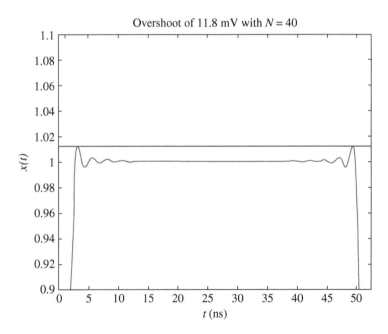

Figure 1.15   Peak overshoot with $N = 40$.

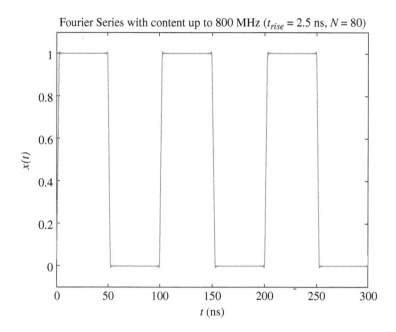

Figure 1.16   Signal reconstruction with $N = 80$.

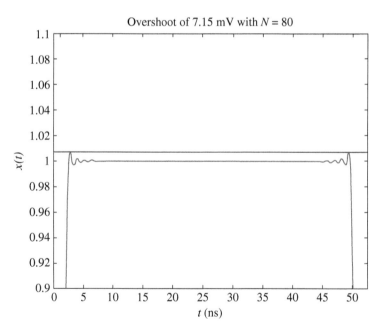

Figure 1.17   Peak overshoot with $N = 80$.

1.4 Effect of the Rise Time, Signal Amplitude, Fundamental Frequency, and Duty Cycle on the Signal Spectrum

1.4.1 Effect of the Rise Time

Returning to Figure 1.10, we make another important observation: the pulses having short rise/fall times have larger high-frequency content than do pulses with long rise/fall times. This is illustrated in Figure 1.18.

Note that the low-frequency spectrum is not affected by the rise time. This is shown in Figures 1.19 and 1.20, which show the first 6 odd harmonics for the signals with the rise time of $t_{r1} = 5$ ns and $t_{r2} = 2.5$ ns, respectively.

Note that the magnitudes of the corresponding harmonics are virtually identical. This result agrees with the plot shown in Figure 1.18.

Next, let's compare the high-frequency content. This is shown in Figure 1.21.

At 250 MHz (Marker 1) the signal with rise time of 5 ns has a frequency content of 67.14 dBμV, while the signal with rise time of 2.5 ns has a frequency content of 77.71 dBμV.

At 490 MHz (Marker 2) the values are 34.53 dBμV, vs. 37.37 dBμV. Clearly, the signal with a shorter rise time has a higher high-frequency content.

1.4.2 Effect of the Signal Amplitude

The effect of the signal amplitude on the frequency content of a trapezoidal signal is shown in Figure 1.22 [Adamczyk, 2017b].

As can be seen, reducing the signal amplitude reduces the frequency content over the entire frequency range. This is verified by the measurement shown in Figure 1.23, which shows the spectral content for two signals: one with an amplitude of $A_1 = 0.5$ V, and the other with an amplitude of $A_2 = 1.0$ V.

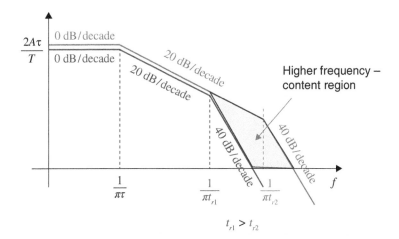

Figure 1.18 Impact of a rise time on the high-frequency content.

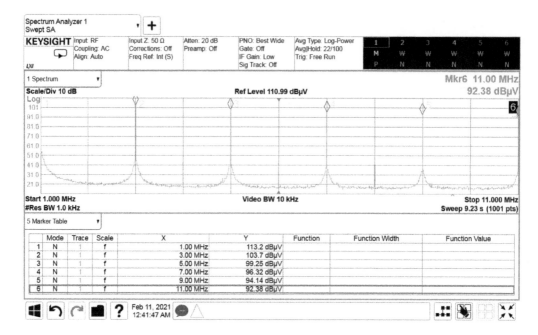

Figure 1.19   Low-frequency spectrum, $t_{r1} = 5$ ns.

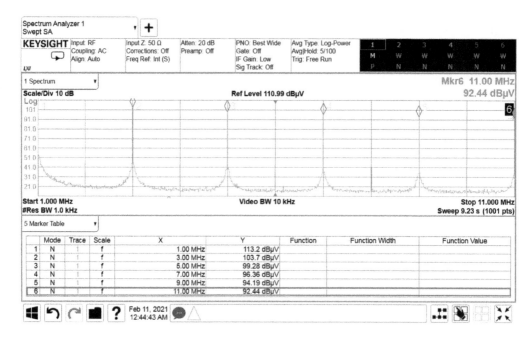

Figure 1.20   Low-frequency spectrum, $t_{r2} = 2.5$ ns.

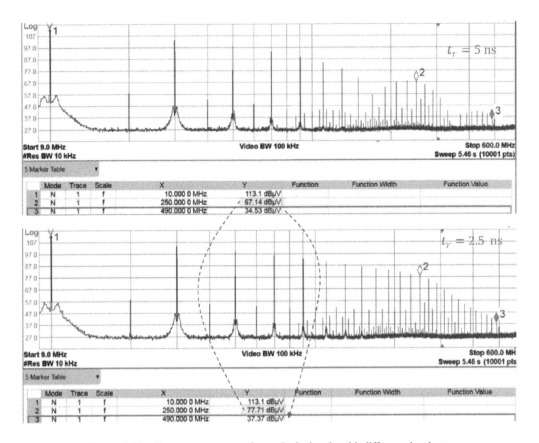

Figure 1.21   Frequency content of two clock signals with different rise times.

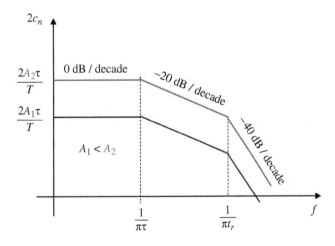

Figure 1.22   Effect of the signal amplitude.

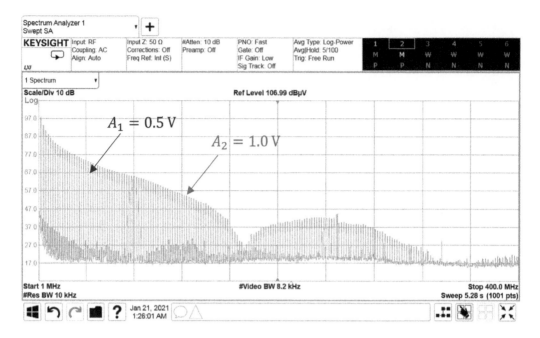

Figure 1.23   Effect of the signal amplitude.

The result shown verifies that reducing the signal amplitude reduces the frequency content over the entire frequency range. Let's look at the first 6 odd harmonics of the two signals shown in Figure 1.23.

Figure 1.24 shows the results for the amplitude $A_1 = 0.5\,$V, while Figure 1.25 shows the results for the amplitude $A_2 = 1.0\,$V.

Note that the magnitude of each harmonic of the signal with the amplitude $A_2 = 1.0\,$V is about 6 dB higher than the corresponding magnitude of the signal with the amplitude $A_1 = 0.5\,$V. This is to be expected as the amplitudes differ by a factor of 2 which corresponds to 6 dB.

1.4.3 Effect of the Fundamental Frequency

Note that the first break frequency on the spectrum bounds is at

$$f = \frac{1}{\pi\tau} \tag{1.54}$$

Since the duty cycle D is related to the period T, and thus the fundamental frequency f_0, we have

$$D = \frac{\tau}{T} = \tau f_0 \tag{1.55}$$

or

$$\tau = \frac{D}{f_0} \tag{1.56}$$

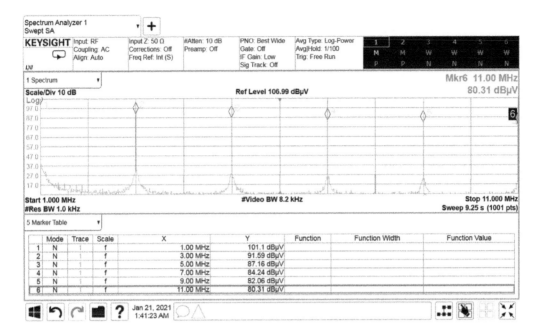

Figure 1.24 First 6 odd harmonics for the amplitude $A_1 = 0.5\,\text{V}$.

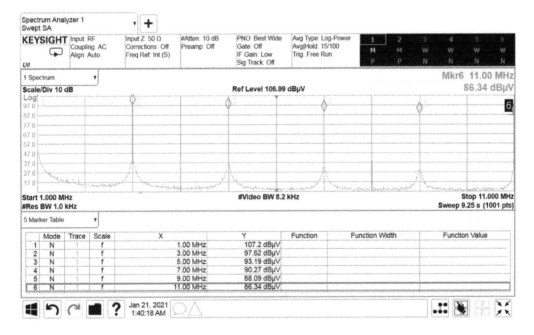

Figure 1.25 First 6 odd harmonics for the amplitude $A_2 = 1.0\,\text{V}$.

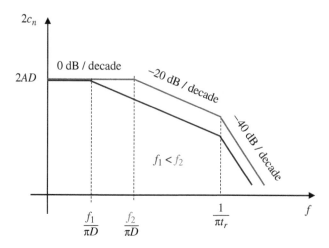

Figure 1.26   Effect of the fundamental frequency while maintaining the duty cycle.

Thus, the first break point on the frequency spectrum is at frequency

$$f = \frac{1}{\pi \tau} = \frac{f_0}{\pi D} \tag{1.57}$$

The effect of reducing the fundamental frequency while maintaining the same duty cycle on the frequency content of signal is shown in Figure 1.26.

Reducing the fundamental frequency (while maintaining the duty cycle) reduces the high-frequency spectral content of the waveform, but does not affect the low-frequency content.

Let the duty cycle $D = 0.5$, $f_1 = 1\,\text{MHz}$, and $f_2 = 2\,\text{MHz}$. Then

$$\frac{f_1}{\pi D} = \frac{1 \times 10^6}{\pi \times 0.5} = 636.6\,\text{kHz} \tag{1.58}$$

$$\frac{f_2}{\pi D} = \frac{2 \times 10^6}{\pi \times 0.5} = 1.27\,\text{MHz} \tag{1.59}$$

The first harmonic of the 2 MHz signal is above these values. Thus, according to Figure 1.26, all the harmonics of the 2 MHz signal should be higher than the harmonics of the 1 MHz signal at the corresponding frequencies.

This is verified by the measurement shown in Figure 1.27.

1.4.4 Effect of the Duty Cycle

Finally, the effect of reducing the duty cycle while maintaining the fundamental frequency is shown in Figure 1.28.

Let the duty cycle $f_0 = 1\,\text{MHz}$, $A = 0.5\,\text{V}$, $D_1 = 0.4$, and $D_2 = 0.8$. Then

$$\frac{f_0}{\pi D_1} = \frac{1 \times 10^6}{\pi \times 0.4} = 796\,\text{kHz} \tag{1.60}$$

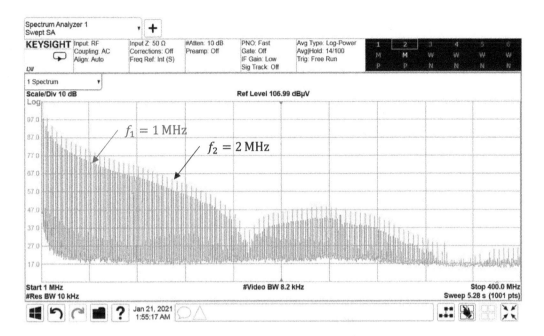

Figure 1.27 Effect of the fundamental frequency while maintaining the duty cycle.

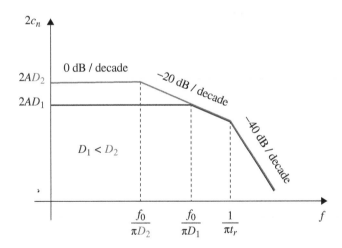

Figure 1.28 Effect of the duty cycle while maintaining the fundamental frequency.

$$\frac{f_0}{\pi D_2} = \frac{1 \times 10^6}{\pi \times 0.8} = 398 \, \text{kHz} \tag{1.61}$$

Both of these values are below the first harmonic of 1 MHz. Thus, according to Figure 1.28, the spectrum of each signal has the same frequency bounds.

This is verified by the measurement shown in Figure 1.29.

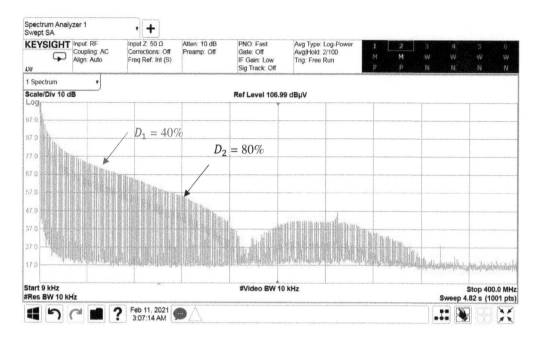

Figure 1.29   Effect of the duty cycle while maintaining the fundamental frequency.

1.5 Laboratory Exercises

1.5.1 Spectrum of a Digital Clock Signal

Objective: This laboratory examines the spectrum of a digital clock signal and determines the impact of signal parameters on the spectrum. Follow this exercise and redo the measurements/calculations with your own parameters.

1.5.2 Laboratory Equipment and Supplies

Spectrum analyzer: KEYSIGHT CXA Signal Analyzer N9000B (or equivalent), shown in Figure 1.30.
  Function generator: Tektronix AFG3252 (or equivalent), shown in Figure 1.31.

RG58C coaxial cable, shown in Figure 1.32.
BNC Male Plug to BNC Male Plug, length 4 feet
Supplier: Pomona Electronics, Part Number: 2249-C
Digi-Key Part Number: 501-1019-ND

Connector adapter, shown in Figure 1.33.
N Plug Male Pin to BNC Jack, Female Socket
Supplier: Amphenol, Part Number: 31-216
Digi-Key Part Number: ARF1096-ND

Figure 1.30 Laboratory equipment: spectrum analyzer.

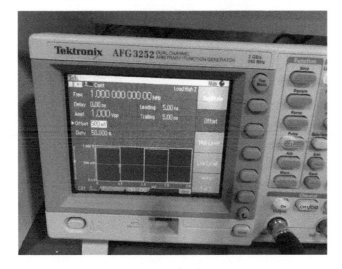

Figure 1.31 Laboratory equipment: function generator.

1.5.3 Measured Spectrum vs. Calculated Spectrum

Objective: Compare the measured and calculated values of the low-frequency harmonics.

Connect the function generator to the spectrum analyzer with the coaxial cable and the adapter, as shown in Figure 1.34.

Initial function generator settings: Function = pulse, frequency = 1 MHz, amplitude = 1 Vpp ($load = 50\,\Omega$, offset = 500 mV, duty cycle = 50%, rise time = 20 ns, fall time = 20 ns).

Figure 1.32   Laboratory equipment: coaxial cable.

Figure 1.33   Laboratory equipment: connector adapter.

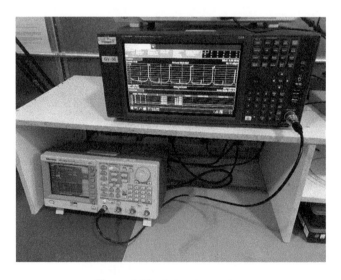

Figure 1.34   Laboratory setup.

	Mode	Trace	Scale	X	Y	Function	Function Width	Function Value
1	N	1	f	1.009 MHz	113.2 dBµV			
2	N	1	f	3.008 MHz	103.6 dBµV			
3	N	1	f	5.007 MHz	99.04 dBµV			
4	N	1	f	7.006 MHz	95.91 dBµV			
5	N	1	f	9.005 MHz	93.44 dBµV			
6	N	1	f	11.004 MHz	91.33 dBµV			

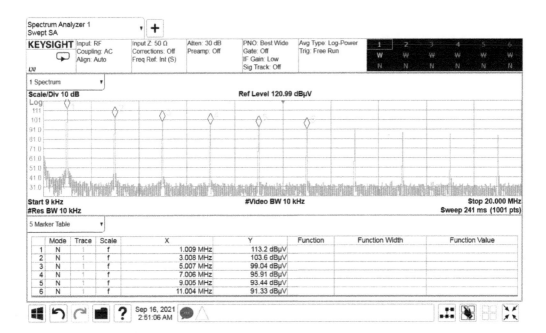

Figure 1.35 Low-frequency spectrum of the signal with $t_r = 20\,\text{ns}$.

Initial spectrum analyzer settings: FREQ: start frequency = 9 kHz, stop frequency = 400 MHz, AMPTD: amplitude units dBµV, reference level 117 dBµV, BW: resolution bandwidth = 10 kHz, VB: video bandwidth = 10 kHz, Trace: max hold.

Figure 1.35 shows the spectrum of the signal with the markers displaying the magnitudes of the first six odd harmonics.

Matlab script, shown in Figure 1.36, can be used to calculate the values of these harmonics according to Eq. (1.46).

In Eq. (1.46) the rise time corresponds to the increase from 0% to 100% of the final value. The rise time set by the function generator corresponds to the increase from 10% to 90% of the final value.

To account for this, the rise time used in the calculations was set to 25 ns.

Figure 1.37 (first two columns) shows the measured and calculated values of the first six odd harmonics.

The measured values are the rms values; the calculated ones are not. To account for that we need to add 3 dB to the measured values as shown in column 3. Note the remarkable agreement between the calculated and (adjusted) measured values.

Next, let's change the rise time to 40 ns (50 ns for the calculated values). Figure 1.38 shows the spectrum of the signal with the markers displaying the magnitudes of the first six odd harmonics.

Figure 1.39 (first two columns) shows the measured and calculated values of the first six odd harmonics.

```
% Lab 1 - Spectrum of a Signal
%Set amplitude value
A = 1;
%Set Peroid T = 1/f0
T = 1/(1000000);
%Set Duty Cycle t/T = D
D = 0.5;
%set rise/fall time
tr = 25*10^(-9);

for n = 1:11

    % The magnitude of the nth harmonic is
given by:

Cn = 2*A*(D)*abs(sin(n*pi*D)/(n*pi*D))*abs((sin
(n*pi*tr/T))/(n*pi*tr/T));
    % Results are displayed for the given n
with the value of the coefficeint
    % in volts and also dBμV:
    disp(n)
    disp(Cn)
    disp(20*log10(Cn/(10^-6)))
end
```

Figure 1.36 Matlab script.

Harmonic	Measured value dBμV	Calculated value dBμV	Measured value + 3 dBμV
1	113.2	116.0687	116.2
3	103.6	106.4547	106.6
5	99.04	101.8738	102.04
7	95.91	98.7336	98.91
9	93.44	96.2570	96.44
11	91.33	94.1410	94.33

Figure 1.37 Calculated and measured values of the harmonics $t_r = 20$ ns.

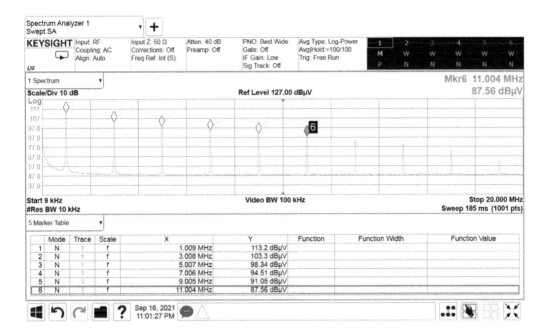

Figure 1.38   Spectrum of the signal with $t_r = 40\,\mathrm{ns}$.

Harmonic	Measured value dBµV	Calculated value dBµV	Measured value + 3 dBµV
1	113.2	116.0419	116.2
3	103.3	106.2113	106.3
5	98.34	101.1861	101.34
7	94.51	97.3489	97.51
9	91.05	93.8779	94.05
11	87.56	90.3919	90.56

Figure 1.39   Calculated and measured values of the harmonics, $t_r = 40\,\mathrm{ns}$.

1.5.4 Effect of the Rise Time

Objective: Determine the impact of the rise on the low-, mid-, and high-frequency content of the spectrum.

Figure 1.40 compares the low-frequency content of the two signals.

Next, let's compare the mid-frequency content of the signals with the rise and fall time equal to $t_r = 20ns$ and $t_r = 40ns$, respectively.

This is shown in Figure 1.41 for $t_r = 20\,\mathrm{ns}$, and in Figure 1.42 for $t_r = 40\,\mathrm{ns}$.

Harmonic	Measured value dBμV $t_r = 20$ ns	Measured value dBμV $t_r = 40$ ns	Δ dBμV
1	113.2	113.2	0
3	103.6	103.3	0.3
5	99.04	98.34	0.7
7	95.91	94.51	1.4
9	93.44	91.05	2.39
11	91.33	87.56	3.77

Figure 1.40   Comparison of the low-frequency content of the two signals.

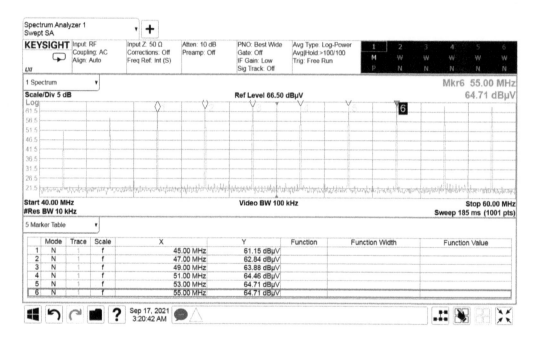

Figure 1.41   Mid-frequency spectrum of the signal with $t_r = 20$ ns.

Figure 1.43 compares the mid-frequency content of the two signals.

We note that the signal with the shorter rise time has a higher mid-frequency content. Note that the difference between the magnitudes is larger than it was for the low-frequency signals.

Let's complete the set by taking the measurements for the high-frequency harmonics.

Figure 1.44 shows the spectrum of the signal with the markers displaying the magnitudes of the high-frequency odd harmonics for the rise time of $t_r = 20$ ns. Figure 1.45 shows the corresponding spectrum for the rise time of $t_r = 40$ ns.

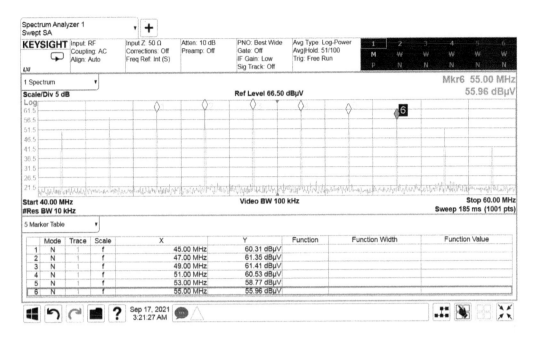

Figure 1.42 Mid-frequency spectrum of the signal with $t_r = 40\,\text{ns}$.

Harmonic	Measured value dBμV $t_r = 20\,\text{ns}$	Measured value dBμV $t_r = 40\,\text{ns}$	Δ dBμV
45	61.15	60.31	0.84
47	62.84	61.35	1.49
49	63.88	61.41	2.47
51	64.46	60.53	3.93
53	64.71	58.77	5.94
55	64.71	55.96	8.75

Figure 1.43 Comparison of the mid-frequency content of the two signals.

Figure 1.46 compares the high-frequency content of the two signals.

Again, the signal with the shorter rise time has a higher high-frequency content. Note that the difference between the magnitudes is, in general, larger than it was for the mid-frequency signals.

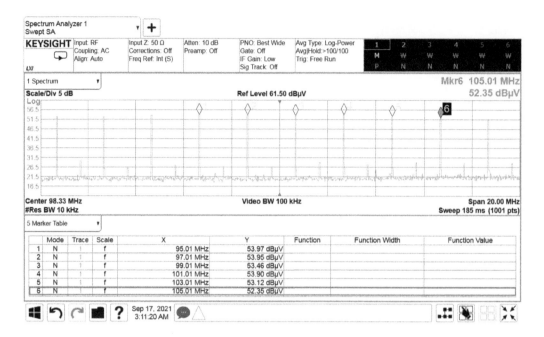

Figure 1.44 High-frequency spectrum of the signal with $t_r = 20$ ns.

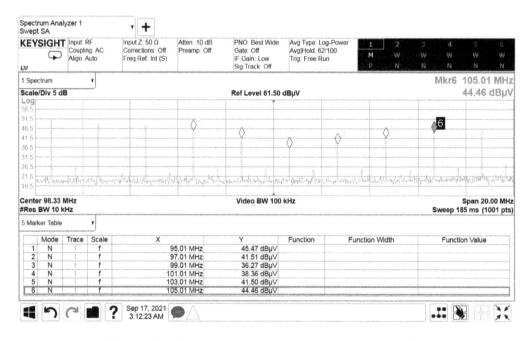

Figure 1.45 High-frequency spectrum of the signal with $t_r = 40$ ns.

Harmonic	Measured value dBμV $t_r = 20$ ns	Measured value dBμV $t_r = 40$ ns	Δ dBμV
95	53.97	45.47	8.5
97	53.95	41.51	12.44
99	53.46	36.27	17.19
101	53.90	38.36	15.54
103	53.12	41.50	11.62
105	52.35	44.46	7.89

Figure 1.46   Comparison of the high-frequency content of the two signals.

1.5.5 Effect of the Signal Amplitude

Objective: Determine the impact of the signal amplitude on the low-, mid-, and high-frequency content of the spectrum.

Let's keep the rise and times at $t_r = t_f = 20$ ns and change the signal amplitude to 0.5 Vpp, offset = 250 mV.

Figures 1.47–1.49 show the corresponding low-, mid-, and high-frequency spectra, respectively.

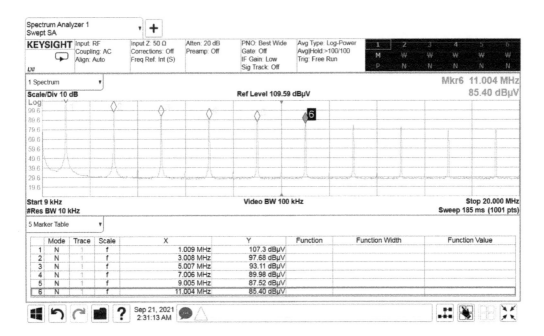

Figure 1.47   Low-frequency spectrum of the signal with $A = 500$ mV.

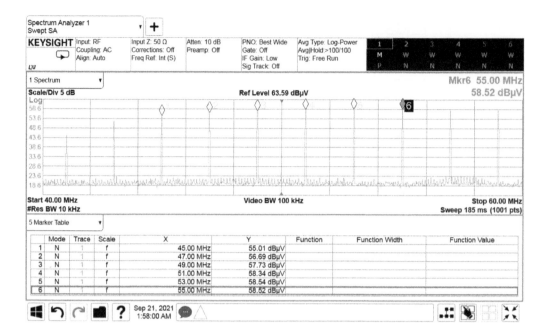

Figure 1.48   Mid-frequency spectrum of the signal with $A = 500\,$mV.

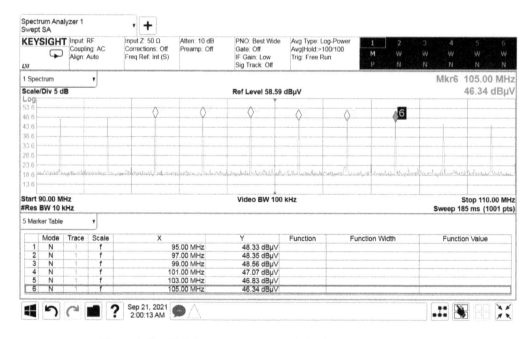

Figure 1.49   High-frequency spectrum of the signal with $A = 500\,$mV.

Harmonic	Measured value dBμV A = 1 V	Measured value dBμV A = 500 mV	Δ dBμV
1	113.2	107.3	5.9
3	103.6	97.68	5.92
5	99.04	93.11	5.93
7	95.91	89.98	5.93
9	93.44	87.52	5.92
11	91.33	85.40	5.93

Figure 1.50   Comparison of the low-frequency spectra for $A = 1$ V and $A = 500$ mV.

Harmonic	Measured value dBμV A = 1 V	Measured value dBμV A = 500 mV	Δ dBμV
45	61.15	55.01	6.14
47	62.84	56.69	6.15
49	63.88	57.33	6.55
51	64.46	58.34	6.12
53	64.71	58.54	6.17
55	64.71	58.52	6.19

Figure 1.51   Comparison of the mid-frequency spectra for $A = 1$ V and $A = 500$ mV.

Figures 1.50–1.52 summarize the results.
Note that the two signals differ by about 6 dBμV across the entire spectrum.

1.5.6 Effect of the Fundamental Frequency

Objective: Determine the impact of the fundamental frequency on the low-, mid-, and high-frequency content of the spectrum.

Let's keep the rise and times at $t_r = t_f = 20$ ns, the amplitude at 0.5 Vpp, and change the fundamental frequency to 2 MHz.

Figures 1.53 and 1.54 show the corresponding low-frequency spectra, for $f_0 = 1$ MHz and $f_0 = 2$ MHz, respectively.

Figure 1.55 compares the results.

Harmonic	Measured value dBμV $A = 1$ V	Measured value dBμV $A = 500$ mV	Δ dBμV
95	53.97	48.33	5.64
97	53.95	48.35	5.6
99	53.46	48.56	4.9
101	53.90	47.07	6.83
103	53.12	46.83	6.29
105	52.35	46.34	6.01

Figure 1.52   Comparison of the high-frequency spectra for $A = 1$ V and $A = 500$ mV.

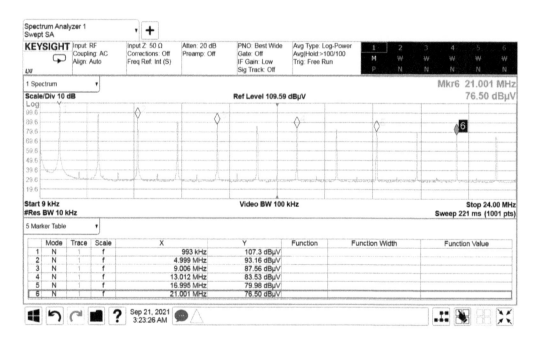

Figure 1.53   Low-frequency spectrum of the signal with $f_0 = 1$ MHz.

Note that the signal with the fundamental frequency $f_0 = 2$ MHz has a higher (about 6 dB) low-frequency content.

Figures 1.56 and 1.57 show the mid-frequency spectra, for $f_0 = 1$ MHz and $f_0 = 2$ MHz, respectively.

Figure 1.58 compares the results.

Again, note that the signal with the fundamental frequency $f_0 = 2$ MHz has a higher (about 6 dB) mid-frequency content.

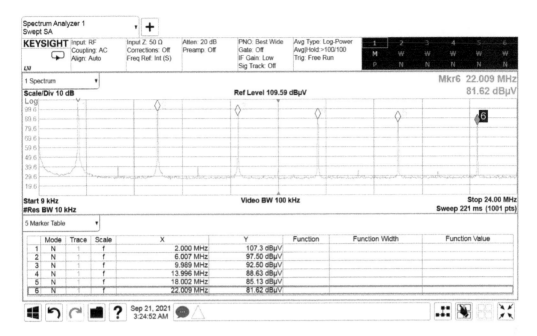

Figure 1.54 Low-frequency spectrum of the signal with $f_0 = 2\,\text{MHz}$.

Harmonic	Measured value dBμV $f_0 = 1$ MHZ	Harmonic	Measured value dBμV $f_0 = 2$ MHZ
1 (@ 1 MHz)	107.3	1 (@ 2 MHz)	107.3
5 (@ 5 MHz)	93.16	3 (@ 6 MHz)	97.50
9 (@ 9 MHz)	87.56	5 (@ 10 MHz)	92.50
13 (@ 13 MHz)	83.53	7 (@ 14 MHz)	88.63
17 (@ 17 MHz)	79.98	9 (@ 18 MHz)	85.13
21 (at 21 MHz)	76.50	11 (@ 22 MHz)	81.62

Figure 1.55 Comparison of the low-frequency spectrum with $f_0 = 1\,\text{MHz}$ and $f_0 = 2\,\text{MHz}$.

Figures 1.59 and 1.60 show the high-frequency spectra, for $f_0 = 1\,\text{MHz}$ and $f_0 = 2\,\text{MHz}$, respectively.

Figure 1.61 compares the results.

Once again, note that the signal with the fundamental frequency $f_0 = 2\,\text{MHz}$ has a higher (about 6 dB) high-frequency content.

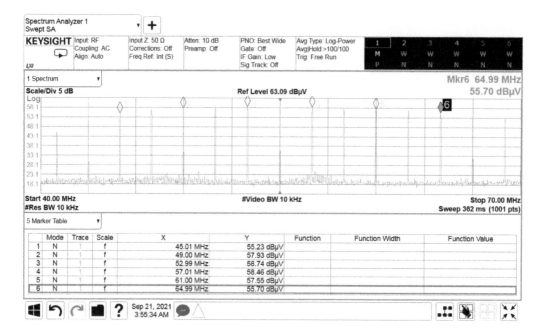

Figure 1.56   Mid-frequency spectrum of the signal with $f_0 = 1\,\text{MHz}$.

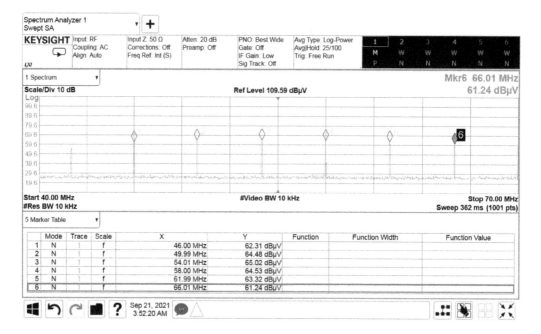

Figure 1.57   Mid-frequency spectrum of the signal with $f_0 = 2\,\text{MHz}$.

Frequency MHz	Measured value dBµV $f_0 = 1$ MHZ	Harmonic	Measured value dBµV $f_0 = 2$ MHZ
45	55.23	46	62.31
49	57.93	50	64.48
53	58.74	54	65.02
57	58.46	58	64.53
61	57.55	62	63.32
65	55.70	66	61.24

Figure 1.58   Comparison of the mid-frequency spectrum with $f_0 = 1$ MHz and $f_0 = 2$ MHz.

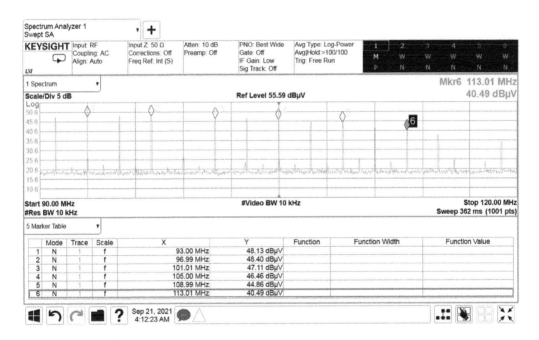

Figure 1.59   High-frequency spectrum of the signal with $f_0 = 1$ MHz.

1.5.7 Effect of the Duty Cycle

Objective: Determine the effect of the duty cycle on the low-, mid-, and high-frequency content of the spectrum.

Let's keep the rise and times at $t_r = t_f = 20$ ns, the amplitude at 0.5 Vpp, and the fundamental frequency at 1 MHz. Let's choose two duty cycle values: $D_1 = 40\%$ and $D_2 = 80\%$.

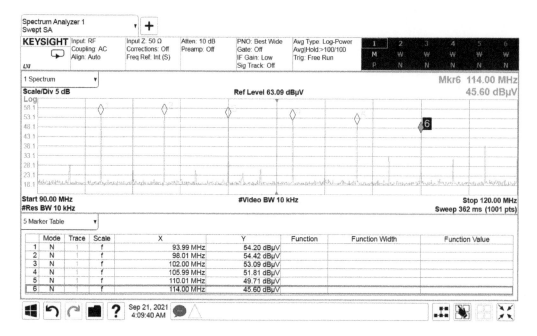

Figure 1.60   High-frequency spectrum of the signal with $f_0 = 2$ MHz.

Frequency MHz	Measured value dBμV $f_0 = 1$ MHZ	Frequency MHz	Measured value dBμV $f_0 = 2$ MHZ
93	48.13	94	54.20
97	48.40	98	54.42
101	47.11	102	53.09
105	46.46	106	51.81
109	44.86	110	49.71
113	40.49	114	45.60

Figure 1.61   Comparison of the high-frequency spectrum with $f_0 = 1$ MHz and $f_0 = 2$ MHz.

Figures 1.62 and 1.63 show the corresponding low-frequency spectra, for $D_1 = 40\%$ and $D_2 = 80\%$, respectively.

Figure 1.64 compares the results.

Observation: The difference between the harmonics' levels is about ±4 dB.

Figures 1.65 and 1.66 show the corresponding mid-frequency spectra, for $D_1 = 40\%$ and $D_2 = 80\%$, respectively.

Figure 1.67 compares the results.

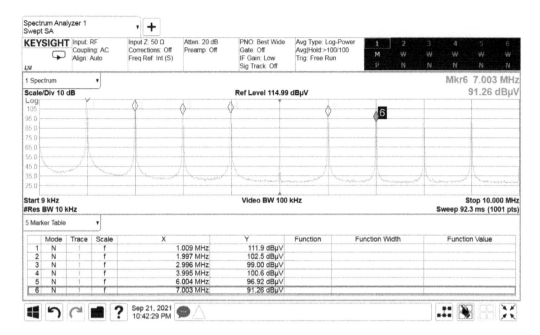

Figure 1.62 Low-frequency spectrum of the signal with $D_1 = 40\%$.

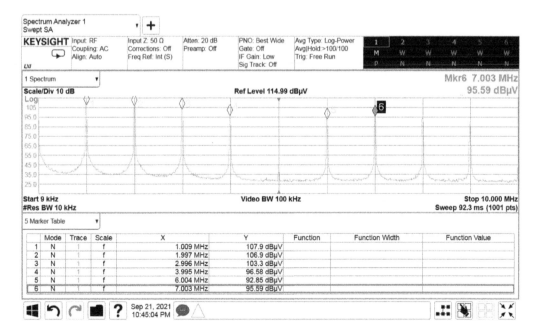

Figure 1.63 Low-frequency spectrum of the signal with $D_2 = 80\%$.

Frequency MHz	Measured value dBμV $D_1 = 40\%$	Measured value dBμV $D_1 = 80\%$	Δ dBμV
1	111.9	107.9	4
2	102.5	106.9	−4.4
3	99	103.3	−4.3
4	100.6	96.58	4.02
6	96.92	92.85	4.07
7	91.26	95.59	−4.33

Figure 1.64 Comparison of the low-frequency spectrum with $D_1 = 40\%$ and $D_2 = 80\%$.

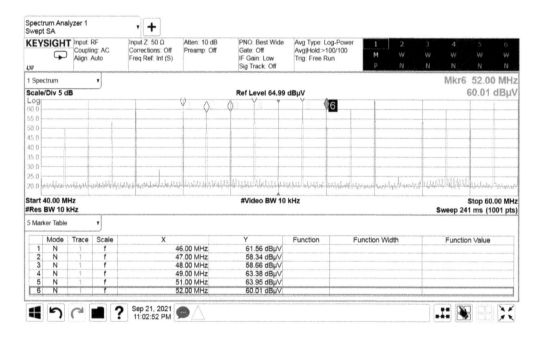

Figure 1.65 Mid-frequency spectrum of the signal with $D_1 = 40\%$.

Observation: Again, the difference between the harmonics' levels is about ±4 dB.

Figures 1.68 and 1.69 show the corresponding high-frequency spectra, for $D_1 = 40\%$ and $D_2 = 80\%$, respectively.

Figure 1.70 compares the results.

Observation: Once again, the difference between the harmonics' levels is about ±4 dB.

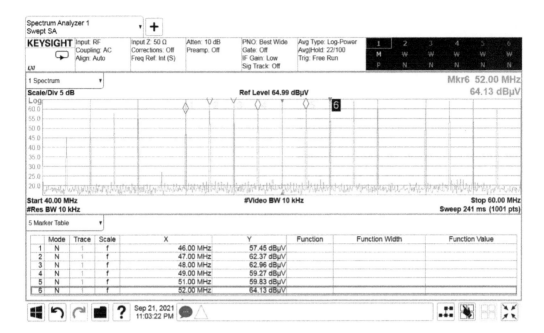

Figure 1.66 Mid-frequency spectrum of the signal with $D_2 = 80\%$.

Frequency MHz	Measured value dBμV $D_1 = 40\%$	Measured value dBμV $D_1 = 80\%$	Δ dBμV
46	61.56	57.45	4.11
47	58.34	62.37	−4.03
48	58.66	62.96	−4.3
49	63.38	59.27	4.11
51	63.95	59.83	4.12
52	60.01	64.13	−4.02

Figure 1.67 Comparison of the mid-frequency spectrum with $D_1 = 40\%$ and $D_2 = 80\%$.

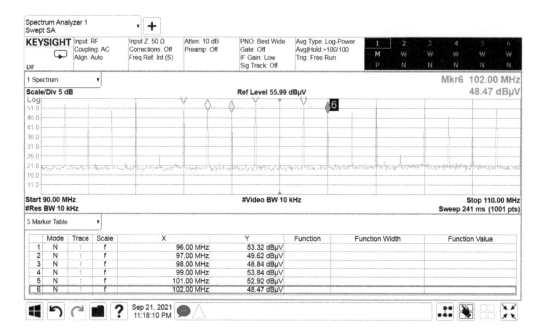

Figure 1.68   High-frequency spectrum of the signal with $D_1 = 40\%$.

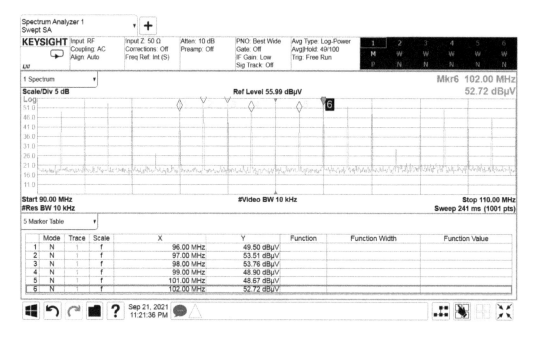

Figure 1.69   High-frequency spectrum of the signal with $D_2 = 80\%$.

Frequency MHz	Measured value dBμV $D_1 = 40\%$	Measured value dBμV $D_1 = 80\%$	Δ dBμV
96	53.32	49.50	3.82
97	49.62	53.51	−3.89
98	48.84	53.76	−4.92
99	53.84	48.90	4.94
101	52.92	48.67	4.25
102	48.47	52.72	−4.25

Figure 1.70 Comparison of the high-frequency spectrum with $D_1 = 40\%$ and $D_2 = 80\%$.

References

Bogdan Adamczyk. *Foundations of Electromagnetic Compatibility*. John Wiley & Sons, Ltd, Chichester, UK, 2017a. ISBN 9781119120810. doi: https://doi.org/10.1002/9781119120810.

Bogdan Adamczyk. Spectra of Digital Clock Signals. *In Compliance Magazine*, April 2017b. URL https://incompliancemag.com/article/spectra-of-digital-clock-signals/.

Bogdan Adamczyk. Concept of a Phasor in Sinusoidal Steady State Analysis. *In Compliance Magazine*, December 2022. URL https://incompliancemag.com/article/concept-of-a-phasor-in-sinusoidal-steady-state-analysis/.

Bogdan Adamczyk and Brian Gilbert. Basic EMC Rules. *In Compliance Magazine*, May 2020. URL https://incompliancemag.com/article/basic-emc-rules/.

Bogdan Adamczyk and Jim Teune. EMC Units in Measurements and Testing. *In Compliance Magazine*, July 2019. URL https://incompliancemag.com/article/emc-units-in-measurements-and-testing/.

Clayton R. Paul. *Introduction to Electromagnetic Compatibility*. John Wiley & Sons, Inc., Hoboken, New Jersey, 2nd edition, 2006. ISBN 978-0-471-75500-5.

Chapter 2: EM Coupling Mechanisms

Coupling mechanisms discussed in this chapter are explained using the concept of electrically short structures. To define such structures we need to relate the wavelength to the physical dimensions of the system. This is done next.

2.1 Wavelength and Electrical Dimensions

We begin by introducing a mathematical concept of a wave and then showing that the time-varying EM fields propagate as waves.

2.1.1 Concept of a Wave

Consider a function of time t and space z, $f(z,t)$, with its argument given by

$$f(z,t) = f\left(t - \frac{z}{v}\right) \tag{2.1}$$

Then,

$$f(z + \Delta z, t + \Delta t) = f\left(t + \Delta t - \frac{z + \Delta z}{v}\right) \tag{2.2}$$

Equation (2.2) is valid for any Δz and any Δt. Thus, we could choose any relationship between the two deltas and the new equation would still be valid. Let's choose this relationship to be

$$\Delta z = v\Delta t \tag{2.3}$$

Then Eq. (2.2) becomes

$$f(z + \Delta z, t + \Delta t) = f\left(t + \Delta t - \frac{z + v\Delta t}{v}\right) = f\left(t + \Delta t - \frac{z}{v} - \Delta t\right) \tag{2.4}$$

or

$$f(z + \Delta z, t + \Delta t) = f(z,t) \tag{2.5}$$

Therefore, after a time Δt, the function f retains the same value at a point that is $\Delta z = v\Delta t$ away from the previous position in space (defined by z), as shown in Figure 2.1.

This means that any function of the form $f(t - \frac{z}{v})$ represents a wave traveling in the positive z direction, with a velocity

$$v = \frac{\Delta z}{\Delta v} \tag{2.6}$$

Principles of Electromagnetic Compatibility: Laboratory Exercises and Lectures, First Edition. Bogdan Adamczyk.
© 2024 John Wiley & Sons Ltd. Published 2024 by John Wiley & Sons Ltd.
Companion website: www.wiley.com/go/principlesofelectromagneticcompatibility

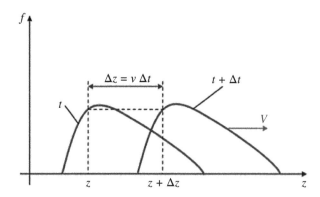

Figure 2.1   Wave propagating in the positive z direction with a velocity v.

Similarly, it can be shown that any function of the form $g(t + \frac{z}{v})$ represents a wave traveling in the negative z direction, as the time advances.

2.1.2 Uniform Plane EM Wave in Time Domain

The time variations of the magnetic (H) and electric fields (E) give rise to the space variations of the electric and magnetic fields, respectively. This interdependence of the space and time variations gives rise to the electromagnetic wave propagation.

The two fields are related by Maxwell's equations (in source-free medium)

$$\nabla \times \boldsymbol{E} = -\mu \frac{\partial \boldsymbol{H}}{\partial t} \tag{2.7}$$

$$\nabla \times \boldsymbol{H} = \sigma \boldsymbol{E} + \epsilon \frac{\partial \boldsymbol{E}}{\partial t} \tag{2.8}$$

where μ, σ, and ϵ are the permeability, conductivity, and permittivity of a medium, respectively.

In general, the electric and magnetic fields have three nonzero components, each being a function of all three coordinates and time. That is,

$$\boldsymbol{E} = [E_x(x,y,z,t), E_y(x,y,z,t), E_z(x,y,z,t)] \tag{2.9}$$

$$\boldsymbol{H} = [H_x(x,y,z,t), H_y(x,y,z,t), H_z(x,y,z,t)] \tag{2.10}$$

We will focus on a simple and very useful type of wave: *uniform plane wave* (see Section 11.1). Under the uniformity in the plane assumption, if the E field points in the $+x$ direction (usual designation) then Maxwell's equations show that the H field is pointing in the $+y$ direction, and [Adamczyk, 2017b]

$$\boldsymbol{E} = [E_x(z,t), 0, 0] \tag{2.11}$$

$$\boldsymbol{H} = [0, H_y(z,t), 0] \tag{2.12}$$

This is shown in Figure 2.2.

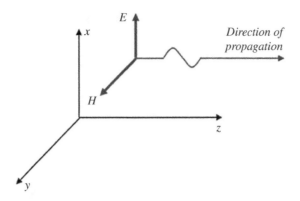

Figure 2.2   Uniform plane EM wave.

The fields propagate as waves in the positive +z direction. Under the uniformity in the *xy* plane assumption, Eqs. (2.69) and (2.70) for a lossless medium ($\sigma = 0$) become [Adamczyk, 2017b]

$$\frac{\partial^2 E_x(z,t)}{\partial z^2} = \mu\epsilon\frac{\partial^2 E_x(z,t)}{\partial t^2} \tag{2.13}$$

$$\frac{\partial^2 H_y(z,t)}{\partial z^2} = \mu\epsilon\frac{\partial^2 H_y(z,t)}{\partial t^2} \tag{2.14}$$

and their general solution, in a lossless medium is,

$$E_x(z,t) = Af\left(t - \frac{z}{v}\right) + Bg\left(t + \frac{z}{v}\right) \tag{2.15}$$

$$H_y(z,t) = \frac{A}{\eta}f\left(t - \frac{z}{v}\right) - \frac{B}{\eta}g\left(t + \frac{z}{v}\right) \tag{2.16}$$

where

$$\eta = \sqrt{\frac{\mu}{\epsilon}} \tag{2.17}$$

is the intrinsic impedance of a (lossless) medium, and A and B are constants.

We recognize the functions f and g, as waves propagating in +z and −z directions, respectively, with a velocity of propagation equal to

$$v = \frac{\Delta z}{\Delta t} = \frac{1}{\sqrt{\mu\epsilon}} \tag{2.18}$$

2.1.3 Uniform Plane EM Wave in Frequency Domain

In Section 2.1.2, we described the wave equations in a lossless medium for an arbitrary time variations. When the time variations are sinusoidal, the wave equations (in a simple medium) become [Adamczyk, 2017b]

$$\frac{d^2\hat{E}_x}{dz^2} = \hat{\gamma}^2 \hat{E}_x(z) \tag{2.19}$$

$$\frac{d^2\hat{H}_y}{dz^2} = \hat{\gamma}^2 \hat{H}_y(z) \tag{2.20}$$

where

$$\hat{\gamma} = \sqrt{j\omega\mu(\sigma + j\omega\epsilon)} \tag{2.21}$$

is the *propagation constant* of the medium. The general solution of Eqs. (2.19) and (2.20) is

$$\hat{E}_x = E_m^+ e^{-\hat{\gamma}z} + E_m^- e^{\hat{\gamma}z} \tag{2.22}$$

$$\hat{H}_y = \frac{E_m^+}{\hat{\eta}} e^{-\hat{\gamma}z} - \frac{E_m^-}{\hat{\eta}} e^{\hat{\gamma}z} \tag{2.23}$$

where

$$\hat{\eta} = \sqrt{\frac{j\omega\mu}{\sigma + j\omega\epsilon}} = \frac{j\omega\mu}{\hat{\gamma}} = \eta\angle\theta_\eta \tag{2.24}$$

is the *intrinsic impedance* of the medium. The propagation constant is often expressed in terms of its real and imaginary parts as

$$\hat{\gamma} = \alpha + j\beta \tag{2.25}$$

where α is the attenuation constant and β is the phase constant. The complex intrinsic impedance is often expressed in an exponential form as

$$\hat{\eta} = \eta e^{j\theta_\eta} \tag{2.26}$$

Then the solution in Eqs. (2.22) and (2.23) can be written as

$$\hat{E}_x = \hat{E}_m^+ e^{-\alpha z} e^{-j\beta z} + \hat{E}_m^- e^{\alpha z} e^{j\beta z} \tag{2.27}$$

$$\hat{H}_y = \frac{\hat{E}_m^+}{\eta} e^{-\alpha z} e^{-j\beta z} e^{-j\theta_\eta} - \frac{\hat{E}_m^-}{\eta} e^{\alpha z} e^{j\beta z} e^{-j\theta_\eta} \tag{2.28}$$

Often, the undetermined complex constants can be expressed as

$$\hat{E}_m^+ = E_m^+ \angle 0 = E_m^+ \tag{2.29}$$

$$\hat{E}_m^- = E_m^- \angle 0 = E_m^- \tag{2.30}$$

Then, the solutions in Eqs. (2.27) and (2.28) become

$$\hat{E}_x = E_m^+ e^{-\alpha z} e^{-j\beta z} + E_m^- e^{\alpha z} e^{j\beta z} \tag{2.31}$$

$$\hat{H}_y = \frac{E_m^+}{\eta} e^{-\alpha z} e^{-j\beta z} - \frac{E_m^-}{\eta} e^{\alpha z} e^{j\beta z} \tag{2.32}$$

The corresponding time-domain solutions, in a *lossless medium*, are [Adamczyk, 2017b] uniform plane wave!time-domain solutions

$$E_x = E_m^+ \cos(\omega t - \beta z) + E_m^- \cos(\omega t + \beta z) \tag{2.33}$$

$$H_y = \frac{E_m^+}{\eta} \cos(\omega t - \beta z) - \frac{E_m^-}{\eta} \cos(\omega t + \beta z) \tag{2.34}$$

Note that

$$\cos(\omega t - \beta z) = \cos \omega \left(t - \frac{\beta}{\omega} z \right) = \cos \omega \left(t - \frac{z}{v} \right) = f \left(t - \frac{z}{v} \right) \tag{2.35}$$

$$\cos(\omega t + \beta z) = \cos \omega \left(t + \frac{\beta}{\omega} z \right) = \cos \omega \left(t + \frac{z}{v} \right) = g \left(t + \frac{z}{v} \right) \tag{2.36}$$

Thus, Eqs. (2.33) and (2.34) represent sinusoidal waves traveling in the +z, and −z directions, respectively!

Figure 2.3 shows a forward propagating EM wave in a lossless medium.

The wavelength λ is related to the velocity of propagation and frequency by

$$\lambda = \frac{v}{f} \tag{2.37}$$

The phase constant β is related to λ by

$$\beta = \frac{2\pi}{\lambda} \tag{2.38}$$

Multiplying both sides by z we obtain

$$\beta z = 2\pi \frac{z}{\lambda} \tag{2.39}$$

We refer to z as a *physical length* and to $\frac{z}{\lambda}$ as the *electrical length*.

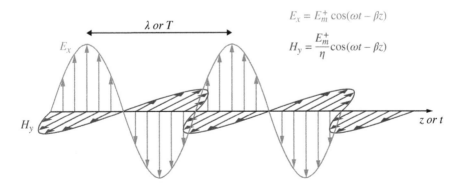

Figure 2.3   Sinusoidal EM wave in a lossless medium.

The time domain solutions in Eqs. (2.33) and (2.34) can now be written in terms of the electrical length as

$$E_x = E_m^+ \cos\left(\omega t - 2\pi\frac{z}{\lambda}\right) + E_m^- \cos\left(\omega t + 2\pi\frac{z}{\lambda}\right) \tag{2.40}$$

$$H_y = \frac{E_m^+}{\eta}\cos\left(\omega t - 2\pi\frac{z}{\lambda}\right) - \frac{E_m^-}{\eta}\cos\left(\omega t + 2\pi\frac{z}{\lambda}\right) \tag{2.41}$$

The definition of electrical length leads to the concept of the *electrically short structure*. Physical object is electrically short [Adamczyk and Gilbert, 2020] if its electrical length

$$\frac{z}{\lambda} \leq \frac{1}{10} \tag{2.42}$$

or equivalently, if its physical length

$$z \leq \frac{\lambda}{10} \tag{2.43}$$

If the physical object is electrically short then the lumped-parameter circuit models are an adequate representation of that object.

The EMC interference problem discussed in Section 2.2 is explained under the assumption of the electrically short structures, just defined.

2.2 EMC Interference Problem

Every EMC interference problem consists of three basic parts, shown in Figure 2.4:

- the interference source,
- the coupling mechanism,
- the susceptible system

There are three ways to prevent (minimize) interference:

- suppress the interference (noise) at the source,
- make the coupling mechanism inefficient,
- make the receptor circuit system less susceptible.

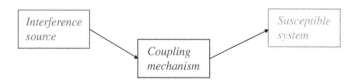

Figure 2.4   EMC interference problem.

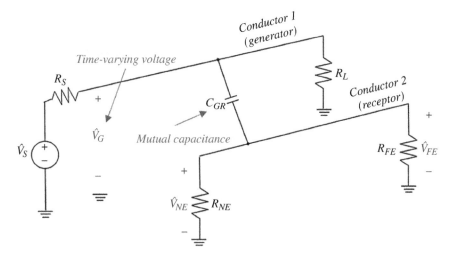

Figure 2.5   Capacitive coupling between circuits.

This chapter focuses on the coupling mechanisms between the interference source and the susceptible system.

There are four basic EMC coupling mechanisms:

- electric field coupling (capacitive coupling),
- magnetic field coupling (inductive coupling),
- common-impedance coupling (conducted coupling),
- electromagnetic field coupling (EM wave coupling).

For *capacitive coupling* to occur there must be:

1. a source of the electric field, i.e. a time-varying voltage dV_G/dt
2. a way to couple this field to another circuit, i.e. mutual capacitance C_{GR}.

This is shown in Figure 2.5.

For *inductive coupling* to occur there must be:

1. a source of the magnetic field, i.e. a time-varying current dI_G/dt
2. a way to couple this field to another circuit, i.e. mutual inductance L_{GR}.

This is shown in Figure 2.6.

The capacitive and inductive coupling may occur when the generator and receptor circuits are in the *near field* of each other. We often refer to the capacitive and inductive coupling occurring simultaneously, as *crosstalk* between circuits.

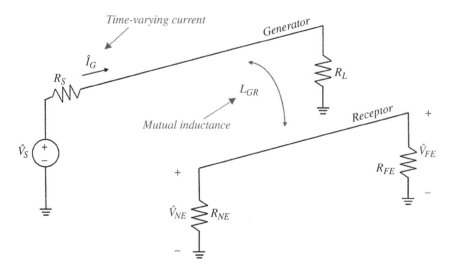

Figure 2.6   Inductive coupling between circuits.

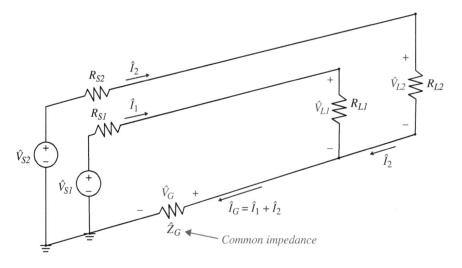

Figure 2.7   Common-impedance coupling between circuits.

Note: The circuit models for the capacitive and inductive coupling are derived for the case of *weak* coupling. The signals in the generator circuit induce signals in the receptor circuit but the induced signals in the receptor circuit do not induce signals back in the generator circuit.

For the *common-impedance coupling* to occur, two circuits must share a current path (with a non-negligible common impedance). This is shown in Figure 2.7.

The electromagnetic wave coupling may occur when the generator, i.e. the radiating circuit, and the receptor, i.e. the susceptible circuit, are in the *far field* of each other. This is shown in Figure 2.8.

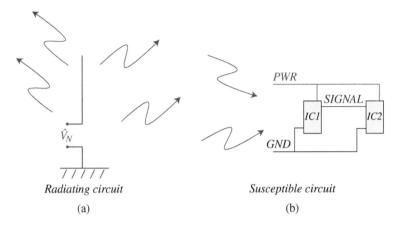

Figure 2.8 EM wave coupling between a radiating circuit (a) and the susceptible circuit (b).

2.3 Capacitive Coupling

The basic model for the capacitive coupling between two conductors (for *electrically short* structures) is shown in Figure 2.9.

$\hat{V}_G$ is the source of interference (generator circuit), C_{GR} is the mutual capacitance between the generator and the receptor circuits, $\hat{V}_{NE}$ is the near-end capacitively induced noise voltage (in the receptor circuit) across the near-end resistance R_{NE}, and $\hat{V}_{FE}$ is the far-end capacitively induced noise voltage across the far-end resistance R_{FE}.

Note: Since, in this section, we are considering only the capacitive coupling (that is, we are not including the inductive coupling) the near- and far-end resistances are in parallel,

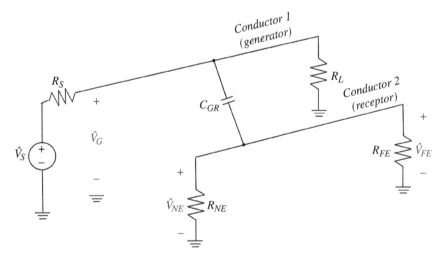

Figure 2.9 Basic capacitive-coupling circuit model.

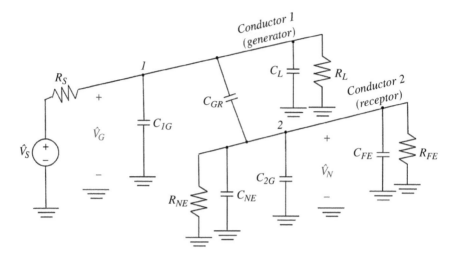

Figure 2.10   Capacitive-coupling circuit model with capacitances.

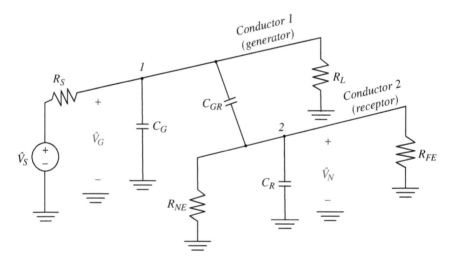

Figure 2.11   Simplified circuit model.

and thus, the near- and far-end voltages are equal, and subsequently will be denoted simply by $\hat{V}_N$.

More detailed circuit model, showing all possible capacitances, is shown in Figure 2.10.

C_{1G} is the capacitance between the generator circuit and ground, C_L is the load capacitance in the generator circuit, C_{2G} is the capacitance between the receptor circuit and ground, C_{NE} is the near-end capacitance in the receptor circuit, and C_{FE} is the far-end capacitance in the receptor circuit.

A simplified circuit model is shown in Figure 2.11.

Figure 2.12 shows the circuit representation of the model shown in Figure 2.11.

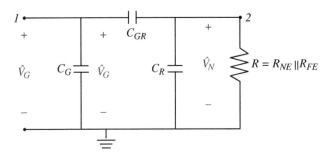

Figure 2.12 Circuit representation.

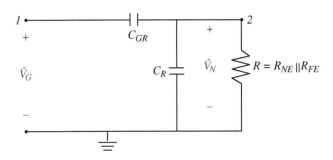

Figure 2.13 Equivalent circuit representation.

C_G is the total capacitance between the generator circuit and ground, and C_R is the total capacitance between the receptor circuit and ground.

Note that the capacitance C_G has no effect on the noise voltage $\hat{V}_N$ leading to the equivalent circuit representation shown in Figure 2.13.

The parallel combination of R and C_R in Figure 2.13 results in an impedance of

$$R\|C_R = \frac{\frac{R}{1+j\omega C_R}}{R + \frac{1}{j\omega C_R}} = \frac{R}{j\omega R C_R + 1} \tag{2.44}$$

The voltage divider produces

$$\hat{V}_N = \frac{\frac{R}{j\omega R C_R+1}}{\frac{1}{j\omega C_{GR}} + \frac{R}{j\omega R C_R+1}} \tag{2.45}$$

or

$$\hat{V}_N = \frac{j\omega R C_{GR}}{j\omega R(C_{GR} + C_R) + 1} \tag{2.46}$$

When

$$j\omega R(C_{GR} + C_R) \gg 1 \tag{2.47}$$

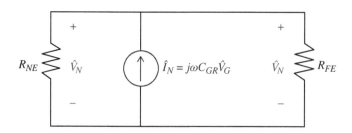

Figure 2.14 Capacitive coupling modeled as a shunt current source.

or equivalently the frequency ω satisfies the inequality (which is generally true for electrically short structures)

$$\omega \ll \frac{1}{R(C_{GR} + C_R)} \tag{2.48}$$

then the noise voltage induced in the receptor circuit approximately equals to

$$\hat{V}_N = j\omega R C_{GR} \hat{V}_G \tag{2.49}$$

Equation (2.49) can be alternatively written as

$$\hat{V}_N = R(j\omega C_{GR}\hat{V}_G) = R\hat{I}_N \tag{2.50}$$

Equation (2.50) shows that the capacitive coupling can be modeled as a shunt current source in the receptor circuit (for electrically short structures). This is shown in Figure 2.14. Figure 2.14 reveals that the noise voltage $\hat{V}_N$ can be reduced by [Ott, 2009]:

— lowering the shunt resistance, R,
— reducing the frequency ω present in the generator signal,
— lowering the amplitude of the generator signal $\hat{V}_G$,
— decreasing the mutual capacitance C_{GR}.

The mutual capacitance C_{GR} can be reduced by:

— moving the conductors further apart,
— changing their orientation,
— shielding.

2.3.1 Shielding to Reduce Capacitive Coupling

The generator–receptor circuit configuration, without a shield, is shown in Figure 2.15.

Let's place a *grounded* metallic shield over the receptor circuit, as shown in Figure 2.16. *Note*: If the shield is not grounded, it has no effect on the capacitive coupling [Ott, 2009].

Since in any practical circuit the receptor conductor extends beyond the shield, we still have the mutual capacitance, C_{GR}, between the generator and receptor circuits, as well as the capacitance, C_R, between the receptor circuit and ground.

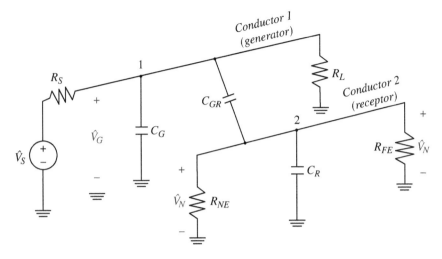

Figure 2.15 Capacitive coupling – receptor circuit without a shield.

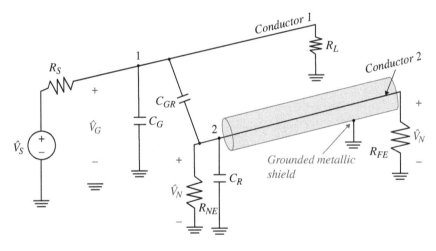

Figure 2.16 Capacitive coupling – receptor circuit with a shield.

Since the shield is metallic, it introduces additional capacitances, C_{GS}, between the generator and the shield, and C_{RS}, between the receptor and the shield. Since the shield is grounded, the capacitance between it and ground is zero. All relevant capacitances are shown in Figure 2.17.

The equivalent circuit representation of the shielded configuration is shown in Figure 2.18.

Following the derivations similar to the unshielded case results in the noise voltage given by

$$\hat{V}_N = \frac{j\omega R C_{GR}}{j\omega R(C_{GR} + C_R + C_{RS}) + 1} \tag{2.51}$$

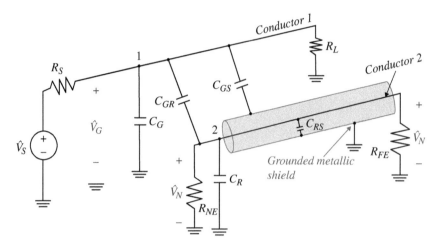

Figure 2.17   Additional capacitances introduced by the shield.

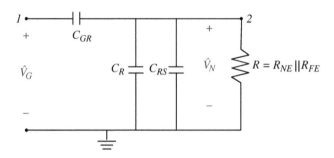

Figure 2.18   Equivalent circuit representation.

When

$$j\omega R(C_{GR} + C_R + C_{RS}) \ll 1 \tag{2.52}$$

or equivalently the frequency ω satisfies the inequality (which is generally true for electrically short structures)

$$\omega \ll \frac{1}{R(C_{GR} + C_R + C_{RS})} \tag{2.53}$$

then the noise voltage induced in the receptor circuit approximately equals to

$$\hat{V}_N = j\omega R C_{GR} \hat{V}_G \tag{2.54}$$

Comparison of Eq. (2.49) with Eq. (2.54) reveals that they look exactly the same! The difference is that the mutual capacitance, C_{GR}, is greatly reduced when the grounded shield is placed over the receptor circuit.

Final conclusion: When a shield over the receptor circuit is implemented to reduce capacitive coupling between circuits, it must be grounded at least at one location (for electrically short structures).

2.4 Inductive Coupling

Consider the generator and receptor circuits shown in Figure 2.19.

The time-varying current i_G flowing in the generator circuit gives rise to the time-varying magnetic field $\mathbf{H}_G$. The time-varying magnetic field $\mathbf{H}_G$, in turn, creates the time-varying magnetic flux Ψ_{GR} that crosses the adjacent receptor circuit.

This time-varying flux crossing the receptor circuit induces a voltage $v_{ind}(t)$ in the receptor circuit, as shown in Figure 2.20.

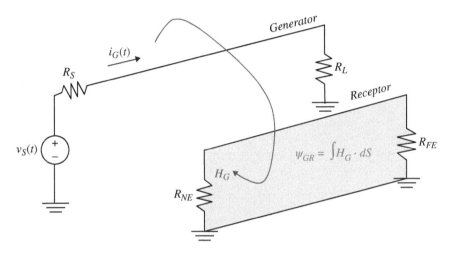

Figure 2.19   Inductive-coupling circuit model.

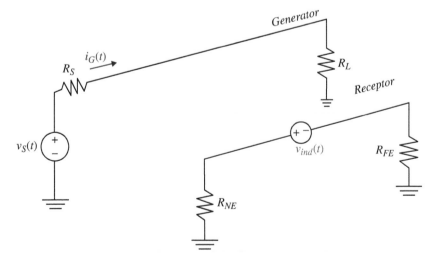

Figure 2.20   Induced voltage in the receptor circuit.

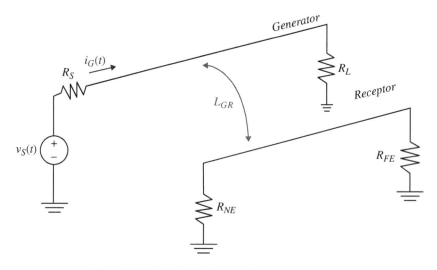

Figure 2.21   Mutual inductance between the circuits.

According to Faraday's law, the induced voltage is given by

$$v_{ind}(t) = \frac{d\Psi_{GR}}{dt} \tag{2.55}$$

To obtain a circuit model of this induced voltage we introduce the concept of the *mutual inductance* L_{GR} between the generator and the receptor circuit. This is shown in Figure 2.21.

The mutual inductance L_{GR} is defined as the ratio of the magnetic flux Ψ_{GR} crossing the receptor circuit to the current i_G, in the generator circuit, that gave rise to this flux.

$$L_{GR} = \frac{\Psi_{GR}}{i_G} \tag{2.56}$$

Using Eq. (2.56) we can express the magnetic flux as

$$\Psi_{GR} = L_{GR}i_G \tag{2.57}$$

Combining Eqs. (2.55) and (2.57) we obtain the time-domain induced voltage v_{ind} as

$$v_{ind}(t) = L_{GR}\frac{di_G}{dt} \tag{2.58}$$

In sinusoidal steady state, Eq. (2.58) becomes

$$\hat{V}_{ind} = j\omega L_{GR}\hat{I}_G \tag{2.59}$$

The frequency domain, inductive-coupling circuit model is shown in Figure 2.22.

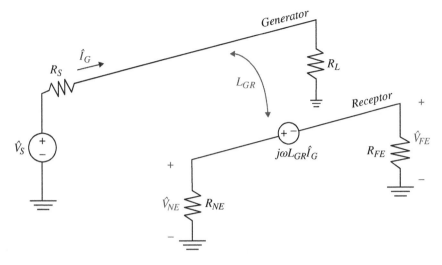

Figure 2.22   Frequency-domain circuit model.

The induced voltage will have the effect of creating the near-end, $\hat{V}_{NE}$, and the far-end, $\hat{V}_{FE}$, noise voltages in the receptor circuit. Using the voltage divider, these voltages are given by

$$\hat{V}_{NE} = \frac{R_{NE}}{R_{NE} + R_{FE}} j\omega L_{GR}\hat{I}_G \tag{2.60}$$

$$\hat{V}_{FE} = -\frac{R_{FE}}{R_{NE} + R_{FE}} j\omega L_{GR}\hat{I}_G \tag{2.61}$$

If the termination resistances in the receptor circuits are fixed, then noise voltages $\hat{V}_{NE}$ and $\hat{V}_{FE}$ can be reduced by [Ott, 2009]:

- reducing the frequency ω present in the generator signal,
- lowering the amplitude of the generator signal $\hat{I}_G$,
- decreasing the mutual inductance L_{GR}.

The mutual inductance L_{GR} can be reduced by:

- reducing the area of the receptor circuit,
- changing the conductors orientation,
- shielding.

2.4.1 Shielding to Reduce Inductive Coupling

The generator–receptor circuit configuration, without a shield, is shown in Figure 2.23.

Let's place a non-magnetic metallic shield, *grounded at both ends*, over the receptor circuit, as shown in Figure 2.24. *Note*: If the shield is not grounded at both ends, it has no effect on the inductive coupling [Ott, 2009].

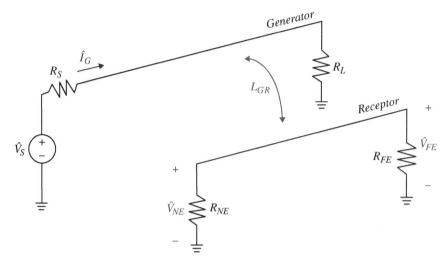

Figure 2.23   Inductive coupling – receptor circuit without a shield.

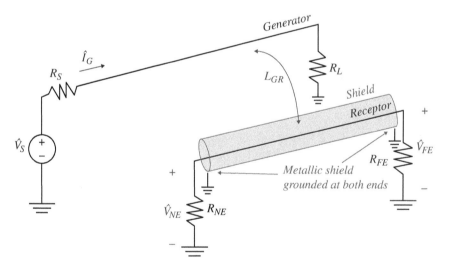

Figure 2.24   Inductive coupling – receptor circuit with a shield.

The addition of the shield introduces additional inductances, L_{GS}, between the generator and the shield, and L_{RS}, between the receptor and the shield. This is shown in Figure 2.25.

Since the shield is non-magnetic, it has no effect on the magnetic properties of the medium between the generator and the receptor circuit. Thus, the magnetic flux produced by the current $\hat{I}_G$ in the generator circuit will cross the receptor-ground circuit, in the same manner as it did when the shield was not present. This is shown in Figure 2.26.

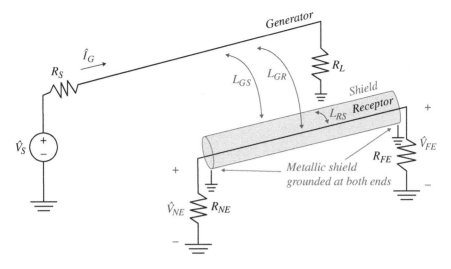

Figure 2.25   Additional mutual inductances.

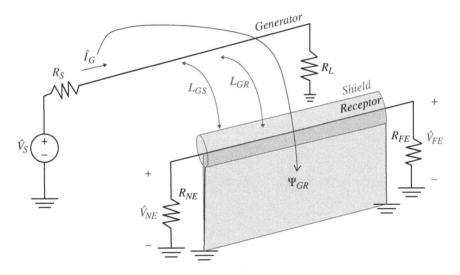

Figure 2.26   Magnetic flux crossing the receptor-ground circuit.

This time-varying flux induces a voltage in the receptor circuit in the same manner as it did when the shield was not present. This is shown in Figure 2.27.

The magnetic flux produced by the current $\hat{I}_G$ in the generator circuit also crosses the shield-ground circuit, as shown in Figure 2.28.

This magnetic flux induces a voltage in the shield circuit, as shown in Figure 2.29. R_{SH} is the shield resistance and L_{SH} is the shield inductance. The voltage induced in the shield circuit produces a shield current $\hat{I}_S$ as shown.

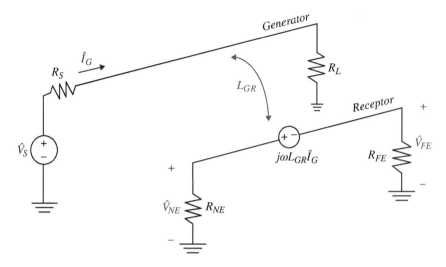

Figure 2.27   Voltage induced in the receptor circuit.

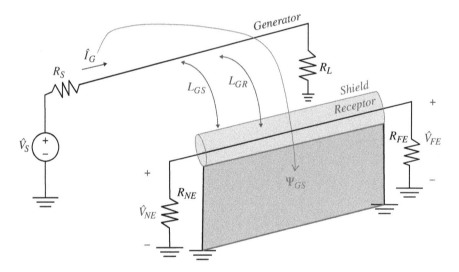

Figure 2.28   Magnetic flux crossing the shield-ground circuit.

The shield current gives rise to the magnetic flux Ψ_{SR} that crosses the shield-receptor circuit, as shown in Figure 2.30.

This flux, in turn, induces another voltage source in the receptor circuit, as shown in Figure 2.31.

The near- and far-end voltages induced in the receptor circuit are now given by

$$\hat{V}_{NE} = \frac{R_{NE}}{R_{NE} + R_{FE}} j\omega(L_{GR}\hat{I}_G - L_{RS}\hat{I}_S) \tag{2.62}$$

$$\hat{V}_{FE} = -\frac{R_{FE}}{R_{NE} + R_{FE}} j\omega(L_{GR}\hat{I}_G - L_{RS}\hat{I}_S) \tag{2.63}$$

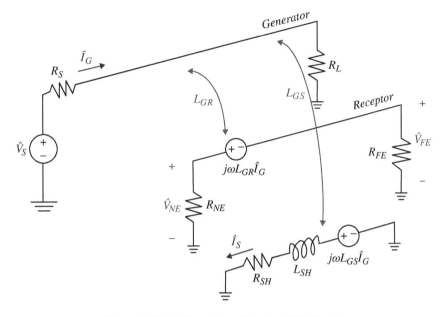

Figure 2.29   Voltage induced in the shield circuit.

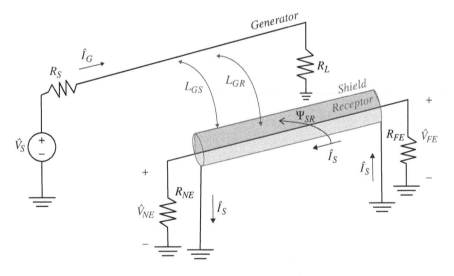

Figure 2.30   Magnetic flux crossing the shield-receptor circuit.

Equations (2.62) and (2.63) clearly show that the shield current $\hat{I}_S$ reduced the voltages induced in the receptor circuit. Note that the shield must be grounded at both ends in order to provide a path for this current. If the shield is not grounded or grounded at one end only, the current cannot flow on the shield and the shield has no effect on the noise voltages induced in the receptor circuit.

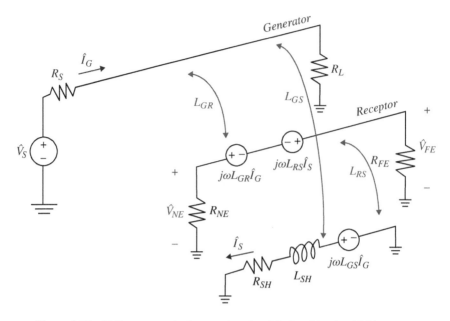

Figure 2.31 Voltage source in the receptor circuit induced by the shield current.

2.5 Crosstalk Between PCB Traces

When two circuits are in the vicinity of one another, a signal propagating in one circuit can induce a signal in another circuit, due to capacitive (electric field) and inductive (magnetic field) coupling between the circuits. This phenomenon is referred to as *crosstalk*. An example is of such arrangement is shown in Figure 2.32 [Adamczyk, 2017a].

Two PCB traces in a microstrip configuration are separated from each other by distance s and from the ground plane (which is return conductor for both) by a distance d. The first trace (generator conductor) is driven by a time-varying voltage source V_S with the impedance R_S, and terminated by a load resistor R_L. The second trace (receptor) is terminated by the load resistors, R_{NE} and R_{FE}, on the near end and the far end, respectively. This arrangement can be modeled by the circuit shown in Figure 2.33.

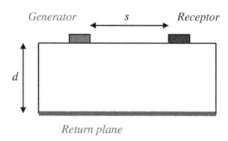

Figure 2.32 Microstrip line PCB configuration.

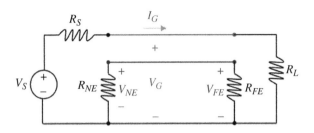

Figure 2.33 Circuit model of a microstrip line.

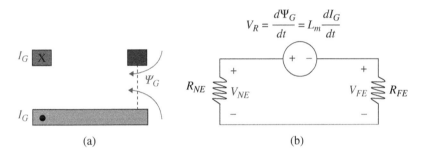

(a)   (b)

Figure 2.34 Inductive coupling between the circuits (a) field model and (b) circuit model.

The signal present on the generator line, V_G and I_G induces the near-end and far-end crosstalk-coupled voltages, V_{NE} and V_{FE}.

The current on the generator line, I_G, creates a magnetic field that results in a magnetic flux Ψ_G crossing the loop of the receptor circuit, as shown in Figure 2.34a [Adamczyk, 2017b].

If this flux is time varying, then according to Faraday's law, it induces a voltage V_R in the receptor circuit. The circuit model of this field phenomenon is represented by a mutual inductance, L_m, and is shown in Figure 2.34b.

Using the voltage divider we obtain the near- and far-end induced voltages as

$$V_{NE}(t) = \frac{R_{NE}}{R_{NE} + R_{FE}} L_m \frac{dI_G}{dt} \tag{2.64}$$

$$V_{FE}(t) = -\frac{R_{FE}}{R_{NE} + R_{FE}} L_m \frac{dI_G}{dt} \tag{2.65}$$

Similarly, the voltage between the two conductors of the generator circuit, V_G, has an associated with it charge separation, which creates the electric field lines, some of which terminate on the conductors of the receptor circuit as shown in Figure 2.35a.

If this charge (voltage) varies with time, it induces a current in the receptor circuit. The circuit model of this field phenomenon is represented by a mutual capacitance, C_m, and is shown in Figure 2.35b.

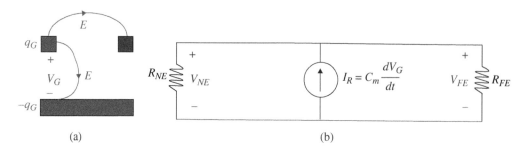

Figure 2.35 Capacitive coupling between the circuits (a) field model and (b) circuit model.

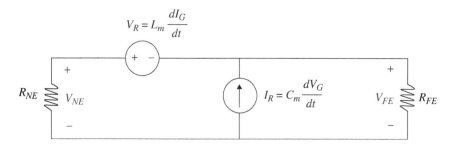

Figure 2.36 Inductive and capacitive coupling circuit model.

Using the current divider and Ohm's law, we obtain the near- and far-end induced voltages as

$$V_{NE}(t) = \frac{R_{NE}R_{FE}}{R_{NE} + R_{FE}} C_m \frac{dV_G}{dt} \tag{2.66}$$

$$V_{FE}(t) = \frac{R_{NE}R_{FE}}{R_{NE} + R_{FE}} C_m \frac{dV_G}{dt} \tag{2.67}$$

Superposition of these two types of coupling results in the circuit model shown in Figure 2.36.

The total induced voltages, by superposition, are given by

$$V_{NE}(t) = \frac{R_{NE}}{R_{NE} + R_{FE}} L_m \frac{dI_G}{dt} + \frac{R_{NE}R_{FE}}{R_{NE} + R_{FE}} C_m \frac{dV_G}{dt} \tag{2.68}$$

$$V_{FE}(t) = -\frac{R_{FE}}{R_{NE} + R_{FE}} L_m \frac{dI_G}{dt} + \frac{R_{NE}R_{FE}}{R_{NE} + R_{FE}} C_m \frac{dV_G}{dt} \tag{2.69}$$

If the circuit is electrically small at the highest significant frequency of interest then the generator voltage and current can be obtained from the circuit shown in Figure 2.37.

Then,

$$V_G \simeq \frac{R_L}{R_S + R_L} V_S(t) \tag{2.70}$$

$$I_G \simeq \frac{V_S(t)}{R_S + R_L} \tag{2.71}$$

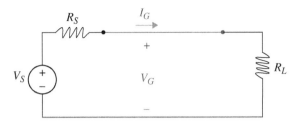

Figure 2.37 Electrically short generator circuit model.

Substituting Eqs. (2.70) and (2.71) into Eqs. (2.68) and (2.69) results in

$$V_{NE}(t) = \left[\underbrace{\frac{R_{NE}}{R_{NE} + R_{FE}} L_m \frac{1}{R_S + R_L}}_{\text{Inductive Coupling}} + \underbrace{\frac{R_{NE}R_{FE}}{R_{NE} + R_{FE}} C_m \frac{R_L}{R_S + R_L}}_{\text{Capacitive Coupling}} \right] \frac{dV_S(t)}{dt} \qquad (2.72)$$

$$V_{FE}(t) = \left[-\underbrace{\frac{R_{FE}}{R_{NE} + R_{FE}} L_m \frac{1}{R_S + R_L}}_{\text{Inductive Coupling}} + \underbrace{\frac{R_{NE}R_{FE}}{R_{NE} + R_{FE}} C_m \frac{R_L}{R_S + R_L}}_{\text{Capacitive Coupling}} \right] \frac{dV_S(t)}{dt} \qquad (2.73)$$

Note that the induced crosstalk voltages are proportional to the mutual inductance and capacitance between the two circuits and the derivative of the source voltage.

The crosstalk circuit model in the frequency domain is shown in Figure 2.38.

From this equivalent circuit in the frequency domain, or directly from Eqs. (2.72) and (2.73) we obtain the near-end and far-end phasor crosstalk voltages as

$$V_{NE}(t) = \left[\underbrace{\frac{R_{NE}}{R_{NE} + R_{FE}} L_m \frac{1}{R_S + R_L}}_{\text{Inductive Coupling}} + \underbrace{\frac{R_{NE}R_{FE}}{R_{NE} + R_{FE}} C_m \frac{R_L}{R_S + R_L}}_{\text{Capacitive Coupling}} \right] j\omega \hat{V}_S \qquad (2.74)$$

$$V_{FE}(t) = \left[-\underbrace{\frac{R_{FE}}{R_{NE} + R_{FE}} L_m \frac{1}{R_S + R_L}}_{\text{Inductive Coupling}} + \underbrace{\frac{R_{NE}R_{FE}}{R_{NE} + R_{FE}} C_m \frac{R_L}{R_S + R_L}}_{\text{Capacitive Coupling}} \right] j\omega \hat{V}_S \qquad (2.75)$$

Note that the crosstalk induced voltages increase at a rate of 20 dB/decade with frequency.

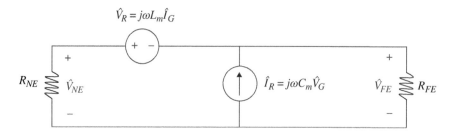

Figure 2.38   Inductive and capacitive coupling circuit model in the frequency domain.

2.6 Common-Impedance Coupling

For common-impedance coupling to occur, two circuits must share a current path (with a non-negligible impedance) [Adamczyk, 2017b]. Before we discuss common-impedance coupling let's consider a couple of scenarios where common-impedance coupling does not occur.

Consider the circuit shown in Figure 2.39.

The current flows from the source to the load along the forward path and returns to the source through a zero-impedance ground, or return path.

The voltage at the load (with respect to the reference ground) is

$$\hat{V}_L = R_L \hat{I} \tag{2.76}$$

Next, let's consider the case where the return path has a non-zero impedance, as shown in Figure 2.40.

Now the voltage at the load (with respect to the ground node) is

$$\hat{V}_L = R_L \hat{I} + \hat{Z}_G \hat{I} \tag{2.77}$$

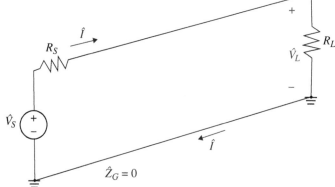

Figure 2.39   Current returns to the source through a zero-impedance ground path.

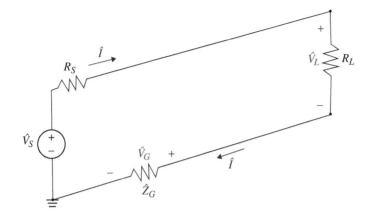

Figure 2.40 Current returns to the source through a non-zero impedance ground.

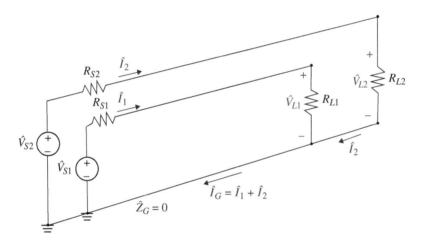

Figure 2.41 Two circuits share a zero-impedance return path.

Obviously the ground impedance, $\hat{Z}_G$, affects the value of the load voltage, but no other circuit influences this value or is impacted by this ground impedance – there is no common-impedance coupling (since there is no other circuit to be coupled to).

Next, consider the situation shown in Figure 2.41 where two circuits share the return path with zero impedance.

The voltages at the loads are

$$\hat{V}_{L1} = R_{L1}\hat{I}_1 \tag{2.78}$$

$$\hat{V}_{L2} = R_{L2}\hat{I}_2 \tag{2.79}$$

Even though both circuits share the return path, the load voltage of circuit 1, $\hat{V}_{L1}$, is not affected by the return current of circuit 2, $\hat{I}_2$; similarly, the load voltage of circuit 2, $\hat{V}_{L2}$, is

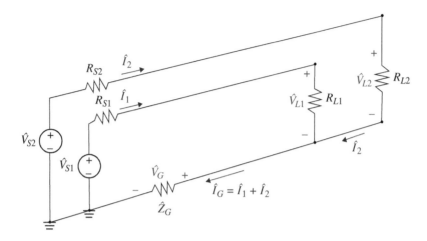

Figure 2.42 Common-impedance coupling circuit.

not affected by the return current of circuit 1, $\hat{I}_1$. There is no impedance coupling between the circuits (since there is no common impedance shared by both circuits).

Finally, consider the situation shown in Figure 2.42 where two circuits share the return path with a non-zero impedance.

The voltages at the loads are

$$\hat{V}_{L1} = R_{L1}\hat{I}_1 + \hat{Z}_G(\hat{I}_1 + \hat{I}_2) \tag{2.80}$$

$$\hat{V}_{L2} = R_{L2}\hat{I}_2 + \hat{Z}_G(\hat{I}_1 + \hat{I}_2) \tag{2.81}$$

Now the load voltage of circuit 1, $\hat{V}_{L1}$, is affected by the return current of circuit 2, $\hat{I}_2$; similarly, the load voltage of circuit 2, $\hat{V}_{L2}$, is affected by the return current of circuit 1, $\hat{I}_1$.

This type of coupling is called the *common-impedance coupling*.

Common-impedance coupling becomes an EMC problem when two or more circuits share a common path (common ground in this example) and one or more of the following conditions exist [Adamczyk and Teune, 2018]: (i) a high-impedance ground (at high frequency: too much inductance; at low frequency: too much resistance), (ii) a large ground current, and (iii) a very sensitive, low-noise margin circuit, sharing the return path with other circuits.

Note: A similar situation occurs when the circuits share a common forward path.

2.7 Laboratory Exercises

2.7.1 Crosstalk Between PCB Traces

Objective: This laboratory exercise examines the crosstalk between PCB traces and investigates the impact of signal parameters and PCB topology on the induced noise voltages. It also examines the impact of a guard trace acting as a shield to reduce the induced noise voltages.

Note: This laboratory requires a custom PCB [Aldridge, 2023], for which the Altium files, together with the Bill of Material (BOM), are provided.

2.7.1.1 Laboratory Equipment and Supplies

Function generator: Tektronix AFG3252 (or equivalent), shown in Figure 2.43.
  Oscilloscope: Tektronix MDO3104 (or equivalent), shown in Figure 2.44.
  PCB: Custom PCB – shown in Figure 2.45.

RG58C coaxial cable assembly, shown in Figure 2.46. BNC Male Plug to BNC Male Plug,
  length 4 feet
Supplier: Pomona Electronics, Part Number: 2249-C
Digi-Key Part Number: 501-1019-ND

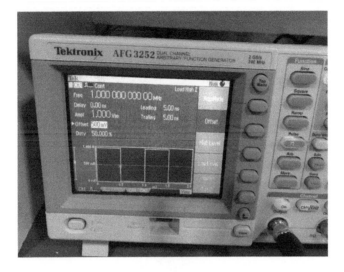

Figure 2.43   Laboratory equipment: function generator.

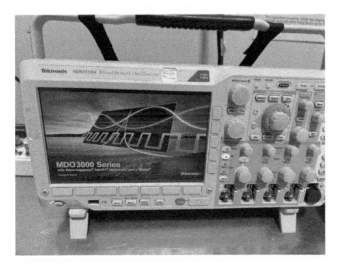

Figure 2.44   Laboratory equipment: oscilloscope.

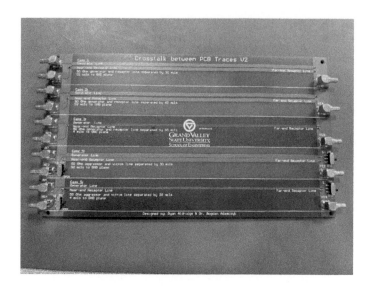

Figure 2.45   Laboratory equipment: custom PCB.

Figure 2.46   Laboratory equipment: coaxial cable.

Coaxial cable assembly (3 sets), shown in Figure 2.47. BNC Male Plug to SMB Male Plug,
   length 3 feet
Supplier: Amphenol RF, Product Number: 095-850-236-036
Digi-Key Part Number: 115-095-850-236-036-ND

Adapter Coaxial Connector BNC Plug, Male Pin To BNC Jack, Female Socket:
Digi-Key Part Number ACX1064-ND, shown in Figure 2.48.

2.7.1.2 Laboratory Procedure

Connect the function generator, PCB, and the oscilloscope, as shown in Figure 2.49.
   Details of the connections for each Case: Attach the Adapter Coaxial *T* Connector to
the output of the function generator. Connect one side of the adapter to channel 1 of
the oscilloscope. Connect the other side of the adapter to the near-end generator-circuit

Figure 2.47 Laboratory equipment: coaxial cable assembly.

Figure 2.48 Laboratory equipment: adapter coaxial connector.

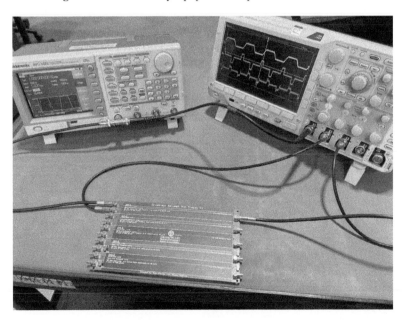

Figure 2.49 Laboratory setup.

connector on the PCB. Connect the near end of the receptor PCB connector to channel 2 of the oscilloscope. Connect the far end of the receptor PCB connector to channel 3 of the oscilloscope.

Initial Function Generator settings: Function = pulse, frequency = 1 MHz, amplitude = 5 Vpp (offset = 2.5 V, duty cycle = 50%, rise time = 100 ns, fall time = 200 ns).

Note: Tabulate all results as you proceed through the sequence of steps.

Step 1: Make the connections corresponding to Case 1. Measure the receptor circuit crosstalk induced voltages at the near end and far end, on the rising and falling edge of the generator signal. Decrease the rise time to 50 ns and the fall time to 100 ns. Repeat the measurements. Comment on the results.

Step 2: Make the connections corresponding to Case 2. Set the rise time = 100 ns, fall time = 200 ns. Repeat the measurements from previous step. Comment on the results.

Step 3: Make the connections corresponding to Case 3. Set the rise time = 100 ns, fall time = 200 ns. Repeat the measurements from previous step. Comment on the results.

Step 4: (a) Make the connections corresponding to Case 4. Set the near end and the far end; set the switches to OPEN. Set the rise time = 100 ns, fall time = 200 ns. Repeat the measurements from previous step. Compare your results to those from Step 1 and comment on the results.

(b) At near end, move the switch to GND, at far end leave the switch in the OPEN position. Repeat the measurements from Step 4a. Comment on the guard trace impact.

(c) At near end, move the switch to OPEN, at far end move the switch to GND. Repeat the measurements from Step 4a. Comment on the guard trace impact.

(d) At near end, move the switch to GND, at far end leave the switch in the GND position. Repeat the measurements from Step 4a. Comment on the guard trace impact.

Step 5: (a) Make the connections corresponding to Case 5. Repeat the measurements from previous step. Comment on the results.

References

Bogdan Adamczyk. Crosstalk Reduction between PCB Traces. *In Compliance Magazine*, March 2017a. URL https://incompliancemag.com/article/crosstalk-reduction-between-pcb-traces/.

Bogdan Adamczyk. *Foundations of Electromagnetic Compatibility*. John Wiley & Sons, Ltd, Chichester, UK, 2017b. ISBN 9781119120810. doi: https://doi.org/10.1002/9781119120810.

Bogdan Adamczyk and Brian Gilbert. Basic EMC Rules. *In Compliance Magazine*, May 2020. URL https://incompliancemag.com/article/basic-emc-rules/.

Bogdan Adamczyk and Jim Teune. Common-Impedance Coupling Between Circuits. *In Compliance Magazine*, January 2018. URL https://incompliancemag.com/article/common-impedance-coupling-between-circuits/.

Ryan Aldridge. Crosstalk PCB Designer, 2023.

Henry W. Ott. *Electromagnetic Compatibility Engineering*. John Wiley & Sons, Inc., Hoboken, New Jersey, 2009. ISBN 978-0-470-18930-6.

Chapter 3: Non-Ideal Behavior of Passive Components

In this chapter we investigate the impedance of the three standard passive circuit components (R, L, and C) as well as the frequency response of a PCB trace. It is shown [Adamczyk, 2019b], that a PCB trace can be modeled as a resonant RLC network just like the standard components, and effectively constitutes the fourth circuit component whose frequency behavior directly affects the impedance of the resistors, inductors, and capacitors on a PCB.

Actual circuit components exhibit non-ideal effects that can be modeled by augmenting the ideal model with a parasitic resistance, inductance, and capacitance. The parasitics for three standard components (R, L, and C) considered in this section come from the component itself and not from the connecting traces. Parasitics associated with a PCB trace considered in this section come from the trace itself and not from the other circuit components.

3.1 Resonance in RLC Circuits

The concept of resonance is of paramount importance in EMC. Resonance (and antiresonance) manifests itself in just about any EMC measurement. This section discusses the fundamental circuit background underlying the resonance study and presents several EMC examples of resonant circuits. These include the non-ideal models of a capacitor, ferrite bead, resistor, and an inductor.

In circuit courses, the study of resonance is usually limited to the two classical second-order circuits, series, and parallel RLC configurations. These circuits, shown in Figure 3.1, contain a single lumped capacitor and a single lumped inductor connected either "purely" in series or "purely" in parallel.

Actual circuits, in addition to the intentional discrete reactive components, contain several distributed parasitic inductances and capacitances. Even though these circuits are non-linear and of a higher order than two, the study of the basic RLC configurations provides an insight into the more complex topologies and their behavior. Let's begin with a series RLC circuit.

3.1.1 "Pure" Series Resonance – Non-Ideal Capacitor Model

Consider the series RLC resonant circuit shown in Figure 3.2.

Since the study of resonance is performed in the sinusoidal steady-state, the voltage and current, in Figure 3.2, are shown in the phasor forms [Adamczyk, 2022], and the component values are replaced by their impedances.

Principles of Electromagnetic Compatibility: Laboratory Exercises and Lectures, First Edition. Bogdan Adamczyk.
© 2024 John Wiley & Sons Ltd. Published 2024 by John Wiley & Sons Ltd.
Companion website: www.wiley.com/go/principlesofelectromagneticcompatibility

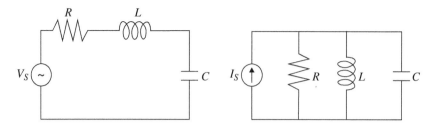

Figure 3.1   The "classical" series and parallel *RLC* resonant circuits.

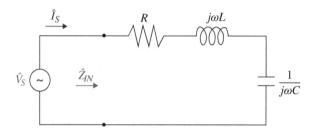

Figure 3.2   The series *RLC* resonant circuit.

In order to introduce the concept of resonance [Adamczyk, 2021], let's calculate the input impedance to the circuit.

$$\hat{Z}_{IN} = \frac{\hat{V}_S}{\hat{I}_S} = R + j\omega L + \frac{1}{j\omega C} \tag{3.1}$$

or

$$\hat{Z}_{IN} = \frac{\hat{V}_S}{\hat{I}_S} = R + j\omega L - j\frac{1}{\omega C} = R + j\left(\omega L - \frac{1}{\omega C}\right) \tag{3.2}$$

When

$$\omega L - \frac{1}{\omega C} = 0 \tag{3.3}$$

the input impedance is purely real

$$\hat{Z}_{IN} = R \tag{3.4}$$

This happens at the frequency

$$\omega_0 = \frac{1}{\sqrt{LC}} \tag{3.5}$$

From the circuit theory we recognize this frequency as the undamped natural frequency.

Let's look at the consequences of the fact that the input impedance is purely real at that frequency. The voltage and current phasors can be expressed in terms of their magnitudes and angles as

$$\hat{V}_S = V\angle\theta_v \tag{3.6}$$

$$\hat{I}_S = I\angle\theta_i \tag{3.7}$$

Consequently, the input impedance in Eq. (3.1) can be written as

$$\hat{Z}_{IN} = \frac{\hat{V}_S}{\hat{I}_S} = \frac{V\angle\theta_v}{I\angle\theta_i} = \frac{V_S}{I_S}\angle(\theta_v - \theta_i) \qquad (3.8)$$

Since, at ω_0, the input impedance is purely real, it follows that

$$\theta_v - \theta_i = 0 \qquad (3.9)$$

or

$$\theta_v = \theta_i \qquad (3.10)$$

Thus, at ω_0, the voltage and current are in phase! We have arrived at the definition of the resonant frequency:

The resonant frequency, ω_r, is a frequency at which the voltage and current phasors are in phase (with respect to the same two terminals of the circuit).

Thus, for the series *RLC* circuit, the resonant frequency is the same as the undamped natural frequency.

$$\omega_r = \omega_0 = \frac{1}{\sqrt{LC}} \qquad (3.11)$$

Note: (i) Not every *RLC* circuit is resonant. (ii) When the circuit is resonant, its resonant frequency, in general, is different from ω_0. (iii) The "classical"series and parallel circuit configurations are always resonant, and their resonant frequency is the same as ω_0.

There are few other interesting phenomena taking place at the resonant frequency. Let's start with the magnitude of the input impedance. From Eq. (3.2) we obtain this magnitude as

$$\left|\hat{Z}_{IN}\right| = Z_{IN} = \sqrt{R^2 + (\omega L - \omega C)^2} \qquad (3.12)$$

At resonant frequency we have

$$\left|\hat{Z}_{IN}\right| = Z_{IN} = \sqrt{R^2 + 0} = R \qquad (3.13)$$

Thus, at resonant frequency, the magnitude of the input impedance is minimum. Let's illustrate this using a circuit model of a non-ideal capacitor [Adamczyk, 2021], and plotting its input impedance. This is shown in Figure 3.3.

At the resonant frequency, the *LC* series combination acts as a short circuit, and the entire source voltage appears across the resistor *R*. Also, at resonant frequency we have

$$\hat{V}_L = V_L\angle 90° \qquad (3.14)$$

$$\hat{V}_C = V_C\angle -90° \qquad (3.15)$$

$$V_L = V_C = V \qquad (3.16)$$

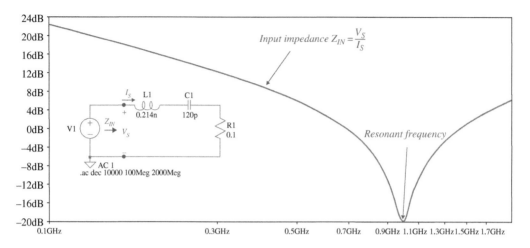

Figure 3.3   Series *RLC* circuit – input impedance is minimal at the resonant frequency.

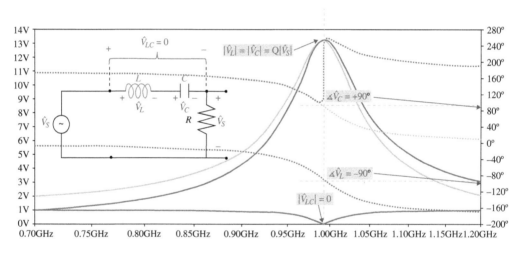

Figure 3.4   Component voltages at resonance.

That is, at resonance, the magnitudes of the capacitor and inductor voltages are equal and opposite in phase. Thus, at resonance,

$$\hat{V}_{LC} = \hat{V}_L + \hat{V}_C = V\angle 90° + V\angle - 90° = 0 \qquad (3.17)$$

This is illustrated in Figure 3.4.

The magnitudes of the inductor and capacitor voltages at resonance are given by

$$\left|\hat{V}_L\right| = \left|\hat{V}_C\right| = Q\left|\hat{V}_S\right| \qquad (3.18)$$

where Q is the quality factor of the circuit [Adamczyk, 2017], given by

$$Q = \frac{1}{R}\sqrt{\frac{L}{C}} \tag{3.19}$$

This factor in a series RLC circuit can be many times larger than 1 (in our case $Q = 13.22$), and thus the capacitor and inductor voltages can exceed the input voltage. This is especially dangerous in high-Q circuits where this voltage might be destructive to a capacitor (inductors handle high voltages much better).

3.1.2 "Pure" Parallel Resonance – Ferrite Bead Model

Consider the parallel RLC resonant circuit shown in Figure 3.5.
Let's calculate the input impedance to this circuit.

$$\frac{1}{\hat{Z}_{IN}} = \frac{1}{R} + \frac{1}{j\omega L} + j\omega C \tag{3.20}$$

Thus

$$\hat{Z}_{IN} = \frac{1}{\frac{1}{R} + \frac{1}{j\omega L} + j\omega C} \tag{3.21}$$

When

$$\omega L - \frac{1}{\omega C} = 0 \tag{3.22}$$

the input impedance is purely real (voltage and current are in phase) and equal to

$$\hat{Z}_{IN} = R \tag{3.23}$$

This happens at the resonant frequency of

$$\omega_r = \omega_0 = \frac{1}{\sqrt{LC}} \tag{3.24}$$

Thus, at resonant frequency, the magnitude of the input impedance is maximum. Let's illustrate this using a circuit model of ferrite bead [Adamczyk, 2021], and plotting its input impedance. This is shown in Figure 3.6.

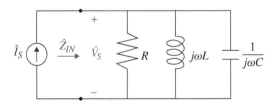

Figure 3.5   The parallel RLC resonant circuit.

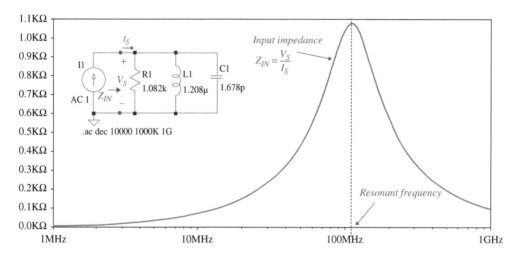

Figure 3.6 Parallel *RLC* circuit – input impedance is maximal at the resonant frequency.

At the resonant frequency, the *LC* series combination acts as an open circuit, and the entire source current flows through the resistor *R*. Also, at resonant frequency we have

$$\hat{I}_L = I_L \angle -90° \qquad (3.25)$$

$$\hat{I}_C = I_C \angle 90° \qquad (3.26)$$

$$I_L = I_C = I \qquad (3.27)$$

That is, at resonance, the magnitudes of the capacitor and inductor currents are equal and opposite in phase. Thus, at resonance,

$$\hat{I}_{LC} = \hat{I}_L + \hat{I}_C = I \angle -90° + I \angle 90° = 0 \qquad (3.28)$$

This is illustrated in Figure 3.7.

The magnitudes of the inductor and capacitor currents at resonance are given by

$$|\hat{I}_L| = |\hat{I}_C| = Q |\hat{I}_S| \qquad (3.29)$$

where *Q* is the quality factor of the circuit [Adamczyk, 2017], given by

$$Q = R \sqrt{\frac{C}{L}} \qquad (3.30)$$

This factor in a parallel *RLC* circuit can be many times larger than 1 (in our case *Q* = 1.27), and thus the capacitor and inductor currents can exceed the input current. This is especially dangerous in high-*Q* circuits where this current might be destructive to an inductor (capacitors handle high currents much better).

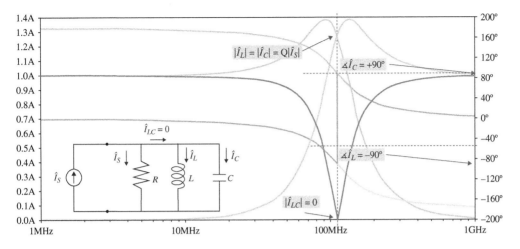

Figure 3.7 Component currents at resonance.

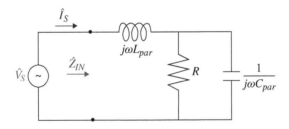

Figure 3.8 "Hybrid" series RLC resonant circuit.

3.1.3 "Hybrid" Series Resonance – Non-Ideal Resistor Model

Consider a variation of the series RLC resonant circuit shown in Figure 3.8.
Let's calculate the input impedance to this circuit.

$$\hat{Z}_{IN} = j\omega L + R \parallel C = j\omega L + \frac{\frac{R}{j\omega C}}{R + j\omega C} = j\omega L + \frac{R}{j\omega RC + 1} \tag{3.31}$$

or

$$\hat{Z}_{IN} = j\omega L + R \parallel C = j\omega L + \frac{\frac{R}{j\omega C}}{R + j\omega C} = j\omega L + \frac{R}{j\omega RC + 1} \tag{3.32}$$

or

$$\hat{Z}_{IN} = \frac{j\omega L(j\omega RC + 1) + R}{j\omega RC + 1} = \frac{R - \omega^2 LRC + j\omega L}{j\omega RC + 1} \tag{3.33}$$

Multiplying the numerator and denominator by the complex conjugate of the denominator, and separating the real and imaginary parts, results in

$$\hat{Z}_{IN} = \frac{R}{1 + \omega^2 R^2 C^2} + j\frac{\omega^3 L R^2 C^2 + \omega L - \omega R^2 C}{1 + \omega^2 R^2 C^2} \tag{3.34}$$

The input impedance is purely real (voltage and current are in phase) when the imaginary part of the impedance is zero, or

$$\omega_r^3 L R^2 C^2 + \omega_r L - \omega_r R^2 C = 0 \tag{3.35}$$

Leading to

$$\omega_r^2 = \frac{R^2 C - L}{L R^2 C^2} = \frac{1}{LC} - \frac{1}{R^2 C^2} \tag{3.36}$$

and therefore the resonant frequency of the circuit is

$$\omega_r = \sqrt{\frac{1}{LC} - \frac{1}{R^2 C^2}} \tag{3.37}$$

Substituting this value into Eq. (3.34) gives the value of the input impedance at resonance.

$$\hat{Z}_{IN}(\omega_r) = \left\{ \frac{R}{1 + \omega^2 R^2 C^2} + j\frac{\omega^3 L R^2 C^2 + \omega L - \omega R^2 C}{1 + \omega^2 R^2 C^2} \right\}_{\omega = \omega_r} \tag{3.38}$$

Since at resonant frequency the input impedance is real, we have

$$\hat{Z}_{IN}(\omega_r) = \left\{ \frac{R}{1 + \omega^2 R^2 C^2} \right\}_{\omega = \omega_r} \tag{3.39}$$

or

$$\hat{Z}_{IN}(\omega_r) = \frac{R}{1 + R^2 C^2 \left(\frac{1}{LC} - \frac{1}{R^2 C^2} \right)} = \frac{R}{1 + \frac{R^2 C}{L} - 1} \tag{3.40}$$

giving the value of the input impedance at resonance as

$$\hat{Z}_{IN}(\omega_r) = \frac{L}{RC} \tag{3.41}$$

Let's return to the resonant frequency of the circuit. From Eq. (3.36) this frequency can alternatively be expressed as

$$\omega_r = \frac{1}{\sqrt{LC}} \sqrt{\frac{R^2 C - L}{R^2 C}} = \omega_0 \sqrt{1 - \frac{L}{R^2 C}} \tag{3.42}$$

When

$$\frac{L}{R^2 C} < 1 \tag{3.43}$$

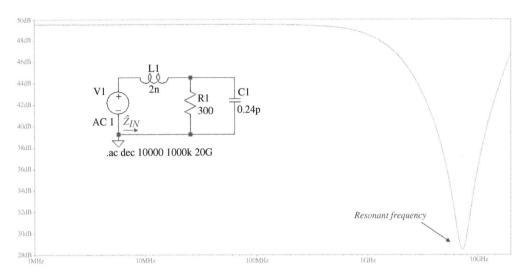

Figure 3.9 "Hybrid" series *RLC* circuit – input impedance of a non-ideal resistor.

the circuit is resonant and its resonant frequency is different from the undamped natural frequency.

To illustrate the circuit behavior at resonance, let's look at the input impedance of a non-ideal resistor. This is shown in Figure 3.9.

Note: The simulation model is Figure 3.9 is used only to show that the hybrid series *RLC* circuit with the component values shown is indeed resonant. This simple model is not valid beyond 2 GHz frequency, and thus the impedance plot and the value of the resonant frequency do not reflect the results that would be obtained from the laboratory measurements.

Component voltages are shown in Figure 3.10.

Again, at resonance the magnitudes of the capacitor and inductor voltages can exceed the magnitude of the input voltage.

3.1.4 "Hybrid" Parallel Resonance – Non-Ideal Inductor Model

Consider a variation of the parallel *RLC* resonant circuit shown in Figure 3.11.

This circuit corresponds to a non-ideal model of an inductor [Adamczyk, 2021]. Let's calculate the input impedance to this circuit.

$$\hat{Y}_{IN} = \frac{1}{\frac{j}{\omega C}} + \frac{1}{R + j\omega L} = j\omega C + \frac{1}{R + j\omega L} = j\omega C + \frac{R - j\omega L}{R^2 + (\omega L)^2} \tag{3.44}$$

or

$$\hat{Y}_{IN} = \frac{R}{R^2 + (\omega L)^2} + j\left[\omega C - \frac{\omega L}{R^2 + (\omega L)^2}\right] \tag{3.45}$$

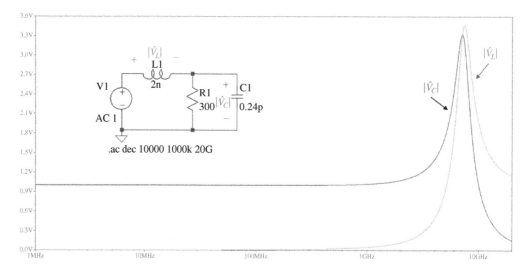

Figure 3.10 Component voltages at resonance.

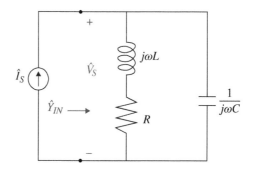

Figure 3.11 "Hybrid" parallel *RLC* resonant circuit.

Resonance occurs when the admittance is real, or

$$\omega_r C - \frac{\omega_r L}{R^2 + (\omega_r L)^2} = 0 \tag{3.46}$$

and thus

$$\omega_r C[R^2 + (\omega_r L)^2] - \omega_r L = 0 \tag{3.47}$$

leading to

$$\omega_r^2 = \frac{L - R^2 C}{L^2 C} = \frac{1}{LC} - \frac{R^2}{L^2} \tag{3.48}$$

and therefore the resonant frequency of the circuit is

$$\omega_r = \sqrt{\frac{1}{LC} - \frac{R^2}{L^2}} \tag{3.49}$$

Substituting this value int Eq. (3.45) gives the value of the input admittance at resonance.

$$\hat{Y}_{IN}(\omega_r) = \left\{ \frac{R}{R^2 + (\omega L)^2} + j\left[\omega C - \frac{\omega L}{R^2 + (\omega L)^2} \right] \right\}_{\omega = \omega_r} \tag{3.50}$$

Since at resonant frequency the input impedance is real, we have

$$\hat{Y}_{IN}(\omega_r) = \left\{ \frac{R}{R^2 + (\omega L)^2} \right\}_{\omega = \omega_r} \tag{3.51}$$

or

$$\hat{Y}_{IN}(\omega_r) = \frac{R}{R^2 + \left(\frac{1}{LC} - \frac{R^2}{L^2} \right) L^2} = \frac{RC}{L} \tag{3.52}$$

giving the value of the input impedance at resonance as

$$\hat{Z}_{IN}(\omega_r) = \frac{L}{RC} \tag{3.53}$$

Let's return to the resonant frequency of the circuit. From Eq. (3.48) this frequency can be alternatively expressed as

$$\omega_r = \frac{1}{\sqrt{LC}} \sqrt{\frac{L - R^2 C}{L}} = \omega_0 \sqrt{1 - \frac{R^2 C}{L}} \tag{3.54}$$

When

$$\frac{R^2 C}{L} < 1 \tag{3.55}$$

the circuit is resonant and its resonant frequency is different from the undamped natural frequency.

3.2 Non-Ideal Behavior of Resistors

3.2.1 Circuit Model and Impedance

Circuit model and the impedance vs. frequency curve for an *ideal* resistor are shown in Figure 3.12.

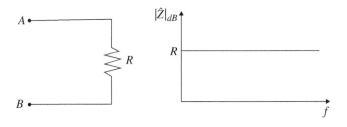

Figure 3.12　Ideal resistor model and impedance curve.

Circuit model of a realistic (non-ideal) resistor and its parasitics (with no traces attached) are shown in Figure 3.13.

Note: This simple model can be used up to the frequency of a few Gigahertz.

Let's obtain the straight-line Bode plots approximations [Adamczyk, 2019a], of the magnitude of the impedance of this resistor vs. frequency, with respect to nodes A and B.

The input impedance to the circuit shown in Figure 3.13 is

$$Z(s) = sL_{par} + R \parallel \frac{1}{sC_{par}} = sL_{par} + \frac{\frac{R}{sC_{par}}}{R + \frac{1}{sC_{par}}}$$

$$= \frac{sL_{par}(1 + sRC_{par}) + R}{1 + sRC_{par}} = \frac{s^2 RC_{par}L_{par} + sL_{par} + R}{1 + sRC_{par}} \tag{3.56}$$

or

$$Z(s) = L_{par} \frac{s^2 + s\frac{1}{RC_{par}} + \frac{1}{L_{par}C_{par}}}{s + \frac{1}{RC_{par}}} \tag{3.57}$$

The magnitude of the impedance is obtained as

$$|Z(s)|_{s=j\omega} = \left| L_{par} \frac{s^2 + s\frac{1}{RC_{par}} + \frac{1}{L_{par}C_{par}}}{s + \frac{1}{RC_{par}}} \right|_{s=j\omega} \tag{3.58}$$

or

$$|Z(j\omega)| = \left| L_{par} \frac{-\omega^2 + j\omega\frac{1}{RC_{par}} + \frac{1}{L_{par}C_{par}}}{j\omega + \frac{1}{RC_{par}}} \right| \tag{3.59}$$

leading to

$$|Z(j\omega)| = R \left| \frac{1 - \frac{\omega^2}{\frac{1}{L_{par}C_{par}}} + j\frac{\omega\frac{1}{RC_{par}}}{\frac{1}{L_{par}C_{par}}}}{1 + j\frac{\omega}{\frac{1}{RC_{par}}}} \right| \tag{3.60}$$

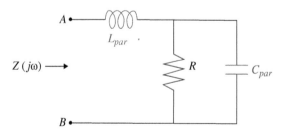

Figure 3.13   Non-ideal resistor model.

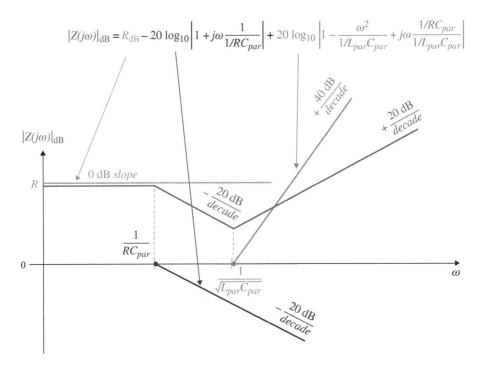

$$|Z(j\omega)|_{dB} = R_{dB} - 20\log_{10}\left|1 + j\omega\frac{1}{1/RC_{par}}\right| + 20\log_{10}\left|1 - \frac{\omega^2}{1/L_{par}C_{par}} + j\omega\frac{1/RC_{par}}{1/L_{par}C_{par}}\right|$$

Figure 3.14 Non-ideal resistor – Bode plots of each factor.

In dB Eq. (3.60) becomes

$$|Z(j\omega)|_{dB} = R_{dB} + \left|1 - \frac{\omega^2}{L_{par}C_{par}} + j\frac{\omega\frac{1}{RC_{par}}}{\frac{1}{L_{par}C_{par}}}\right|_{dB} - \left|1 + j\frac{\omega}{\frac{1}{RC_{par}}}\right|_{dB} \tag{3.61}$$

Bode plot of a constant R is a straight horizontal line in dB at a value R_{dB}. The straight-line approximation to the second term in Eq. (3.61) consists of two asymptotes: 0 dB line until the frequency $\omega = \frac{1}{\sqrt{L_{par}C_{par}}}$ and a +40 dB line beyond that frequency.

The straight-line approximation to the third term in Eq. (3.61) consists of two asymptotes: 0 dB line until the frequency $\omega = \frac{1}{RC_{par}}$ and a −20 dB line beyond that frequency.

The contributions of each circuit component and the resulting plot are shown in Figure 3.14.

We are now ready to construct the magnitude Bode plot of the impedance of a non-ideal resistor. This is shown in Figure 3.15.

3.2.2 Parasitic Capacitance Estimation – Discrete Components

Parasitic capacitance of a resistor can be estimated from its impedance measurements. Figure 3.16 shows a possible measurement setup using a network analyzer.

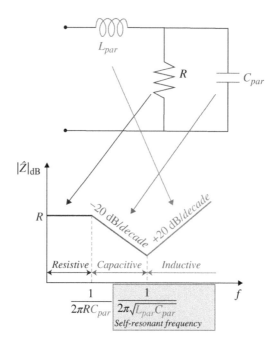

Figure 3.15   Non-ideal resistor – resulting Bode plot of the magnitude,

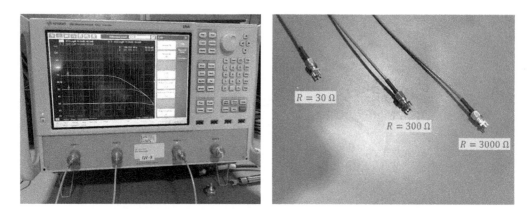

Figure 3.16   Setup for the resistor impedance measurements.

The measurement setup shown in Figure 3.16 is perhaps the simplest measurement setup. The measurements results, however, could be contaminated by the calibration process. To explain this let's look at the details of the resistor–SMA connector assembly, shown in Figure 3.17.

The simplest calibration process often involves an SMA calibration kit like the one shown in Figure 3.18.

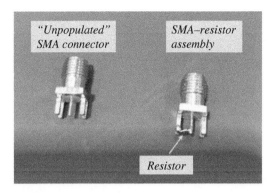

Figure 3.17 Resistor–SMA connector assembly.

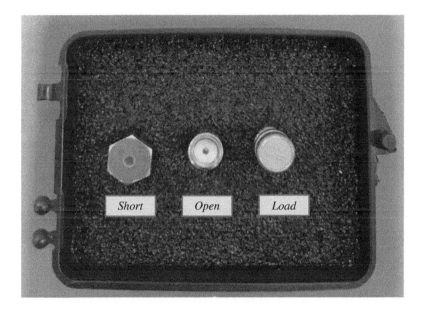

Figure 3.18 SMA calibration kit.

The calibration kit has its own calibration file that is used by the network analyzer during the calibration process. If the calibration were performed with such a kit, then the accuracy of the measurements would be impacted. Additionally, the measurement results represent the impedance of the resistor–SMA connector assembly and not the resistor itself.

To improve the accuracy of the measurements a custom calibration kit, shown in Figure 3.19, could be created.

Note that the SMA connector pins have been shaved off to eliminate their impact on parasitics. Now, a new calibration file would need to be created for this custom calibration kit.

Figure 3.19 Custom calibration kit.

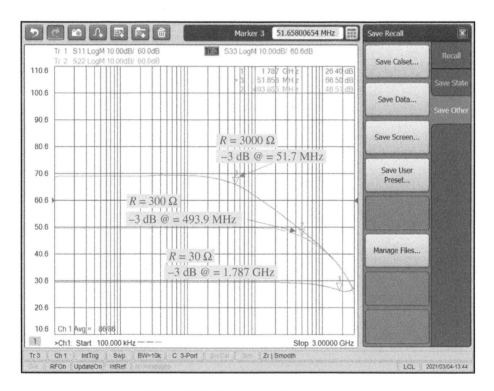

Figure 3.20 Resistor impedance curves.

This makes the seemingly simple measurement process more complicated. For educational purposes one could use the original setup shown in Figures 3.16 and 3.17. Even though the results suffer from the calibration process, they serve the goal of illustrating the non-ideal behavior of components.

The measurement results for such a setup are shown in Figure 3.20.

The markers in Figure 3.20 are placed at the corner frequencies corresponding to the 3 dB points for each curve. These frequencies are equal to

$$f = \frac{1}{2\pi RC_{par}}$$ (3.62)

From Eq. (3.62) we obtain the parasitic capacitance of a resistor as

$$C_{par} = \frac{1}{2\pi fR}$$ (3.63)

Using the results shown in Figure 3.20 the parasitic capacitances for the three resistors are

$$C_{par,30\Omega} = \frac{1}{2\pi \times 1.787 \times 10^9 \times 30} = 1.03\,\text{pF}$$ (3.64)

$$C_{par,300\Omega} = \frac{1}{2\pi \times 493.9 \times 10^6 \times 300} = 1.037\,\text{pF}$$ (3.65)

$$C_{par,3000\Omega} = \frac{1}{2\pi \times 51.7 \times 10^6 \times 3000} = 1.026\,\text{pF}$$ (3.66)

We can use these results, together with the measurements in Figure 3.20, to create the LTSpice circuit models of the resistors.

Figure 3.21 shows the LTspice model and the impedance of a non-ideal $30\,\Omega$ resistor with its parasitics.

Figure 3.22 shows the LTspice model and the impedance of a non-ideal $300\,\Omega$ resistor with its parasitics.

Figure 3.23 shows the LTspice model and the impedance of a non-ideal $3000\,\Omega$ resistor with its parasitics.

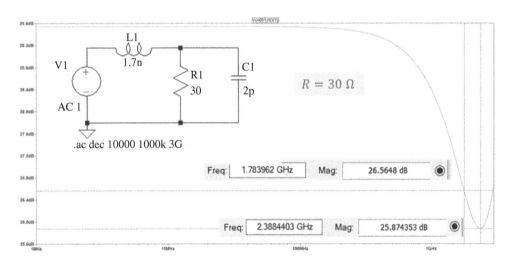

Figure 3.21 $30\,\Omega$ resistor – LTspice model.

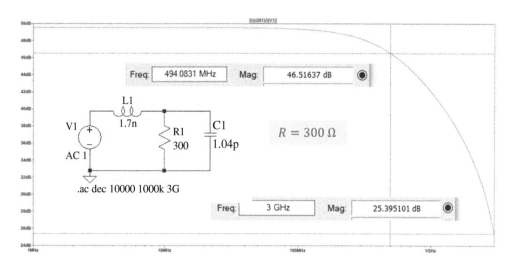

Figure 3.22 300 Ω resistor – LTspice model.

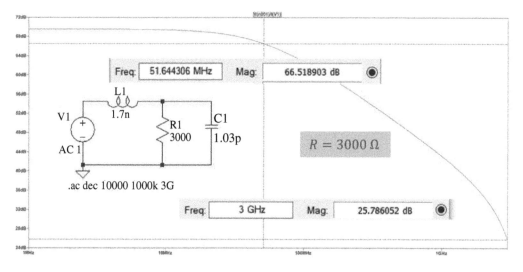

Figure 3.23 3000 Ω resistor – LTspice model.

3.2.3 Parasitic Capacitance Estimation – PCB Components

Alternatively, the parasitic capacitance of a resistor can be estimated from the measurement performed on a PCB populated with resistors of different values [Adamczyk et al., 2022].

Figure 3.24 shows the PCB board used in this study.

Since the calibration traces are of the same length as the traces leading to the components, the connectors and traces are effectively taken out of the measurements. Thus, the measured parasitics come from the component itself and not from the connecting traces.

Note: In order to further improve the accuracy of the results, the new calibration file for this PCB-based calibration kit would need to be created. This, again, makes the measurement process more complicated. For the educational purposes (and to save time and effort) there is little harm from using the calibration file from the original SMA calibration kit (shown in Figure 3.18).

Figure 3.25 shows the measurement setup used to obtain the impedance curves for three different resistor values.

Resistor impedance curves are shown in Figure 3.26.

Each curve shows a 3-dB point corresponding the frequency

$$f = \frac{1}{2\pi R C_{par}} \tag{3.67}$$

From Eq. (3.67) we obtain the parasitic capacitance of a resistor as

$$C_{par} = \frac{1}{2\pi f R} \tag{3.68}$$

The resulting parasitic capacitance values for the three resistors are shown in Figure 3.27.

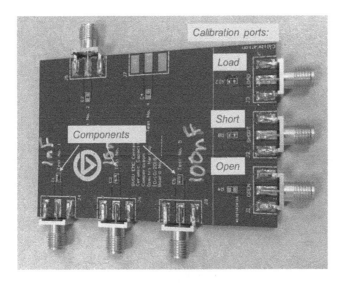

Figure 3.24   PCB and its details.

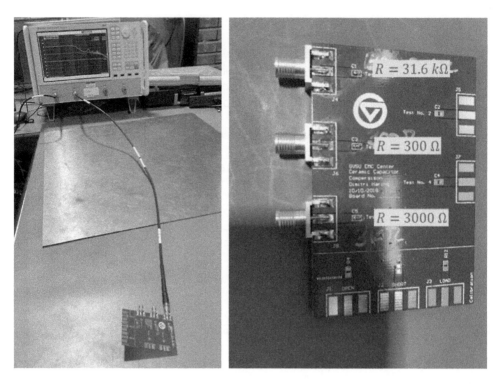

Figure 3.25   Measurement setup – resistor impedance curves.

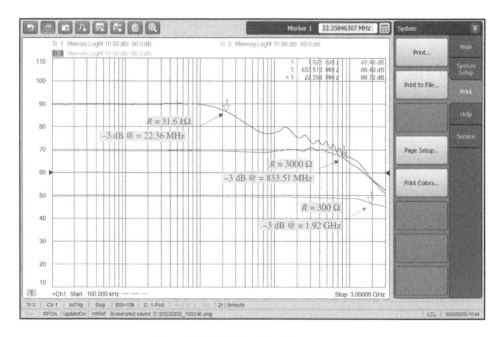

Figure 3.26   Resistor impedance curves.

$R\ (\Omega)$	$f\ (\text{MHz})$	$C_{par}\ (\text{pF})$
30	1921	0.276
300	833.5	0.064
3000	22.36	0.229

Figure 3.27 Resistor – parasitic capacitance.

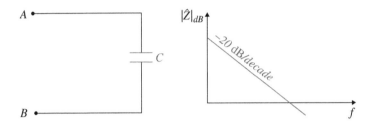

Figure 3.28 Ideal capacitor model and impedance curve.

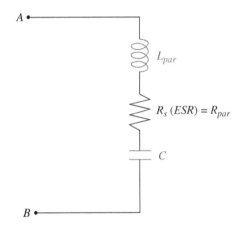

Figure 3.29 Non-ideal capacitor model.

3.3 Non-Ideal Behavior of Capacitors

3.3.1 Circuit Model and Impedance

Circuit model and the impedance vs. frequency curve for an *ideal* capacitor are shown in Figure 3.28.

Circuit model of a realistic (non-ideal) capacitor and its parasitics (with no traces attached) are shown in Figure 3.29.

Note: This simple model can be used up to the frequency of about 1 GHz.

The input impedance to the circuit shown in Figure 3.29 is

$$Z(s) = sL_{par} + R_{par} + \frac{1}{sC} \tag{3.69}$$

or

$$Z(s) = L_{par} \frac{s^2 + s\frac{R_{par}}{L_{par}} + \frac{1}{L_{par}C}}{s} \tag{3.70}$$

The magnitude of the impedance is obtained as

$$|Z(s)|_{s=j\omega} = \left| L_{par} \frac{s^2 + s\frac{R_{par}}{L_{par}} + \frac{1}{L_{par}C}}{s} \right|_{s=j\omega} \tag{3.71}$$

or

$$|Z(j\omega)| = \left| L_{par} \frac{-\omega^2 + j\omega\frac{R_{par}}{L_{par}} + \frac{1}{L_{par}C}}{j\omega} \right| \tag{3.72}$$

or

$$|Z(j\omega)| = \left| \frac{-\omega^2 L_{par} + j\omega R_{par} + \frac{1}{C}}{\omega} \right| \tag{3.73}$$

leading to

$$|Z(j\omega)| = \frac{1}{C} \left| \frac{1 - \frac{\omega^2}{\frac{1}{L_{par}C}} + j\frac{\omega}{\frac{1}{R_{par}C}}}{\omega} \right| \tag{3.74}$$

In dB Eq. (3.74) becomes

$$|Z(j\omega)|_{dB} = \left(\frac{1}{C} \right)_{dB} + \left| 1 - \frac{\omega^2}{\frac{1}{L_{par}C}} + j\frac{\omega}{\frac{1}{R_{par}C}} \right|_{dB} - \omega_{dB} \tag{3.75}$$

The contributions of each circuit component and the resulting plot are shown in Figure 3.30 [Adamczyk, 2019a].

We are now ready to construct the magnitude Bode plot of the impedance of non-ideal capacitor. This is shown in Figure 3.31.

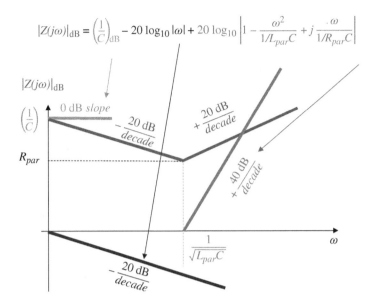

$$|Z(j\omega)|_{dB} = \left(\frac{1}{C}\right)_{dB} - 20\log_{10}|\omega| + 20\log_{10}\left|1 - \frac{\omega^2}{1/L_{par}C} + j\frac{\omega}{1/R_{par}C}\right|$$

Figure 3.30   Non-ideal capacitor – Bode plots of each factor.

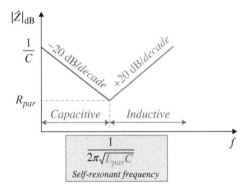

Figure 3.31   Non-ideal capacitor – Bode plot of the magnitude.

3.3.2 Parasitic Inductance Estimation – Discrete Components

Figure 3.32 shows the setup for the impedance measurements of a capacitor.

The measurement results are shown in Figure 3.33.

The markers in Figure 3.20 are placed at the self-resonant frequencies. These frequencies are equal to

$$f_r = \frac{1}{2\pi L_{par}C} \tag{3.76}$$

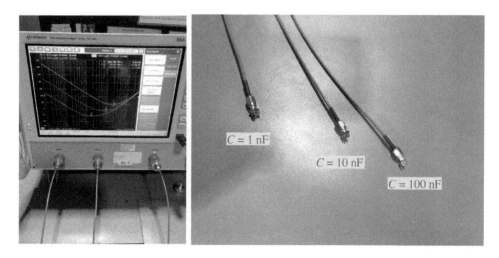

Figure 3.32   Setup for the capacitor impedance measurements.

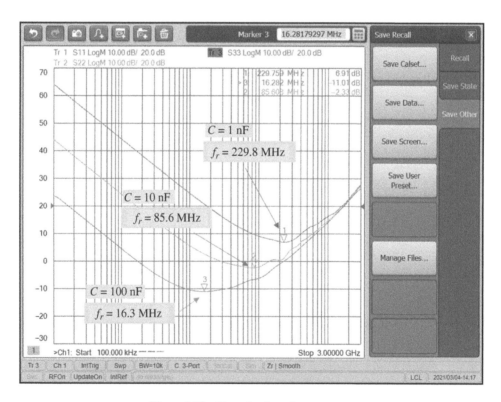

Figure 3.33   Capacitor impedance curves.

From Eq. (3.76) we obtain the parasitic inductance of a capacitor as

$$L_{par} = \frac{1}{4\pi^2 f_r^2 C} \tag{3.77}$$

Using the results shown in Figure 3.33 the parasitic inductances for the three capacitors are

$$L_{par,1\,nF} = \frac{1}{4\pi^2 \times 229.8^2 \times 10^{12} \times 1 \times 10^{-9}} = 0.48\,\text{nH} \tag{3.78}$$

$$L_{par,10\,nF} = \frac{1}{4\pi^2 \times 85.6^2 \times 10^{12} \times 10 \times 10^{-9}} = 0.35\,\text{nH} \tag{3.79}$$

$$L_{par,100\,nF} = \frac{1}{4\pi^2 \times 16.3^2 \times 10^{12} \times 100 \times 10^{-9}} = 0.95\,\text{nH} \tag{3.80}$$

We can use these results, together with the measurements in Figure 3.33, to create the LTSpice circuit models of the capacitors.

Figure 3.34 shows the LTspice model and the impedance of a non-ideal 1 nF capacitor with its parasitics.

Figure 3.35 shows the LTspice model and the impedance of a non-ideal 10 nF capacitor with its parasitics.

Figure 3.36 shows the LTspice model and the impedance of a non-ideal 100 nF capacitor with its parasitics.

3.3.3 Parasitic Inductance Estimation – PCB Components

Figure 3.37 shows the measurement setup used to obtain the impedance curves for three different capacitor values [Adamczyk et al., 2022].

Capacitor impedance curves are shown in Figure 3.38.

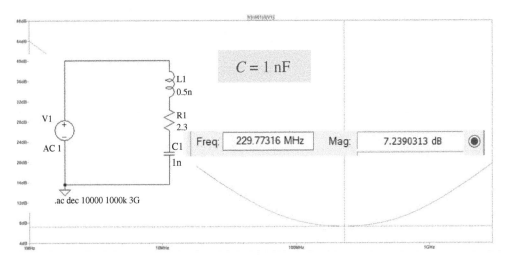

Figure 3.34   1 nf capacitor – LTspice model.

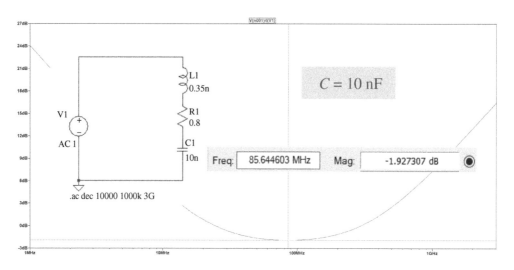

Figure 3.35   10 nF capacitor – LTspice model.

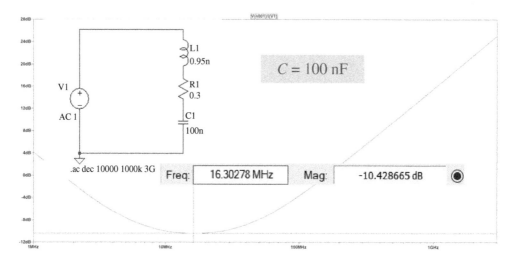

Figure 3.36   100 nF capacitor – LTspice model.

Each curve shows a self-resonant point corresponding to the frequency

$$f_r = \frac{1}{2\pi L_{par}C} \qquad (3.81)$$

From Eq. (3.81) we obtain the parasitic inductance of a capacitor as

$$L_{par} = \frac{1}{4\pi^2 f_r^2 C} \qquad (3.82)$$

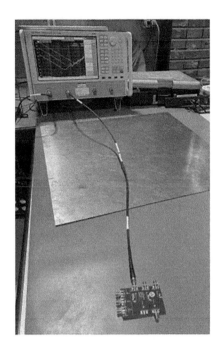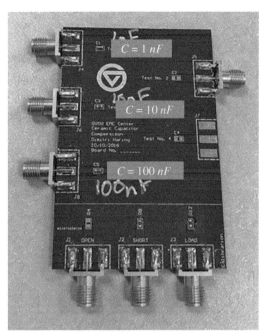

Figure 3.37 Measurement setup – capacitor impedance curves.

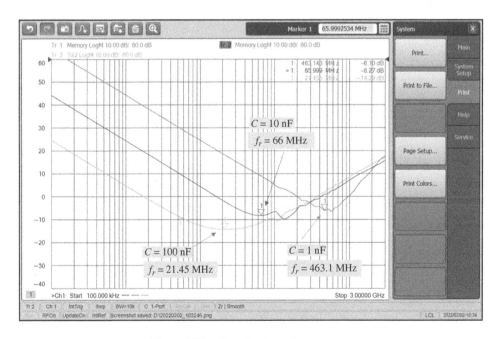

Figure 3.38 Capacitor impedance curves.

C (nF)	f (MHz)	L_{par} (nH)
1	463.1	0.12
10	66	0.58
100	21.4	0.55

Figure 3.39 Capacitor – parasitic inductance.

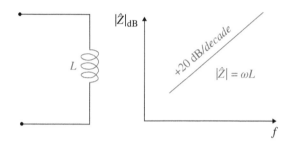

Figure 3.40 Ideal inductor model and impedance curve.

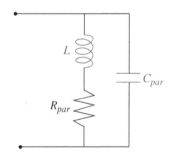

Figure 3.41 Non-ideal inductor model.

The resulting parasitic inductance values for the three capacitors are shown in Figure 3.39.

3.4 Non-Ideal Behavior of Inductors

3.4.1 Circuit Model and Impedance

Circuit model and the impedance vs. frequency curve for an *ideal* inductor are shown in Figure 3.40.

Circuit model of a realistic (non-ideal) inductor and its parasitics (with no traces attached) are shown in Figure 3.41.

Note: This simple model can be used up to the frequency of about 1 GHz.

The input impedance to the circuit shown in Figure 3.41 is

$$Z(s) = (sL + R_{par}) \parallel \frac{1}{sC_{par}} \tag{3.83}$$

or

$$Z(s) = \frac{(sL + R_{par}+)\frac{1}{sC_{par}}}{sL + R_{par} + \frac{1}{sC_{par}}} \tag{3.84}$$

or

$$Z(s) = \frac{sL + R_{par}}{s^2 LC_{par} + sR_{par}C_{par} + 1} \tag{3.85}$$

or

$$Z(s) = \frac{s + \frac{R_{par}}{L}}{s^2 C_{par} + s\frac{R_{par}C_{par}}{L} + \frac{1}{L}} \tag{3.86}$$

leading to

$$Z(s) = \frac{1}{C_{par}} \frac{s + \frac{R_{par}}{L}}{s^2 + s\frac{R_{par}}{L} + \frac{1}{LC_{par}}} \tag{3.87}$$

The magnitude of the impedance is obtained as

$$|Z(s)|_{s=j\omega} = \left| \frac{1}{C_{par}} \frac{s + \frac{R_{par}}{L}}{s^2 + s\frac{R_{par}}{L} + \frac{1}{LC_{par}}} \right|_{s=j\omega} \tag{3.88}$$

or

$$|Z(j\omega)| = \left| \frac{1}{C_{par}} \frac{j\omega + \frac{R_{par}}{L}}{-\omega^2 + j\omega\frac{R_{par}}{L} + \frac{1}{LC_{par}}} \right| \tag{3.89}$$

or

$$|Z(j\omega)| = \frac{1}{C_{par}} \left| \frac{RC_{par} + j\frac{\omega}{\frac{1}{LC_{par}}}}{1 - \frac{\omega^2}{\frac{1}{LC_{par}}} + j\frac{\omega}{\frac{1}{RC_{par}}}} \right| \tag{3.90}$$

leading to

$$|Z(j\omega)| = R_{par} \left| \frac{1 + j\frac{\omega}{\frac{1}{\frac{R_{par}}{L}}}}{1 - \frac{\omega^2}{\frac{1}{LC_{par}}} + j\frac{\omega}{\frac{1}{\frac{R_{par}}{L}}}} \right| \tag{3.91}$$

In dB Eq. (3.91) becomes

$$|Z(j\omega)|_{dB} = R_{par,dB} + \left|1 + j\frac{\omega}{\frac{1}{\frac{R_{par}}{L}}}\right|_{dB} - \left|1 - \frac{\omega^2}{\frac{1}{LC_{par}}} + j\frac{\omega}{\frac{1}{R_{par}C_{par}}}\right|_{dB} \qquad (3.92)$$

The contributions of each circuit component and the resulting plot are shown in Figure 3.42 [Adamczyk, 2019a].

We are now ready to construct the magnitude Bode plot of the impedance of non-ideal inductor. This is shown in Figure 3.43.

3.4.2 Parasitic Capacitance Estimation – Discrete Components

Figure 3.44 shows the setup for the impedance measurements of an inductor.

The measurement results are shown in Figure 3.45.

The markers in Figure 3.45 are placed at the self-resonant frequencies. These frequencies are equal to

$$f_r = \frac{1}{2\pi LC_{par}} \qquad (3.93)$$

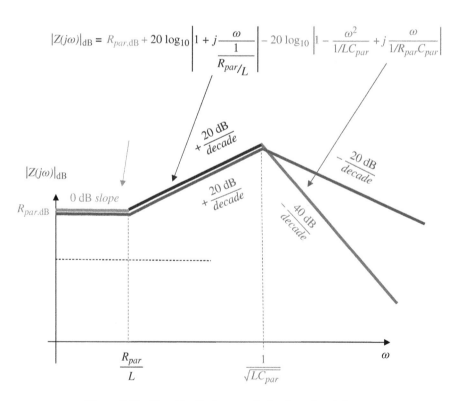

Figure 3.42   Non-ideal inductor – Bode plots of each factor.

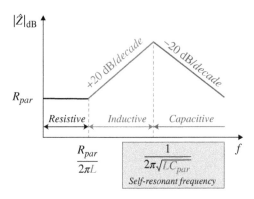

Figure 3.43 Non-ideal inductor – Bode plot of the magnitude.

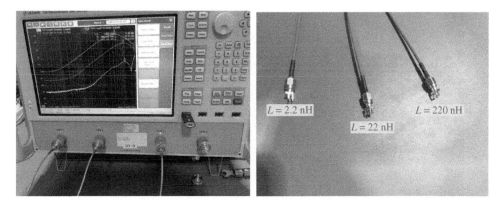

Figure 3.44 Setup for the inductor impedance measurements.

From Eq. (3.93) we obtain the parasitic capacitance of an inductor as

$$C_{par} = \frac{1}{4\pi^2 f_r^2 L} \tag{3.94}$$

Using the results shown in Figure 3.45 the parasitic capacitances for the three inductors are

$$C_{par,2.2\,nH} = \frac{1}{4\pi^2 \times 1000^2 \times 10^{12} \times 2.2 \times 10^{-9}} = 11.5\,pF \tag{3.95}$$

$$C_{par,22\,nH} = \frac{1}{4\pi^2 \times 905^2 \times 10^{12} \times 22 \times 10^{-9}} = 1.4\,pF \tag{3.96}$$

$$C_{par,220\,nH} = \frac{1}{4\pi^2 \times 381^2 \times 10^{12} \times 220 \times 10^{-9}} = 0.79\,pF \tag{3.97}$$

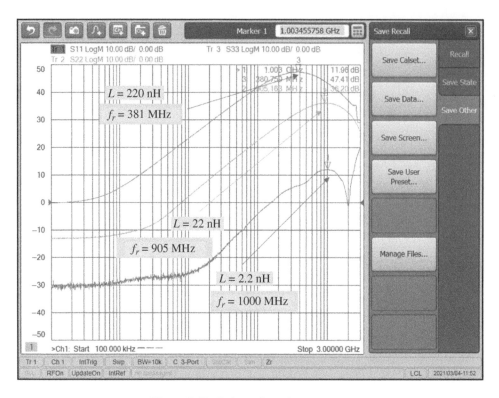

Figure 3.45   Inductor impedance curves.

We can use these results, together with the measurements in Figure 3.45, to create the LTSpice circuit models of the inductors.

Figure 3.46 shows the LTspice model and the impedance of a non-ideal 2.2 nH inductor with its parasitics.

Figure 3.47 shows the LTspice model and the impedance of a non-ideal 22 nH inductor with its parasitics.

Figure 3.48 shows the LTspice model and the impedance of a non-ideal 220 nH inductor with its parasitics.

3.4.3 Parasitic Capacitance Estimation – PCB Components

Figure 3.49 shows the measurement setup used to obtain the impedance curves for three different inductor values [Adamczyk et al., 2022].

Inductor impedance curves are shown in Figure 3.50.

Each curve shows a self-resonant point corresponding to the frequency

$$f_r = \frac{1}{2\pi L C_{par}}$$

(3.98)

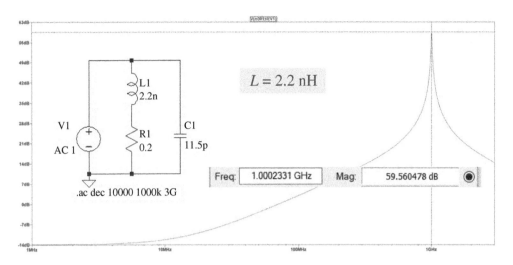

Figure 3.46   2.2 nH inductor – LTspice model.

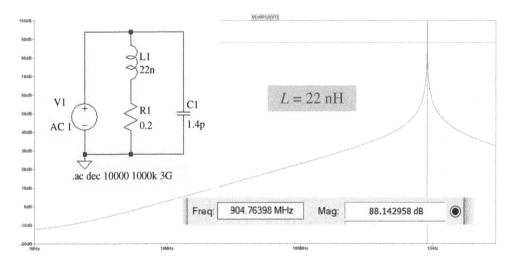

Figure 3.47   22 nH inductor – LTspice model.

From Eq. (3.98) we obtain the parasitic capacitance of an inductor as

$$C_{par} = \frac{1}{4\pi^2 f_r^2 L} \tag{3.99}$$

The resulting parasitic capacitance values for the three inductors are shown in Figure 3.51.

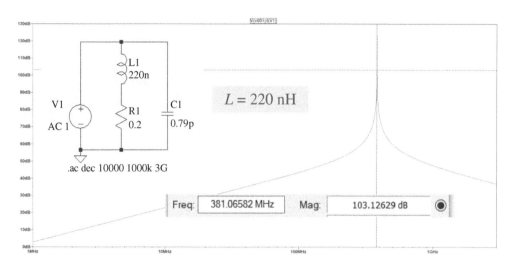

Figure 3.48   220 nH inductor – LTspice model.

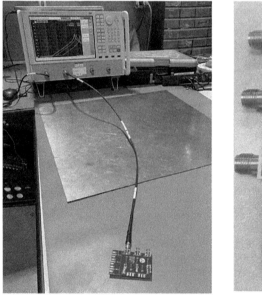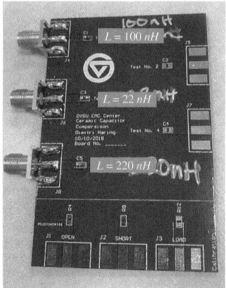

Figure 3.49   Measurement setup – inductor impedance curves.

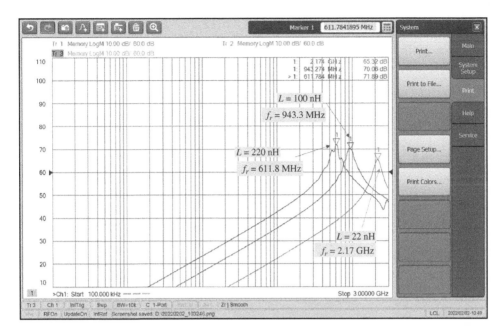

Figure 3.50   Inductor impedance curves.

C (nF)	f (MHz)	L_{par} (nH)
1	463.1	0.12
10	66.0	0.58
100	21.4	0.55

Figure 3.51   Inductor – parasitic capacitance.

3.5 Non-Ideal Behavior of a PCB Trace

3.5.1 Circuit Model and Impedance

Circuit model and the impedance vs. frequency curve for an *ideal* PCB trace are shown in Figure 3.52.

Circuit model of a realistic (non-ideal) PCB trace is shown in Figure 3.53.

Note: This simple model can be used up to the frequency of a few Gigahertz.

Figure 3.54 shows the setup used for impedance measurement of a PCB trace.

Figure 3.55 shows the impedance measurement for the three PCB traces. The short trace has the length of 1–inch, while the long trace is 2-inch long. The "no trace" length is minimal.

Figures 3.56 and 3.57 show the LTspice circuit model and the impedance curve for the short and long traces, respectively.

Figure 3.52   Ideal PCB trace and impedance curve.

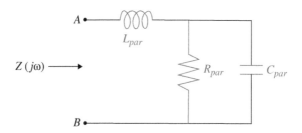

Figure 3.53   Non-ideal PCB trace model.

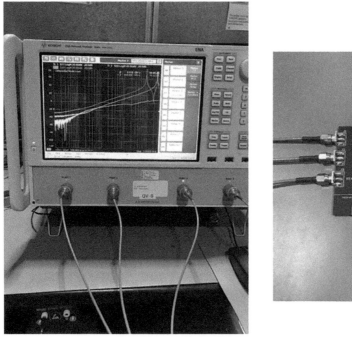

Figure 3.54   Setup for impedance measurement of a PCB trace.

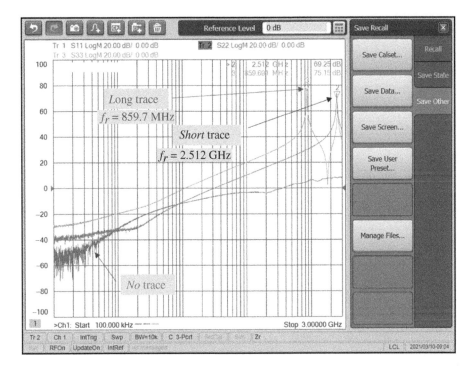

Figure 3.55 PCB trace – impedance measurements.

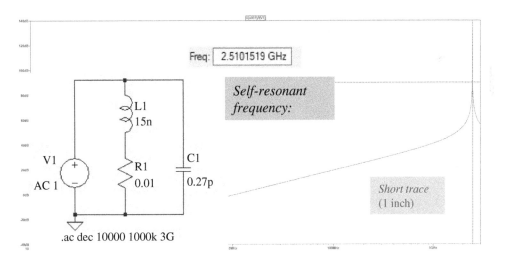

Figure 3.56 Short PCB trace – circuit model and impedance vs. frequency.

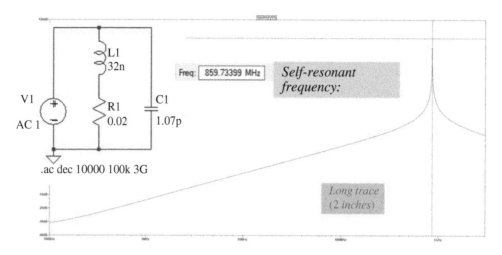

Figure 3.57   Long PCB trace – circuit model and impedance vs. frequency.

3.6 Impact of the PCB Trace Length on Impedance of the Passive Components

In this section we investigate the impact of PCB trace length on the impedance of a resistor, capacitor, and an inductor.

3.6.1 Impedance of a Resistor – Impact of the PCB Trace

Figure 3.58 shows the measurement setup used to examine the impact of PCB trace on the impedance of a resistor. The short trace has a length of 1 inch, while the long trace is 2-inch long.

The measurement results are shown in Figure 3.59.

It is apparent that increasing the trace length causes the self-resonant frequency to shift to the left.

3.6.2 Impedance of a Capacitor – Impact of the PCB Trace

Figure 3.60 shows the measurement setup used to examine the impact of PCB trace on the impedance of a capacitor. The short trace has a length of 1 inch, while the long trace is 2-inch long.

The measurement results are shown in Figure 3.61.

Again, increasing the trace length causes the self-resonant frequency to shift to the left.

3.6.3 Impedance of an Inductor – Impact of the PCB Trace

Figure 3.62 shows the measurement setup used to examine the impact of PCB trace on the impedance of an inductor. The short trace has a length of 1 inch, while the long trace is 2-inch long.

The measurement results are shown in Figure 3.63.

Again, increasing the trace length causes the self-resonant frequency to shift to the left.

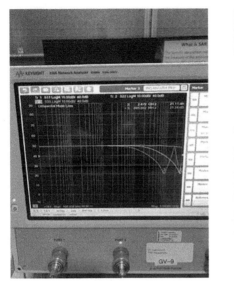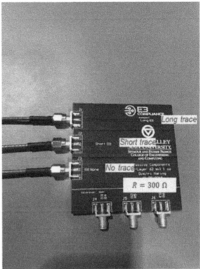

Figure 3.58 Measurement setup – resistor and a PCB trace.

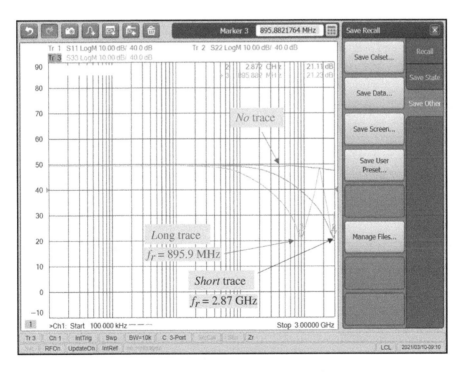

Figure 3.59 Impedance of a resistor and a PCB trace.

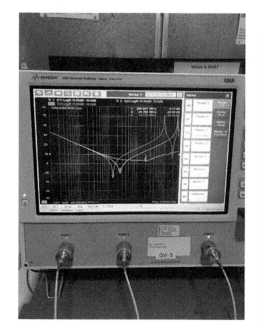 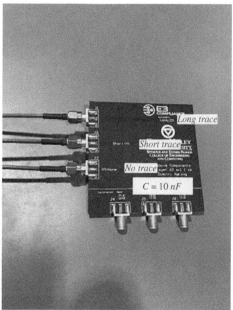

Figure 3.60 Measurement setup – capacitor and a PCB trace.

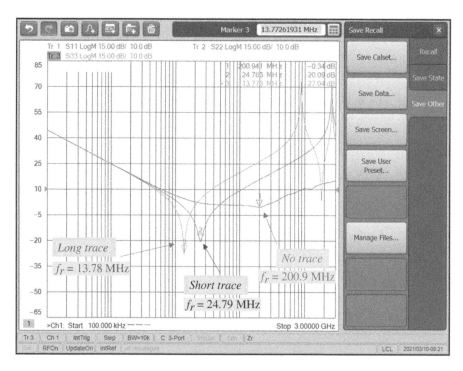

Figure 3.61 Impedance of a Capacitor and a PCB trace.

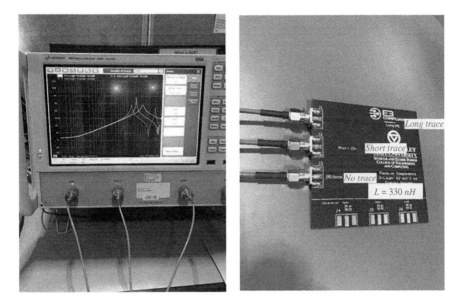

Figure 3.62 Measurement setup – inductor and a PCB trace.

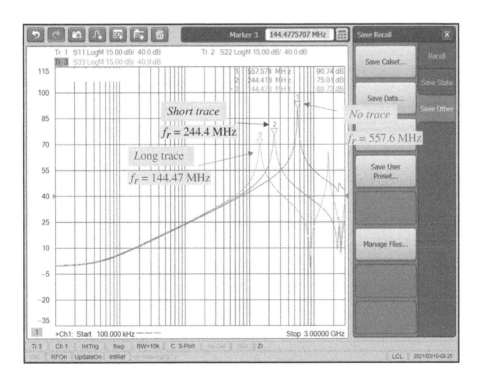

Figure 3.63 Impedance of an inductor and a PCB trace.

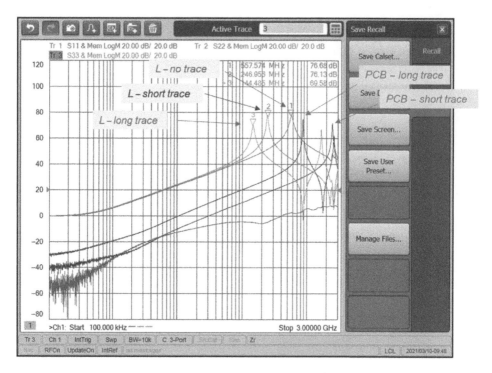

Figure 3.64   Impedance of an inductor vs. the PCB trace.

3.6.4 Impedance of an Inductor vs. Impedance of the PCB Trace

It is very instructive to compare the impedances of the PCB traces vs. those of the inductor. Figure 3.64 shows this comparison.

Figure 3.64 reveals that the impedance curves of the unpopulated PCB traces exhibit the inductive behavior. Effectively, the PCB trace can be thought of as a fourth passive circuit element!

3.7 Laboratory Exercises

3.7.1 Non-Ideal Behavior of Capacitors and Inductors, and Impact of the PCB Trace Length on Impedance

Objective: This laboratory exercise examines the non-ideal behavior of capacitors and inductors. It also investigates the impact of the trace length on impedance of capacitors and inductors.

Note: This laboratory requires a custom PCB [Haring, 2018], for which the Altium files and Bill of Material (BOM) are provided.

3.7.2 Laboratory Equipment and Supplies

Network Analyzer: KEYSIGHT ENA Network Analyzer E5080A (or equivalent), shown in Figure 3.65.

Adapter Coaxial Connector N Plug, Male Pin To SMA Jack, Female Socket 50 Ohm: Digikey PN: ACX1341-ND (or equivalent), shown in Figure 3.66.

SMA Male to SMA Male Cable Assembly, Set of 3: Digikey PN: 3648-P1CA-STMSTM-ST085-36-ND (or equivalent), shown in Figure 3.67.

SMA Connector Jack, Female Socket – Board Edge: Digikey PN: CON-SMA-EDGE-S-ND (or equivalent), shown in Figure 3.68.

Ceramic Capacitor: 150 nF, Digikey PN: 490-11960-1-ND (or equivalent), shown in Figure 3.69.

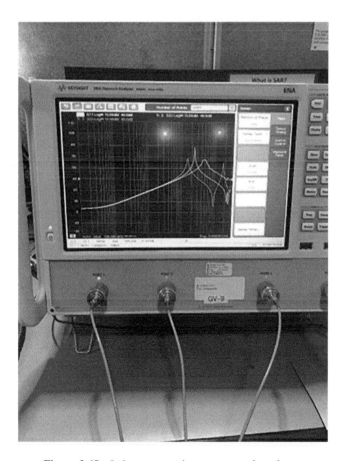

Figure 3.65 Laboratory equipment: network analyzer.

Figure 3.66   Laboratory equipment: N-type to SMA connector.

Figure 3.67   Laboratory equipment: SMA to SMA cable.

Figure 3.68   Laboratory equipment: SMA connector.

Figure 3.69   Laboratory equipment: ceramic capacitor.

Figure 3.70   Laboratory equipment: ceramic inductor.

Figure 3.71   Laboratory equipment: custom PCB.

Ceramic Inductor: 100 nH, Digikey PN: 490-1111-1-ND (or equivalent), shown in Figure 3.70.

PCB: Custom PCB, populated with a capacitor or an inductor – shown in Figure 3.71.

3.7.3 Laboratory Procedure – Non-Ideal Behavior of Capacitors and Inductors

3.7.3.1 Parasitic Inductance of a Capacitor

Step 1 (optional): Create a custom calibration kit like the one shown in Figure 3.19.

Step 2: Connect the SMA cable to the network analyzer and perform the calibration process using either a standard calibration kit (shown in Figure 3.18) or a custom calibration kit (shown in Figure 3.19).

Step 3: Create a set of the capacitor–SMA connector assemblies for a three different values, decades apart, of capacitance (like the one shown in Figure 3.17.

Step 4: Attach the capacitor–SMA assembly to the SMA cable, as shown in Figure 3.32.

Step 5: Perform the impedance measurements for the three different values of capacitance. Place the markers at the self-resonant frequencies (see Figure 3.33). Use the equation

$$L_{par} = \frac{1}{4\pi^2 f_r^2 C} \tag{3.100}$$

to determine the parasitic inductance value for each capacitor.

3.7.3.2 Parasitic Inductance of an Inductor

Step 1: Create a set of the inductor–SMA connector assemblies for a three different values, decades apart, of inductance.
Step 2: Attach the inductor–SMA assembly to the SMA cable.
Step 3: Perform the impedance measurements for the three different values of inductance. Place the markers at the self-resonant frequencies (see Figure 3.45). Use the equation

$$C_{par} = \frac{1}{4\pi^2 f_r^2 L} \tag{3.101}$$

to determine the parasitic capacitance value for each inductor.

3.7.4 Laboratory Procedure – Impact of the PCB Trace Length on Impedance

3.7.4.1 PCB Trace and Capacitor

Step 1: Connect the SMA cable to the calibration ports on the PCB and network analyzer, and perform the calibration process.
Step 2: Connect the three SMA cables to the PCB populated with a capacitor, and network analyzer, as shown in Figure 3.60.
  Place the markers at the self-resonant frequencies and examine the impact of PCB trace on the impedance of a capacitor. Comment on the results.

3.7.4.2 PCB Trace and Inductor

Connect the three SMA cables to the PCB populated with an inductor and network analyzer.
  Place the markers at the self-resonant frequencies and examine the impact of PCB trace on the impedance of an inductor. Comment on the results.

References

Bogdan Adamczyk. *Foundations of Electromagnetic Compatibility*. John Wiley & Sons, Ltd, Chichester, UK, 2017. ISBN 9781119120810. doi: https://doi.org/10.1002/9781119120810.

Bogdan Adamczyk. Basic Bode Plots in EMC Applications – Part II: Examples. *In Compliance Magazine*, May 2019a. URL https://incompliancemag.com/article/basic-bode-plots-in-emc-applications-part-ii-examples/.

Bogdan Adamczyk. Impedance of the Four Passive Circuit Components: R, L, C, and A PCB Trace. *In Compliance Magazine*, January 2019b. URL https://incompliancemag.com/article/impedance-of-the-four-passive-circuit-components-r-l-c-and-a-pcb-trace/.

Bogdan Adamczyk. EMC Resonance Part I: Non-Ideal Passive Components. *In Compliance Magazine*, February 2021. URL https://incompliancemag.com/article/emc-resonance-part-i-non-ideal-passive-components/.

Bogdan Adamczyk. Concept of a Phasor in Sinusoidal Steady State Analysis. *In Compliance Magazine*, December 2022. URL https://incompliancemag.com/article/concept-of-a-phasor-in-sinusoidal-steady-state-analysis/.

Bogdan Adamczyk, Nick Koeller, and Megan Healy. Estimating the Parasitics of Passive Circuit Components. *In Compliance Magazine*, April 2022. URL https://incompliancemag.com/article/estimating-the-parasitics-of-passive-circuit-components/.

Dimitri Haring. RLC PCB Designer, 2018.

Chapter 4: Power Distribution Network

4.1 CMOS Inverter Switching

Let's begin with the CMOS inverter operation as it naturally leads to the impact of the decoupling capacitors. Toward this end, consider a CMOS inverter logic gate in a totem-pole configuration, shown in Figure 4.1 [Adamczyk, 2017].

In a high-speed digital circuits one often encounters the cascaded CMOS configuration shown in Figure 4.2.

A simplified model of this configuration is shown in Figure 4.3.

Let's investigate the operation of this configuration on the low-to-high and high-to-low transition of the input to the first inverter. First, assume that the load capacitors C_{GP} and C_{GN} are initially uncharged. When the input signal $IN = Low$, the upper transistor is ON and the lower is OFF. The current flows though the upper transistor, signal trace, and the capacitor C_{GN} to ground. This is shown in Figure 4.4.

Eventually the capacitor C_{GN} is charged to (approximately) V_{CC} and the current flow stops, as shown in Figure 4.5.

Now, the driver inverter transitions from low to high. Subsequently, the upper transistor turns OFF and the lower transistor turns ON, as shown in Figure 4.6.

At this point we have two sources of current: (i) Current supplied by C_{GN} as it discharges (dashed arrow). (ii) Current supplied by V_{CC} as it charges the upper load capacitor.

The current then flows along the trace toward the driver and through the lower transistor to ground. Eventually the current flow stops and the voltage across capacitor C_{GP} is V_{CC}. This is shown in Figure 4.7.

Now, the driver inverter transitions from high to low. Subsequently, the upper transistor turns ON and the lower transistor turns OFF, as shown in Figure 4.8.

At this point we have two sources of current: (i) Current supplied by C_{GP} as it discharges (dashed arrow). (ii) Current supplied by V_{CC} flowing through the upper transistor, along the trace, and through the lower load capacitor, eventually charging it to V_{CC}.

4.2 Decoupling Capacitors

Now let's turn our attention a typical scenario in digital logic circuits with a dc voltage source, CMOS driver, and a load IC, as shown in Figure 4.9.

When a CMOS gate switches, a current transient is drawn from the power distribution network (PDN). This current flows from the source to the load along the forward path i.e. the power traces and back to the source through the return path i.e. the ground traces.

This time-varying current creates a time-varying magnetic flux that crosses the loop area of the circuit inducing a voltage drop along the traces. We can model this phenomenon by inserting either a voltage source somewhere in the loop, or by inserting

Principles of Electromagnetic Compatibility: Laboratory Exercises and Lectures, First Edition. Bogdan Adamczyk.
© 2024 John Wiley & Sons Ltd. Published 2024 by John Wiley & Sons Ltd.
Companion website: www.wiley.com/go/principlesofelectromagneticcompatibility

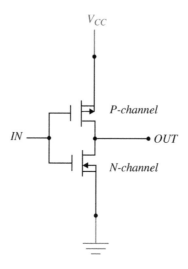

Figure 4.1   CMOS inverter logic gate.

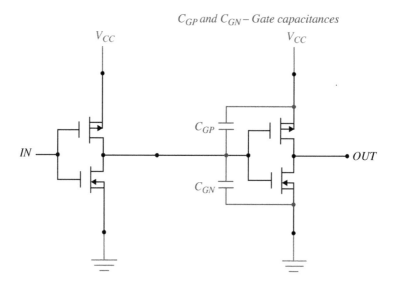

Figure 4.2   Cascaded CMOS configuration.

an inductance somewhere in the loop. Both the induced voltage and the inductance are distributed along the loop. When the loop is electrically small at the frequencies considered we can model these distributed effects as lumped sources or inductances.

In any practical circuit the forward and return paths (horizontal lines in Figure 4.9), are orders of magnitude longer than the length of a path between the power and ground pins (vertical lines in Figure 4.9).

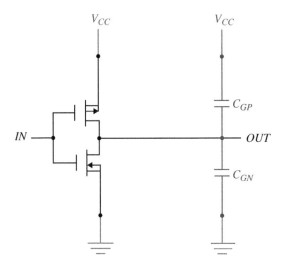

Figure 4.3 Simplified model of a cascaded CMOS configuration.

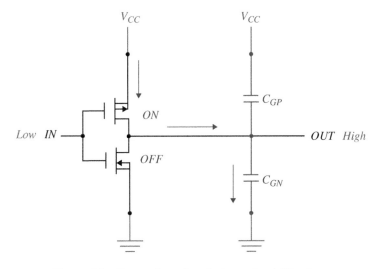

Figure 4.4 Current flow when the input signal *IN* = *Low*.

It is therefore, reasonable to insert the lumped parameters only into the forward and return paths. To model the flux-induced effect we will insert inductances along the power and ground traces. This is shown in Figure 4.10.

This model applies at lower frequencies where the partial inductances of the power and ground traces connecting the ICs themselves can be neglected. At higher frequencies we would augment the model with additional partial and parasitic inductances between the ICs and within the ICs themselves [Adamczyk, 2019].

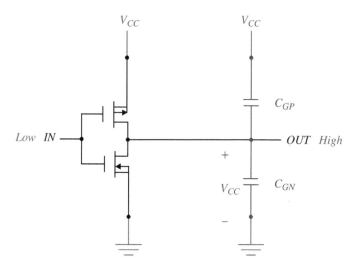

Figure 4.5   Current flow stops when C_{GN} is charged to V_{CC}.

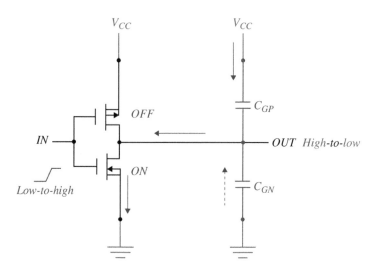

Figure 4.6   Current flow on transition from low to high.

When the driver IC switches, the current is drawn from the source resulting in the voltages V_P and V_G across the power and ground inductances. We often refer to these voltages as a *power rail collapse* and a *ground bounce*, respectively.

During the off-switching time (dc condition) inductances act as short circuits and the voltage V_{IC} between the driver-IC power and ground pins equals the source voltage V_S. During the switching time the voltage V_{IC} no longer equals the source voltage V_S potentially causing signal integrity issues.

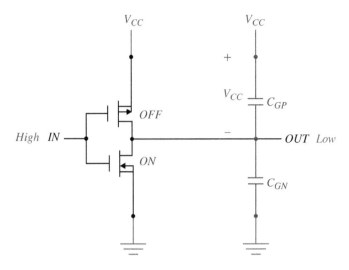

Figure 4.7   Current flow stops when C_{GP} is charged to V_{CC}.

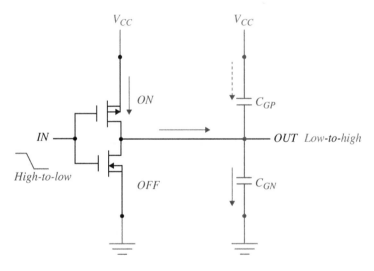

Figure 4.8   Transition from high to low.

This voltage now equals

$$V_{IC} = V_S - (L_P + L_G)\frac{di(t)}{dt} \tag{4.1}$$

Additionally, the transient current flows in a large loop creating an efficient loop antenna.
   Now, let's place a capacitor between the power and ground pins near the switching IC, as shown in Figure 4.11.

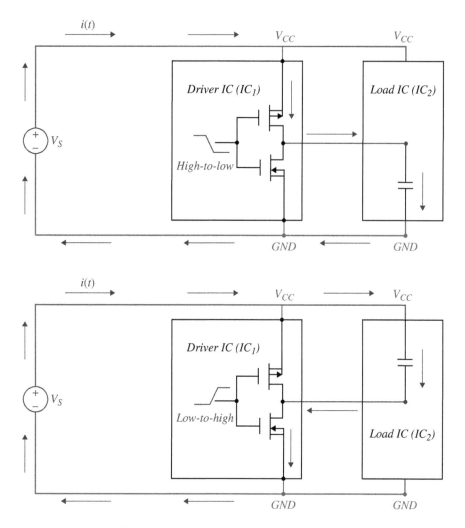

Figure 4.9 CMOS transistors in a high-speed logic circuit.

During the off-switching time this capacitor charges to the source voltage V_S. During the switching time this capacitor ideally supplies the total required current to the load, as shown in Figure 4.11.

In reality it supplies most of the current to the load, thus reducing the current draw from the source and consequently reducing the voltage drops across the inductances.

Assuming the ideal scenario the transient current flows now in a smaller loop, as shown in Figure 4.12. This in turn reduces the radiated emissions.

4.2.1 Decoupling Capacitor Impact – Measurements

The measurement setup used to show the decoupling capacitor impact on signal integrity is shown in Figure 4.13 [Adamczyk, 2019].

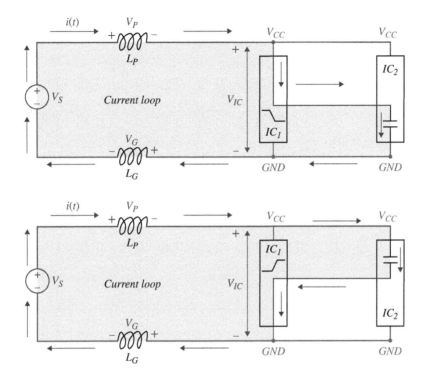

Figure 4.10   Current loop area and partial inductances.

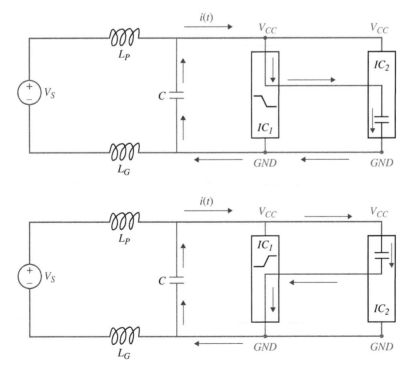

Figure 4.11   Decoupling capacitor placed near the switching IC.

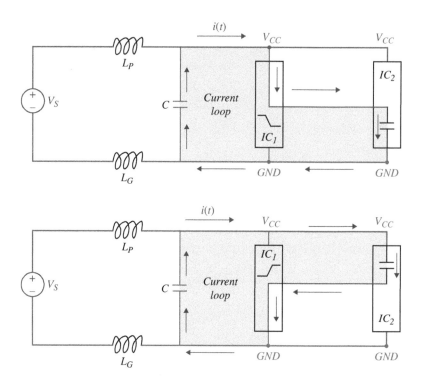

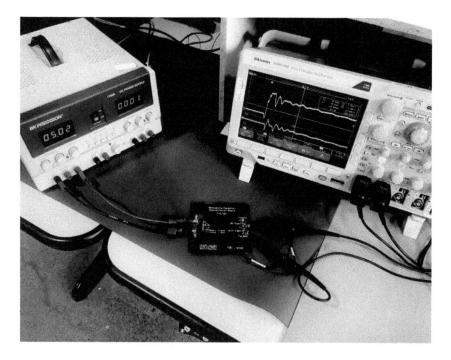

Figure 4.12 Current loop area with a decoupling capacitor.

Figure 4.13 Measurement setup.

Circuit schematic is shown in Figure 4.14, while the board details are shown in Figure 4.15. The board was purposely designed with the long traces to show the negative impact of the associated inductance.

Additional measurement details are shown in Figure 4.16.

Voltage measurements were taken on both the rising edge and the falling edge of the signals, while changing the topology between short and long trace, with or without the decoupling capacitors.

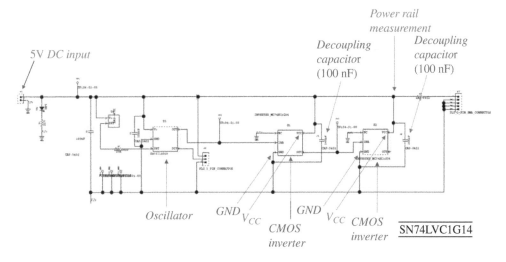

Figure 4.14   Circuit schematic.

Figure 4.15   CMOS inverter board.

Vs: 0 – 5 V, t_r = 10 ns, f = 133 kHz

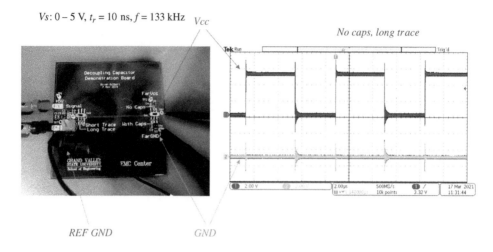

Figure 4.16   CMOS inverter board – measurement nodes.

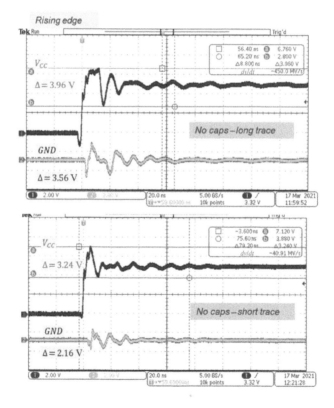

Figure 4.17   Voltage variations at V_{CC} and *GND* pins: rising edge, no decoupling capacitors, short vs. long trace.

Let us look at the voltage variations on the rising edge, first.

Figure 4.17 shows the voltage measurement at the V_{CC} and *GND* pins, on the rising edge, for a short and long traces, with no decoupling capacitors.

Observation: Long trace exhibits larger and longer peak-to-peak voltage variations at both the V_{CC} and *GND* pins.

Figure 4.18 shows the voltage measurement at the V_{CC} and *GND* pins, on the rising edge, for a long trace, with and without the decoupling capacitors.

Observation: Adding the decoupling capacitors reduces the peak-to-peak voltage variations, and their duration, at both the V_{CC} and *GND* pins.

Figure 4.19 shows the voltage measurement at the V_{CC} and *GND* pins, on the rising edge, for a short traces, with and without the decoupling capacitors.

Observation: Adding the decoupling capacitors reduces the peak-to-peak voltage variations, and their duration, at both the V_{CC} and *GND* pins. The voltage variations and their duration are smaller for a short trace than those for a long trace.

Next, let us look at the voltage variations on the falling edge.

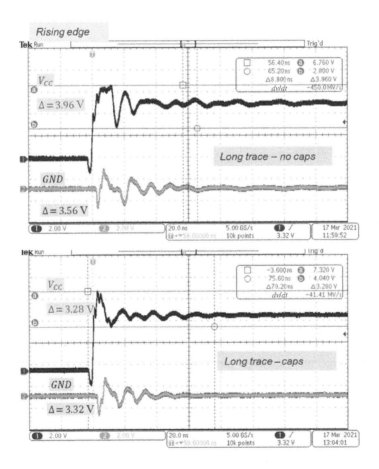

Figure 4.18   Voltage variations at V_{CC} and *GND* pins: rising edge, long trace, no caps. vs. caps.

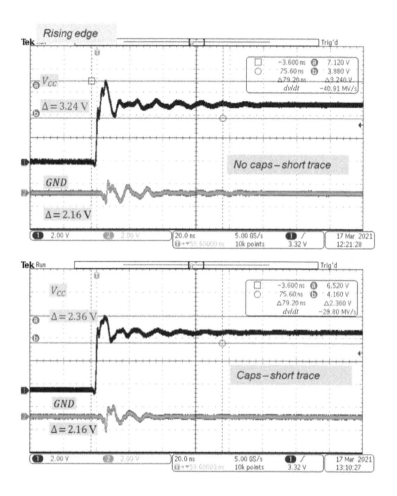

Figure 4.19   Voltage variations at V_{CC} and *GND* pins: rising edge, short trace, no caps. vs. caps.

Figure 4.20 shows the voltage measurement at the V_{CC} and *GND* pins, on the falling edge, for a short and long traces, with no decoupling capacitors.

Observation: Long trace exhibits larger and longer peak-to-peak voltage variations at both the V_{CC} and *GND* pins.

Figure 4.21 shows the voltage measurement at the V_{CC} and *GND* pins, on the falling edge, for a long trace, with and without the decoupling capacitors.

Observation: Adding the decoupling capacitors reduces the peak-to-peak voltage variations, and their duration, at both the V_{CC} and *GND* pins.

Figure 4.22 shows the voltage measurement at the V_{CC} and *GND* pins, on the falling edge, for a short traces, with and without the decoupling capacitors.

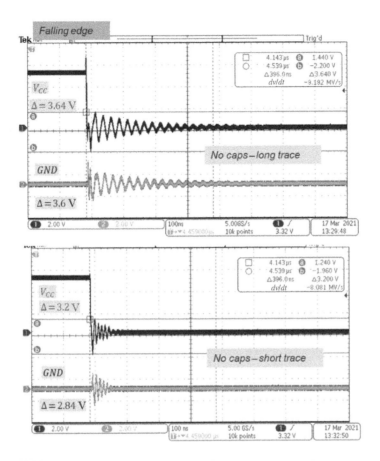

Figure 4.20 Voltage variations at V_{CC} and *GND* pins: falling edge, no decoupling capacitors, short vs. long trace.

Observation: Adding the decoupling capacitors reduces the peak-to-peak voltage variations, and their duration, at both the V_{CC} and *GND* pins. The voltage variations and their duration are smaller for a short trace than those for a long trace.

4.2.2 Decoupling Capacitor Configurations

Rarely, a single decoupling capacitor is used in PDNs; instead, multiple decoupling capacitors are used. Some of the common decoupling approaches include:

1. Multiple capacitors of the same value.
2. Two capacitors of different values spaced decades apart.
3. Three capacitors of different values spaced a decade apart each.

We will investigate these three approaches next [Adamczyk, 2021b].

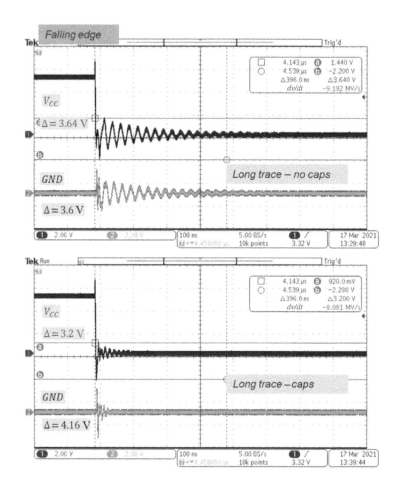

Figure 4.21   Voltage variations at V_{CC} and *GND* pins: falling edge, long trace, no caps. vs. caps.

4.2.2.1 Multiple Capacitors of the Same Value

Consider a network consisting of a 1 nF capacitor with a circuit parasitic inductance of 2.4 nH, and a parasitic resistance of 0.1 Ω. Let's look at the impedance curves of three different networks: single 1 nF capacitor, two 1 nF capacitors, and four 1 nF capacitors, shown in Figure 4.23.

The impedance curves for these networks are shown in Figure 4.24.

The resonant frequency of the series *RLC* circuit is given by [Adamczyk, 2021a]:

$$f_r = f_0 = \frac{1}{2\pi\sqrt{LC}} = \frac{1}{2\pi\sqrt{(2.4 \times 10^{-9})(1 \times 10^{-9})}} = 102.73\,\text{MHz} \tag{4.2}$$

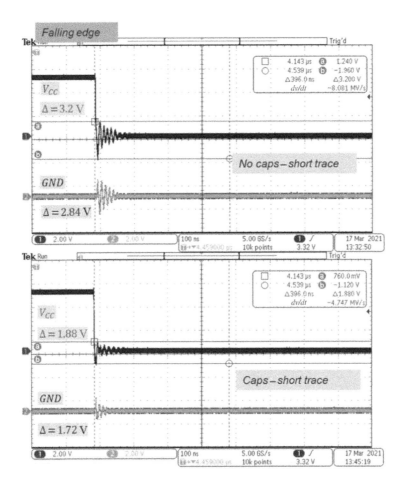

Figure 4.22 Voltage variations at V_{CC} and GND pins: falling edge, short trace, no caps. vs. caps.

and is consistent with the value shown in Figure 4.24. L and C in Eq. (4.2) are the total inductances and capacitances, respectively. These values do not change for different configurations of Figure 4.23 since, at resonance, the capacitor networks are connected in parallel and the total capacitance is and inductance (assuming that the mutual inductances are negligible) are given by [Ott, 2009]:

$$C_{total} = nC, \qquad L_{total} = \frac{L}{n}, \qquad L_{total} \times C_{total} = LC \qquad (4.3)$$

where n is the number of capacitors. The impedance value at resonance does not stay constant and is equal to

$$Z(f_r) = R_{total} = \frac{R}{n} \qquad (4.4)$$

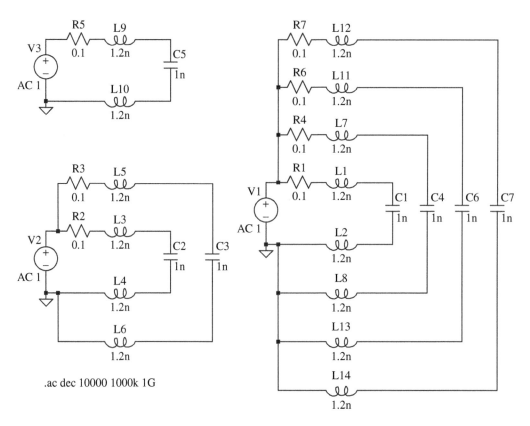

Figure 4.23 Single 1 nF capacitor vs. multiple 1 nF capacitors.

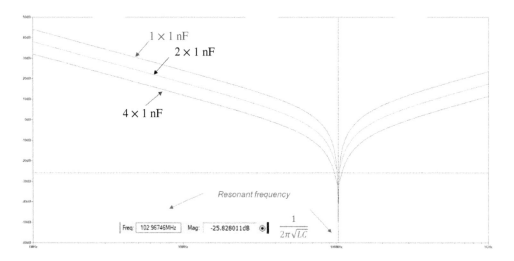

Figure 4.24 Impedance simulation results – capacitors of the same value.

In Adamczyk and Teune [2020b] we measured the impedance a four-layer PCBs populated with one 1 nF, two 1 nF, and four 1 nF capacitors. The PCB boards used are shown in Figure 4.25.

Figure 4.26 shows the impedance curves for the case of a single 1 nF capacitor vs. two 1 nF capacitors, while Figure 4.27. Compares two 1 nF capacitors vs. four 1 nF capacitors.

Note that, in Region 1, the measurement results, shown in Figures 4.26 and 4.27, are consistent with the simulation results of Figure 4.24.

4.2.2.2 Two Capacitors of Different Values Spaced Decades Apart

Next, consider a network with two capacitors, with their values either one or two decades apart, as shown in Figure 4.28.

The impedance curves for these networks are shown in Figure 4.29.

Let $C_1 = 100\,\text{nF}$, $C_2 = 10\,\text{nF}$, $C_3 = 1\,\text{nF}$. Under the condition $C_1 \ll C_3$, the first resonant frequency is approximately [Paul, 2006]:

$$f_1 = \frac{1}{2\pi\sqrt{LC_1}} = \frac{1}{2\pi\sqrt{(2.4\times 10^{-9})(100\times 10^{-9})}} = 10.27\,\text{MHz} \qquad (4.5)$$

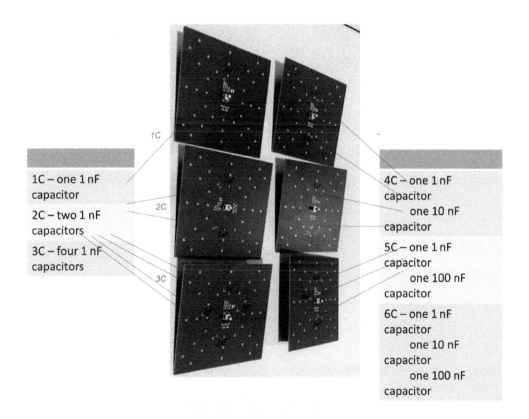

Figure 4.25   PCB boards and capacitor values investigated.

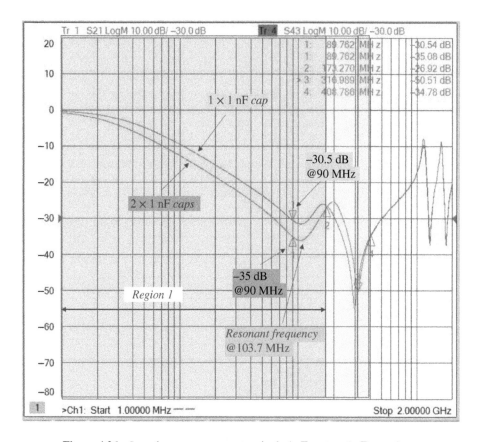

Figure 4.26   Impedance measurement – single 1 nF vs. two 1 nF capacitors.

while under the condition $C_2 \ll C_3$, the second resonant frequency is approximately

$$f_2 = \frac{1}{2\pi\sqrt{LC_2}} = \frac{1}{2\pi\sqrt{(2.4 \times 10^{-9})(10 \times 10^{-9})}} = 32.48\,\text{MHz} \tag{4.6}$$

The anti-resonant frequency can be approximated from

$$f_3 = \frac{1}{2\pi\sqrt{LC_3}} = \frac{1}{2\pi\sqrt{(2.4 \times 10^{-9})(1 \times 10^{-9})}} = 72.64\,\text{MHz} \tag{4.7}$$

Note that these calculated resonant frequencies are consistent with the simulation results in Figure 4.29. Next, let's compare these results with the measured values, shown in Figure 4.30.

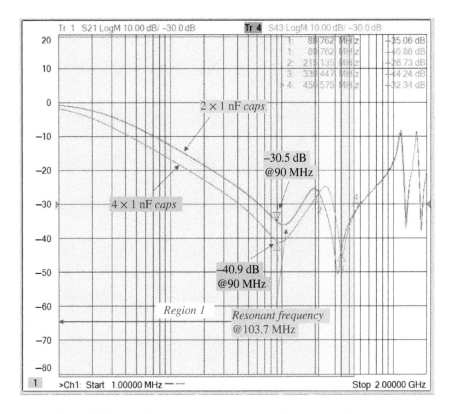

Figure 4.27   Impedance measurement – two 1 nF vs. four 1 nF capacitors.

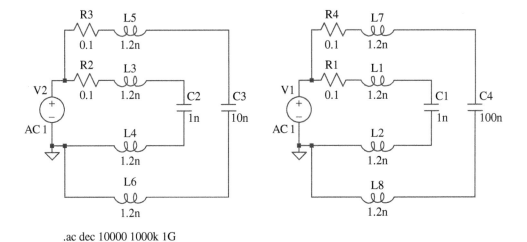

.ac dec 10000 1000k 1G

Figure 4.28   Two capacitors decades apart.

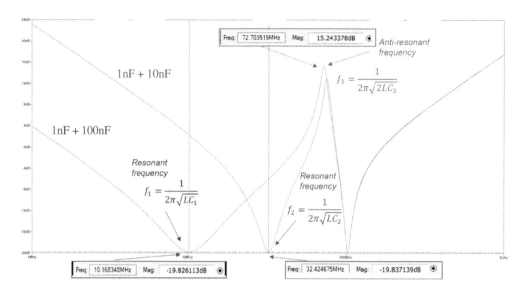

Figure 4.29   Impedance curves – two capacitors decades apart.

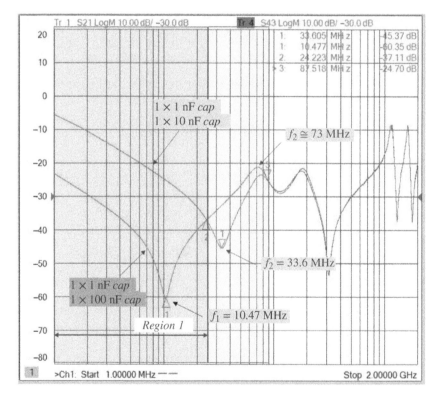

Figure 4.30   Impedance measurement – capacitors decade apart.

Again, the measured resonant frequencies are consistent with the analytical and simulation results.

4.2.2.3 Three Capacitors with Values Decades Apart

Finally, let's consider a network with three capacitors decades apart and compare their impact against four capacitors of the same value, as shown in Figure 4.31.

The impedance curves for these networks are shown in Figure 4.32.

Note that now we have three resonant frequencies and two anti-resonant frequencies. Next, let's compare these results with the measured values, shown in Figure 4.33.

Note that the measured resonant frequencies are consistent with the calculated and simulated values. Also, the measured anti-resonant frequencies agree with the predicted simulated values. It interesting to note that the simulated and measured border frequencies of Region 1 are very close to each other (59.2 MHz vs. 56.8 MHz). Region 1 is the frequency range where the impedance of three capacitors decades apart is lower than the impedance of four capacitors of the same value.

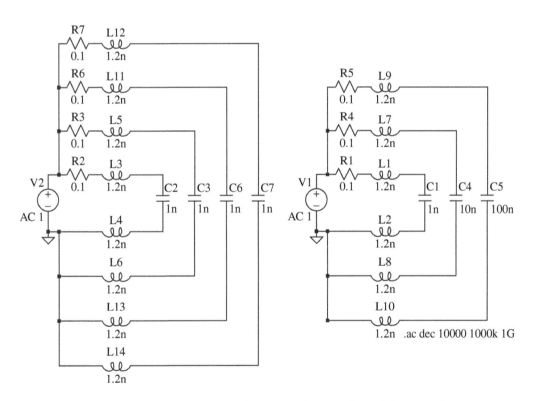

Figure 4.31   Three capacitors decades apart vs. four capacitors of the same values.

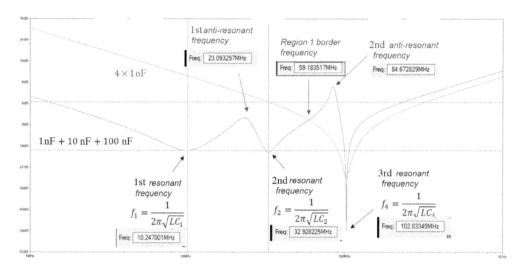

Figure 4.32 Impedance curves – three capacitors decades apart.

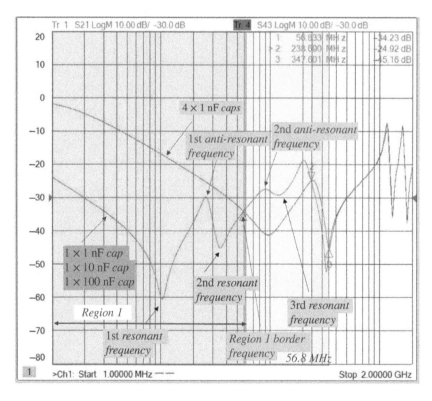

Figure 4.33 Impedance measurement – three capacitors decade apart.

4.3 Decoupling Capacitors and Embedded Capacitance

Studies have shown that when the PCB power and ground planes are closely spaced they provide a low-impedance path for the switching currents in a PDN [Hubing et al., 1995]. The power- and ground-plane pairs also impact the effectiveness of decoupling capacitors. In this section we investigate several PCB configurations in which we vary both the distance between power and ground planes, as well as the placement and values of the decoupling capacitors.

4.3.1 Decoupling Capacitors and Closely vs. Not Closely Spaced Power and Ground Planes

In this section we investigate two four-layer PCBs, shown in with a power- and ground-plane pair spaced 3 mils (Case A) and 30 mils (Case B) apart, respectively [Adamczyk and Teune, 2020a]. The two boards are shown in Figure 4.34.

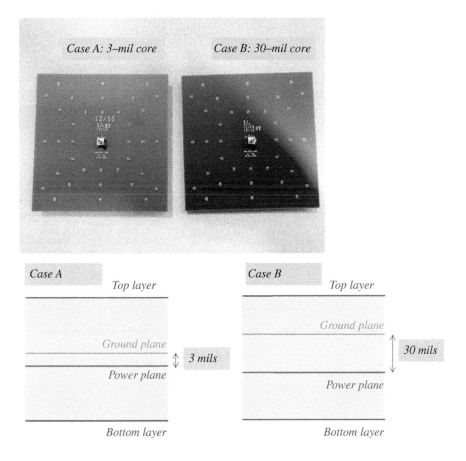

Figure 4.34 Four-layer PCB boards investigated in this study.

The boards are populated with four capacitors of the same value, at various distances from the measurement point.

We measure the impedance of the two boards for the following cases: Case 1 – Bare boards, no decoupling capacitors; Case 2 – Four 1 nF caps at a short distance (1 inch) from the measurement point; Case 3 – Four 1 nF caps at a medium distance (1.5 inch) from the measurement point; Case 4 – Four 1 nF caps at a far distance (2 inches) from the measurement point.

Figure 4.35 shows the details of the capacitor locations on a PCB.

Figure 4.36 summarizes the PCB configurations investigated in this study.

The measurement setup used in this study is shown in Figure 4.37.

Figure 4.38 shows the impedance measurement for Cases 1A and 1B – 3 mil-board vs. 30 mil-board – no decoupling capacitors.

Observations: The impedance of the 3-mil PCB is lower than the impedance of the 30-mil PCB, over the entire frequency range, except for the small region around the resonant frequency of 282 MHz corresponding to the 30-mil board. At 200 MHz the impedance of the 3-mil PCB is 25.4 dB ($60.2 - 34.8 = 25.4$ dB) lower than the impedance of the 30-mil PCB. The 3-mil PCB resonates at 228 MHz.

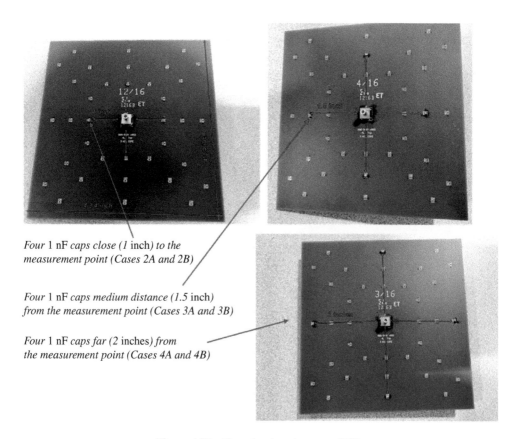

Figure 4.35   Capacitor locations on a PCB.

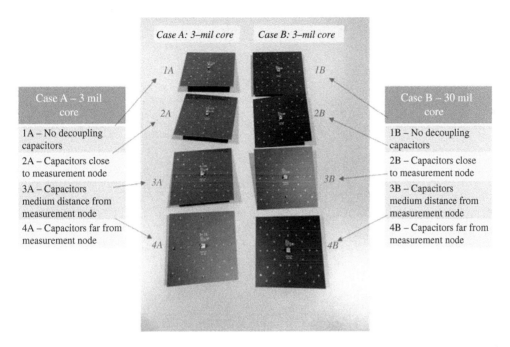

Figure 4.36 Summary of the PCB configurations.

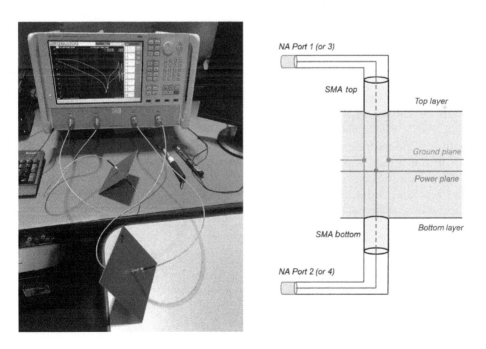

Figure 4.37 Measurement set-up.

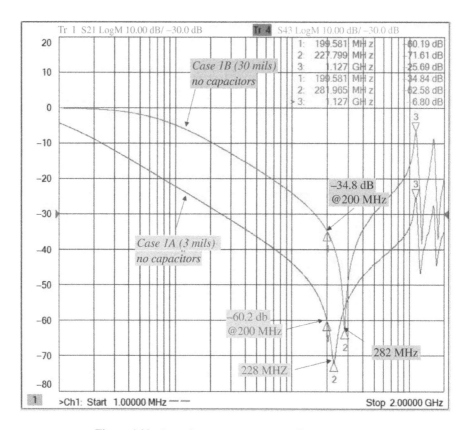

Figure 4.38   Impedance measurements – Cases 1A vs. 1B.

Figure 4.39 shows the impedance measurement for Cases 1A and 2A – (3 mil PCB) no capacitors vs. capacitors at close distance.

Observations: Addition of the capacitors (close to the measurement point) introduces a resonance at 105 MHz and an antiresonance at 138.5 MHz. In Region 1 (1–122 MHz) Case 2A-PCB has a lower impedance (4.2 dB difference at 50 MHz). In Region 2 (122–235 MHz) Case 1A-PCB has a lower impedance. In Region 3 (235–306 MHz) Case2A-PCB has a lower impedance again. Beyond the frequency of 306 MHz the capacitors have no impact on the impedance.

Figure 4.40 shows the impedance measurement for Cases 1B and 2B – (30 mil PCB) no capacitors vs. capacitors at close distance.

Observations: In Region 1 (1–164 MHz) Case 2B-PCB has a lower impedance (17.6 dB difference at 50 MHz). In Region 2 (164–331 MHz) Case 1B-PCB has a lower impedance. In Region 3 (331–496 MHz) Case 2B-PCB has a lower impedance again. Beyond the frequency of 496 MHz the capacitors have no impact on the impedance.

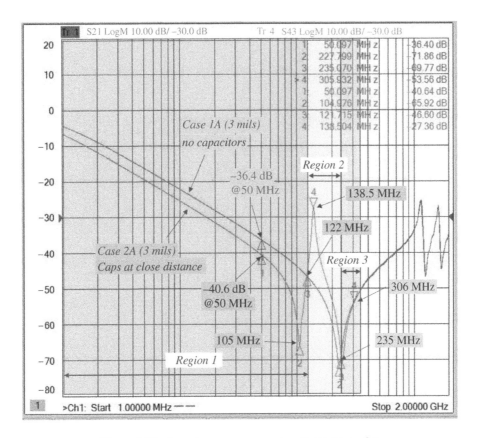

Figure 4.39 Impedance measurements – Cases 1A vs. 2A.

Figure 4.41 summarizes the results for Cases 1A vs. 2A and 1B vs. 2B.

Figure 4.42 shows the impedance measurement for Cases 2A and 3A – (3 mil PCB) capacitors at a close distance vs. capacitors at a medium distance.

Observations: On a 3-mil board, moving the capacitors further away (from 1 to 1.5 inch) from the measurement point has no impact on the board impedance.

Let's see if the same holds for the 30-mil board.

Figure 4.43 shows the impedance measurement for Cases 2B and 3B – (30 mil PCB) capacitors at a close distance vs. capacitors at a medium distance.

Observations: On a 30-mil board, moving the capacitors further away (from 1 to 1.5 inch) from the measurement point shifts the resonant and antiresonant frequencies as marked. Thus, moving the capacitors does impact the impedance, but effect is relatively small (there was virtually no effect for the 3-mil board).

Next, let's move the capacitors even further away (2 inches) from the measurement point. Let's investigate the 3-mil board first.

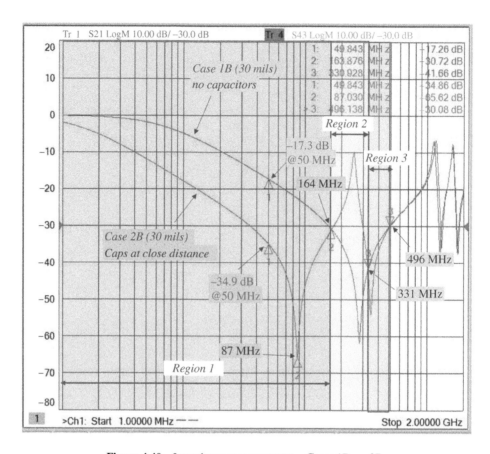

Figure 4.40   Impedance measurements – Cases 1B vs. 2B.

Case	Region 1 (MHz)	Δ @ 50 MHz dB	Region 2 MHz	Region 3 MHz
1A vs. 2A	1–122	4.2	122–235	235–306
1B vs. 2B	1–164	17.6	164–331	331–496

Figure 4.41   Summary: Cases 1A vs. 2A and 1B vs. 2B.

Figure 4.44 shows the impedance measurement for Cases 2A and 4A – (3 mil PCB) capacitors at a close distance vs. capacitors at a far distance.

Observations: On a 3-mil board, moving the capacitors even further away (from 1 inch to 2 inches) from the measurement point still has no impact on the impedance.

Let's see if the same holds for the 30-mil board.

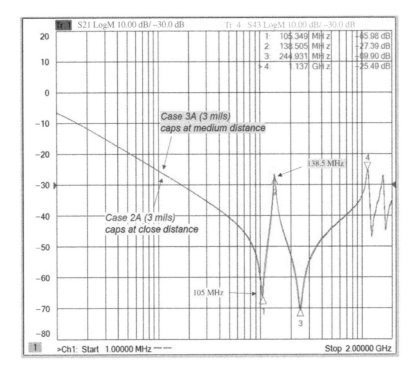

Figure 4.42   Impedance measurements – Cases 2A vs. 3A.

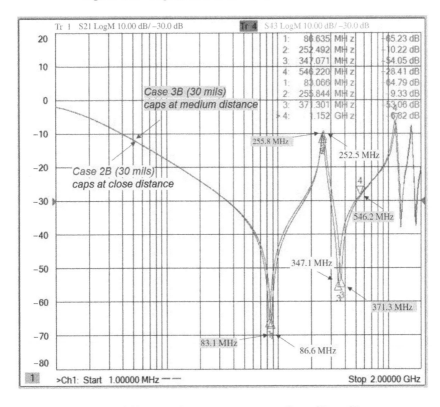

Figure 4.43   Impedance measurements – Cases 2B vs. 3B.

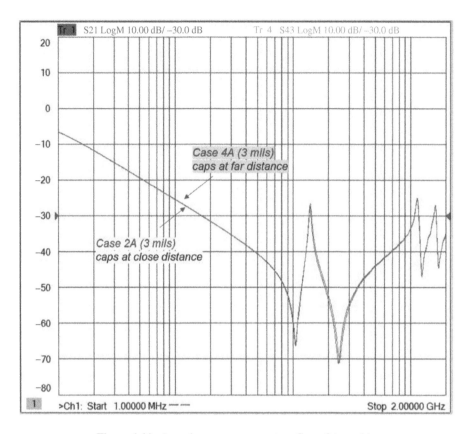

Figure 4.44   Impedance measurements – Cases 2A vs. 4A.

Figure 4.45 shows the impedance measurement for Cases 2B and 4B – (30 mil PCB) capacitors at a close distance vs. capacitors at a far distance.

Observations: On a 30-mil board, moving the capacitors even further away (from 1h to 2 inches) from the measurement point has a large impact (17.5 dB difference at 50 MHz) on the impedance.

This result should look familiar (compare Figure 4.45 with Figure 4.40). Or better yet, let's compare Case 1B (no capacitors) to Case 4B (capacitors 2 inches away). This is shown in Figure 4.46 – Cases 1B vs. 4B – (30 mil PCB) no capacitors vs. capacitors at a far distance.

Observations: On a 30-mil board, the capacitors placed 2 inches away from the mea- surement point virtually have no impact on the board impedance! The impedance curve for the bare board with no capacitors is the same as the one with the capacitors 2 inches away from the measurement point. When the capacitors were placed at 1 or 1.5 inch away from the measurement point, they had large impact on lowering the board impedance. When placed too far from the measurement point (2 inches in this case) they have no impact!

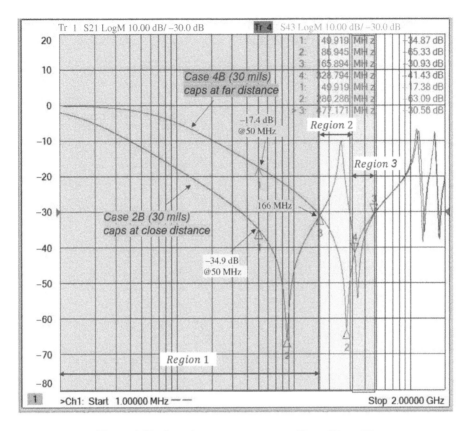

Figure 4.45   Impedance measurements – Cases 2B vs. 4B.

Recall: On a 3-mil board adding the capacitors at a close distance had a relatively small impact (Figure 4.39); moving them 1.5 inch away (Figure 4.42) or 2 inches away (Figure 4.44) had the same small impact. That is, the location of the capacitors did not matter, as long as they were up two 2 inches away.

On a 30 mil-board adding the capacitors at a close distance had a large impact (Figure 4.40); moving them 1.5 inch away had a similar large impact (Figure 4.43), but moving them 2 inches away (Figures 4.45 and 4.46) had no impact on the board impedance. That is, placing the capacitors too far (2 inches) renders them useless as they provide no benefit!

These results are consistent with the established guidelines: (i) When power and ground planes are spaced less than 10 mils apart, the location of decoupling capacitors is NOT critical as long as they are within the general vicinity of the IC. (ii) When power and ground planes are spaced more than 10 mils apart, the location of decoupling capacitors is critical.

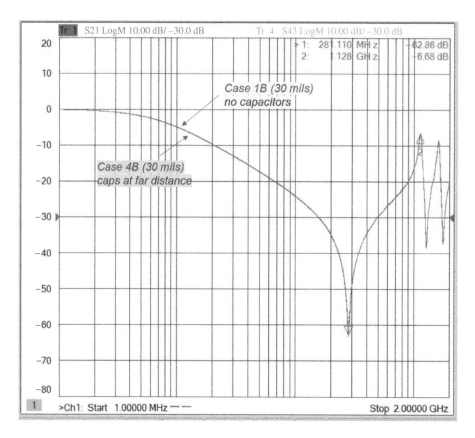

Figure 4.46   Impedance measurements – Cases 1B vs. 4B.

4.3.2 Impact of the Number and Values of the Decoupling Capacitors

The conclusion of Section 4.3.1 was that the location of capacitors was more critical on a 30-mil board than it was on a 3-mil board, leading to the choice of 30-mil boards for the study described next.

In this section we use the 30-mil boards and populate them with multiple capacitors of the same value (Cases 1C, 2C, 3C) as well as with the capacitors of different values, decades apart (Cases 4C, 5C, 6C). The location of the capacitors for all cases in this study is 1 inch away from the measurement point, as shown in Figure 4.47 [Adamczyk and Teune, 2020b].

The cases shown in Figure 4.47 are summarized in Figure 4.48.

Figure 4.49 shows the impedance measurement for Cases 1B and 1C – no decoupling capacitors vs. one 1 nF capacitor.

Observations: In Region 1 (1–136 MHz) Case 1C-PCB has a lower impedance (7.6 dB difference at 90 MHz). In Region 2 (136–291 MHz) Case 1B-PCB has a lower impedance.

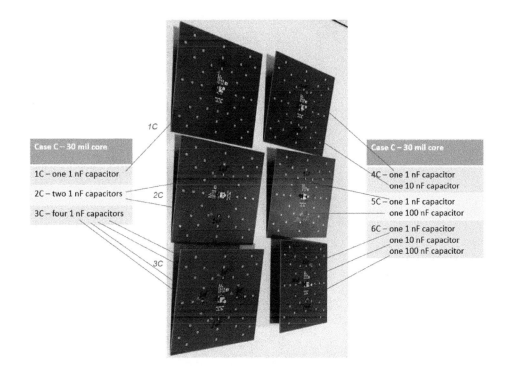

Figure 4.47 Capacitor locations and values.

	Case 1C	Case 2C	Case 3C	Case 4C	Case 5C	Case 6C
Capacitor number and values	1×1 nF	2×1 nF	4×1 nF	1×1 nF 1×10 nF	1×1 nF 1×100 nF	1×1 nF 1×10 nF 1×100 nF

Figure 4.48 Capacitor number and values.

In Region 3 (291–445 MHz) Case 1C-PCB has a lower impedance again. Beyond the frequency of 445 MHz the capacitor has no impact on the impedance.

Figure 4.50 shows the impedance measurement for Cases 1C and 2C – one 1 nF capacitor vs. two 1 nF capacitors.

Observations: The addition of the second 1 nF capacitor extends the Region 1 in frequency from 136 MHz (see Figure 4.49) to 173 MHz. It also lowers the impedance curve in that region – at 90 MHz the impedance is lowered from −30.5 to −35 dB. This is a very desirable result. Region 2 (where the Case 1C PCB outperforms the Case 2C PCB) is shifted to the right and extends from 173 to 317 MHz. Region 3 is also affected and extends now from 317 to 409 MHz. Beyond the frequency of 409 MHz the capacitors have no impact on the impedance.

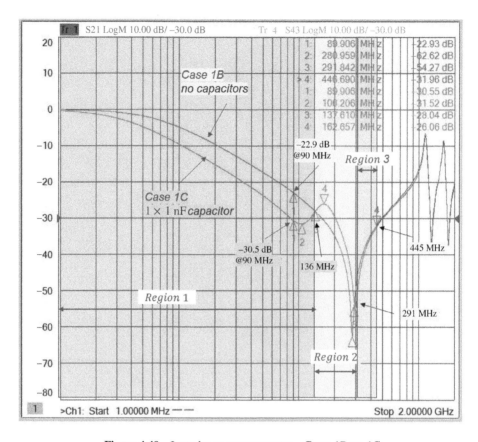

Figure 4.49   Impedance measurements – Cases 1B vs. 1C.

Figure 4.51 shows the impedance measurement for Cases 2C and 3C – two 1 nF capacitors vs. four 1 nF capacitors.

Observations: Increasing the number of 1 nF capacitors from two to four extends the Region 1 further in frequency from 173 MHz (see Figure 4.50) to 215 MHz. It also lowers the impedance curve in that region – at 90 MHz the impedance is lowered from −35 to −40.9 dB. Again, this is a very desirable result.

Region 2 (where the Case 2C PCB outperforms the Case 3C PCB) is shifted to the right and extends from 215 to 339 MHz. Region 3 is also affected and extends now from 339 to 457 MHz. Beyond the frequency of 457 MHz the capacitors have no impact on the impedance.

The major conclusion of this part of the study is that, in Region 1, increasing the number of capacitors of the same value lowers the impedance curve and extends its frequency range.

The results of this study are summarized in Figure 4.52.

Let us compare the measurement results to the LTspice simulations. The simulation circuits are shown in Figure 4.53, while the simulation results are shown in Figure 4.54.

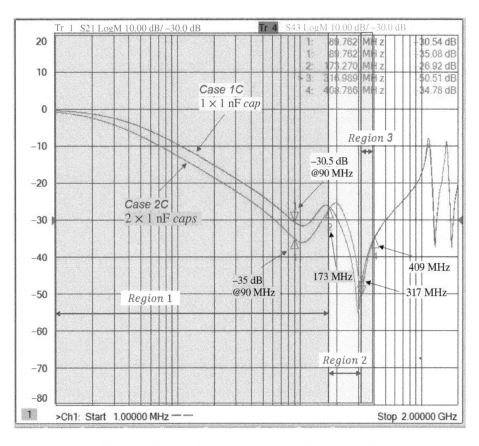

Figure 4.50 Impedance measurements – Cases 1C vs. 2C.

Figure 4.55 shows the corresponding measurement results (presented earlier).

It is quite remarkable that the self-resonant frequency obtained from simulations and measurements closely agrees with the calculated value:

$$f_r = \frac{1}{2\pi\sqrt{LC}} = \frac{1}{2\pi\sqrt{(2.4\times10^{-9})(1\times10^{-9})}} = 102.73\,\text{MHz} \tag{4.8}$$

Next, let's investigate the cases when we use multiple capacitors spaced decades apart in values.

Figure 4.56 shows the impedance measurement for Cases 2C and 4C – two 1 nF capacitors vs. one 1 nF and one 10 nF capacitors.

Observations: The impact of the two capacitors one decade apart (1 and 10 nF) on the impedance curve is quite profound. Region 1 (where the Case 4C PCB has lower impedance) is reduced in frequency and extends from 1 to 58 MHz. Region 2 (where the Case 2C PCB has lower impedance) increases in its span and extends from 58 to 208 MHz.

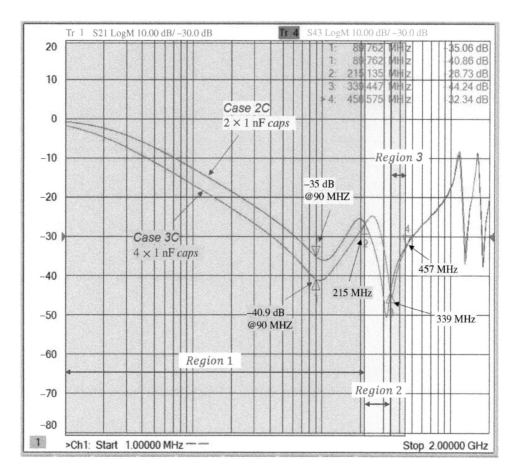

Figure 4.51   Impedance measurements – Cases 2C vs. 3C.

Case	Region 1 (MHz)	Δ @ 90 MHz dB	Region 2 MHz	Region 3 MHz
1B vs. 1C	1–136	7.6	136–291	291–445
1C vs. 2C	1–173	4.5	173–317	317–409
2C vs. 3C	1–215	5.9	215–339	339–457

Figure 4.52   Summary of the results.

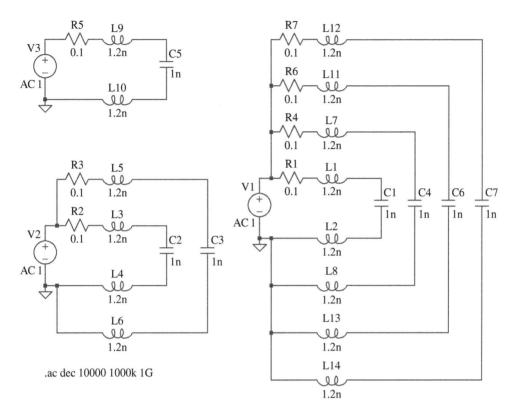

Figure 4.53 Simulation circuits.

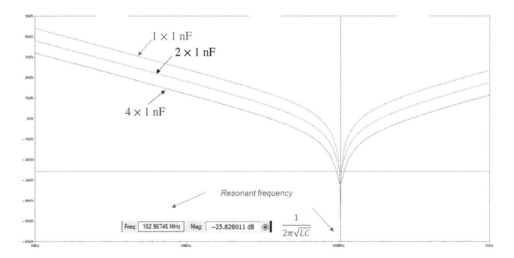

Figure 4.54 Simulation results.

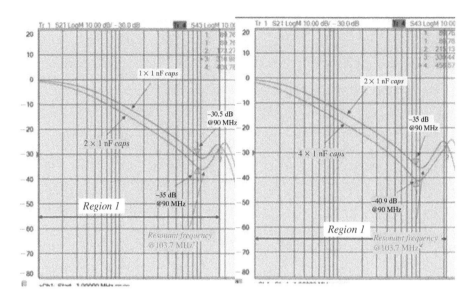

Figure 4.55 Measurement results.

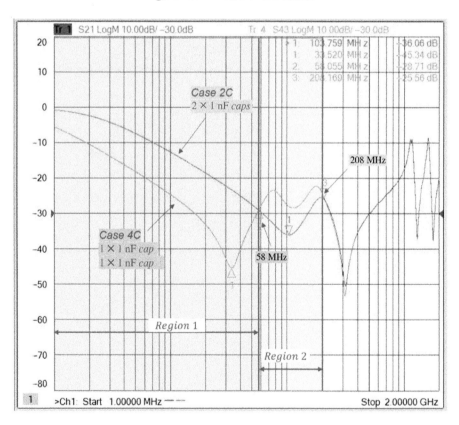

Figure 4.56 Impedance measurements – Cases 2C vs. 4C.

Figure 4.57 shows the impedance measurement for Cases 2C and 5C – two 1 nF capacitors vs. one 1 nF and one 100 nF capacitors.

Observations: Using two capacitors two decades apart (1 and 100 nF) shifts the Region 1 upper frequency to the left, from 58 MHz (see Figure 4.56) to 49 MHz. Region 2 increases in its span and extends from 49 MHz (vs. 58 MHz) to 224 MHz (vs. 208 MHz).

Next, let's compare the Cases 4C (1 and 10 nF) and 5C (1 and 100 nF) directly. Figure 4.58 shows the impedance measurement.

Observations: Using two capacitors two decades apart (Case 5C) vs. two capacitors one decade apart (Case 4C) decreases the impedance in the low frequency range, i.e. Region 1 (1–24 MHz). In Region 2 (24–87 MHz), Case 4C PCB has lower impedance. Beyond the frequency of 87 MHz the impedance of both boards is virtually the same.

The results of three comparisons are summarized in Figure 4.59.

Finally, let's compare a board with four capacitors of the same value (Case 3C: four 1 nF) with the board with three capacitors decades apart in values (Case 6C: 1, 10, 100 nF). The comparison is shown in Figure 4.60.

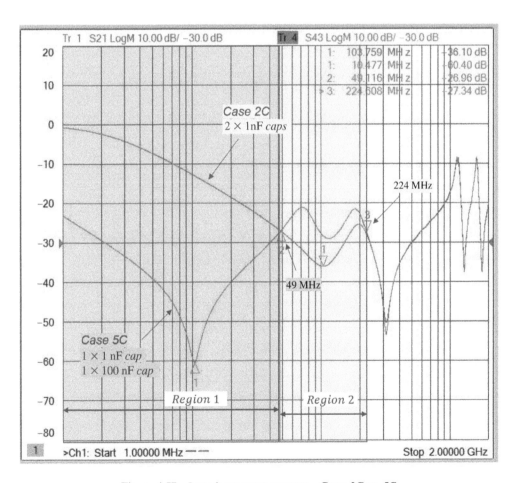

Figure 4.57 Impedance measurements – Cases 2C vs. 5C.

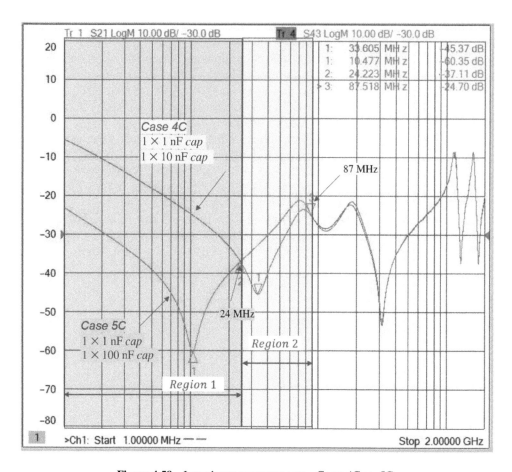

Figure 4.58   Impedance measurements – Cases 4C vs. 5C.

Case	Region 1 (MHz)	Region 2 MHz
2C vs. 4C	1–58	58–208
2C vs. 5C	1–49	49–224
2C vs. 5C	1–24	24–87

Figure 4.59   Summary of the results.

Observations: Using three capacitors decades apart (Case 6C) vs. four capacitors of the same value (Case 3C) decreases the impedance in the low frequency range, i.e. Region 1 (1–57 MHz). In Region 2 (57–215 MHz), Case 3C PCB has lower impedance. In Region 3 (215–347 MHz) the Case 3C PCB has lower impedance. Beyond the frequency of 347 MHz the capacitors have no impact on the impedance.

Again, let us compare the measurement results (for Cases 4C and 5C) to the LTspice simulations. The simulation circuits are shown in Figure 4.61, while the simulation results are shown in Figure 4.62.

Figure 4.63 shows the corresponding measurement results (presented earlier).

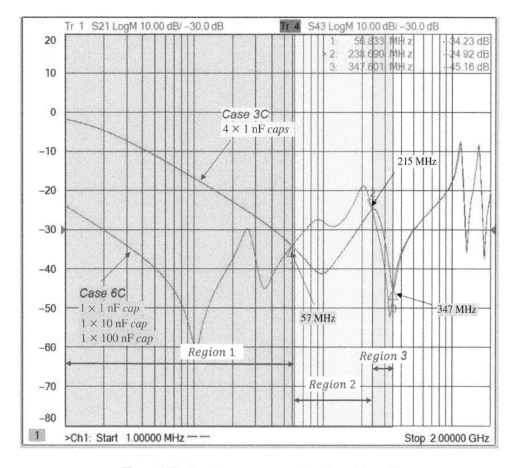

Figure 4.60 Impedance measurements – Cases 3C vs. 6C.

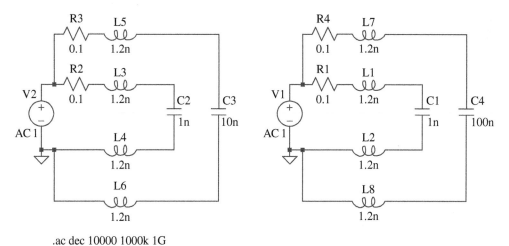

.ac dec 10000 1000k 1G

Figure 4.61 Simulation circuits: Cases 4C and 5C.

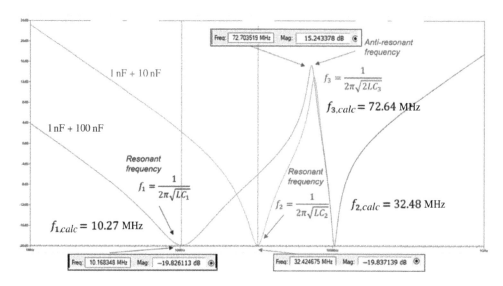

Figure 4.62   Simulation results: Cases 4C and 5C.

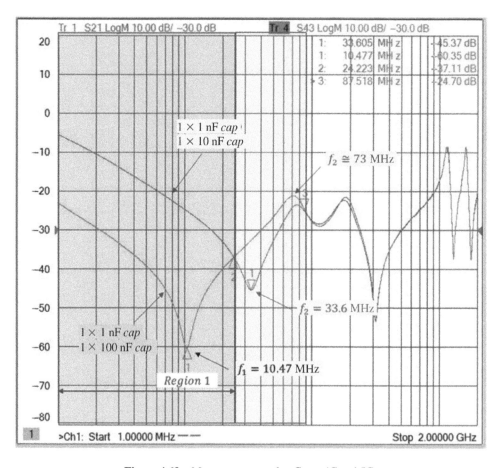

Figure 4.63   Measurement results: Cases 4C and 5C.

The results shown in Figure 4.63 display several resonant and antiresonant frequencies. These can be obtained analytically as [Paul, 2006]:

$$f_1 = \frac{1}{2\pi\sqrt{LC_1}} = \frac{1}{2\pi\sqrt{(2.4 \times 10^{-9})(100 \times 10^{-9})}} = 10.27\,\text{MHz} \qquad (4.9)$$

$$f_2 = \frac{1}{2\pi\sqrt{LC_2}} = \frac{1}{2\pi\sqrt{(2.4 \times 10^{-9})(10 \times 10^{-9})}} = 32.48\,\text{MHz} \qquad (4.10)$$

$$f_3 = \frac{1}{2\pi\sqrt{LC_3}} = \frac{1}{2\pi\sqrt{(2.4 \times 10^{-9})(1 \times 10^{-9})}} = 72.64\,\text{MHz} \qquad (4.11)$$

Again, it is quite remarkable that the self-resonant frequencies obtained from simulations and measurements closely agree with the calculated values.

Finally, let us compare the measurement results (for Cases 3C and 6C) to the LTspice simulations. The simulation circuits are shown in Figure 4.64, while the simulation results are shown in Figure 4.65.

Once again, we observe a close agreement between the measured and simulated values.

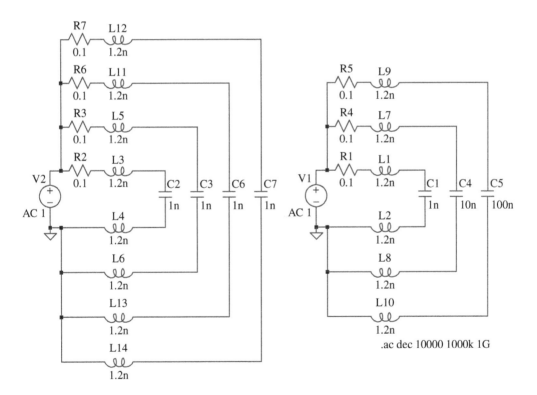

Figure 4.64 Simulation circuits: Cases 3C and 6C.

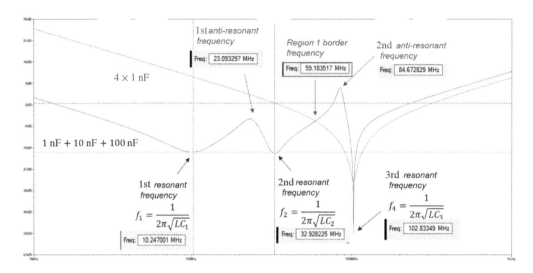

Figure 4.65   Simulation results: Cases 3C and 6C.

4.4 Laboratory Exercises

4.4.1 Decoupling Capacitors

Objective: This laboratory exercise examines the impact of the decoupling capacitors and the trace length on the PDN signal integrity during the CMOS inverter switching.

Note: This laboratory requires a custom PCB [French, 2023], for which the Altium files and BOM are provided.

4.4.1.1 Laboratory Equipment and Supplies

DC Power Supply: BK Precision1760A (or equivalent), shown in Figure 4.66.

Oscilloscope: Tektronix MDO3104 (or equivalent), shown in Figure 4.67.

Banana Cables: Tektronix MDO3104 (or equivalent), shown in Figure 4.68.
Test Lead Banana to Banana 36″
Supplier: Ponoma Electronics
Digi-Key Part Number: 501-1423-ND

PCB: Custom PCB, shown in Figure 4.69.
Circuit schematic for this PCB is shown in Figure 4.70.

The switching frequency of the oscillator (U1) is controlled by the resistor R_1 whose value needs to be in the range $10\,\text{K}\Omega < R_1 < 2\,\text{M}\Omega$. A switch adjacent to U1 can be set to the frequency multiplying factor of 1, 10, or 100.

There are additional two sets of switches. The first set controls the length of the PWR and GND traces. The length of each trace can be set to short, medium, or long.

The second set of switches controls the decoupling capacitor values at each CMOS. These can be set to no capacitor, $0.1\,\mu\text{F}$, or $1\,\mu\text{F}$.

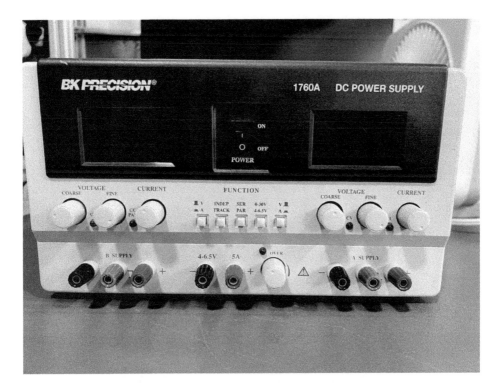

Figure 4.66 Laboratory equipment: DC power supply.

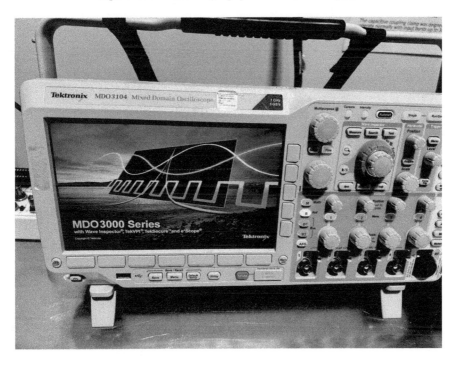

Figure 4.67 Laboratory equipment: oscilloscope.

Figure 4.68   Laboratory equipment: banana cables.

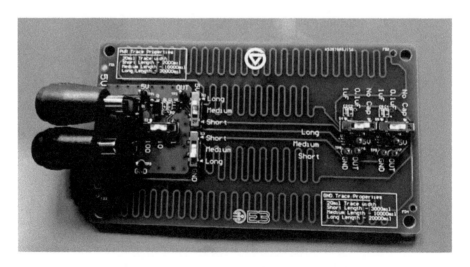

Figure 4.69   Laboratory equipment: custom PCB.

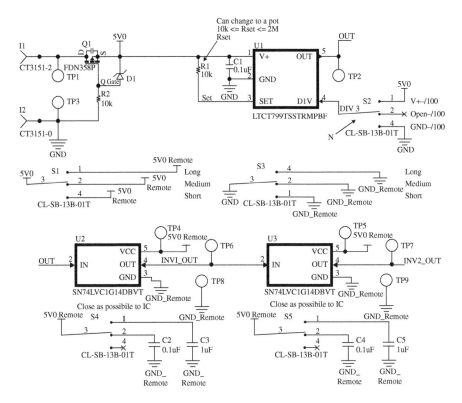

Figure 4.70 Circuit schematic – custom PCB.

4.4.1.2 Laboratory Procedure

Connect the power supply, PCB, and the oscilloscope, as shown in Figure 4.71.

Note: The GND leads of the oscilloscope probes should be connected to the main reference GND node, Test Point 3 (TP3).

With five switches, each having three positions, there is a plentitude of possible measurements.

Connect Channel 1 oscilloscope probe to TP8 (U2 GND). Connect Channel 2 oscilloscope probe to TP4 (U2 VCC). Alternatively, Connect Channel 1 oscilloscope probe to TP9 (U3 GND). Connect Channel 2 oscilloscope probe to TP5 (U3 VCC).

Set the frequency of the oscillator to a desired value. Set the trace lengths and capacitor values, and capture the voltage waveforms on both the rising and falling edges of the oscillator output (measured at TP2). Move the switches to a new position and capture the waveforms again. Use the measurements of Section 4.2.1 as a guideline. Add additional cases as desired.

Tabulate the results and comment on the impact of decoupling capacitors and trace lengths on the VCC and GND voltages.

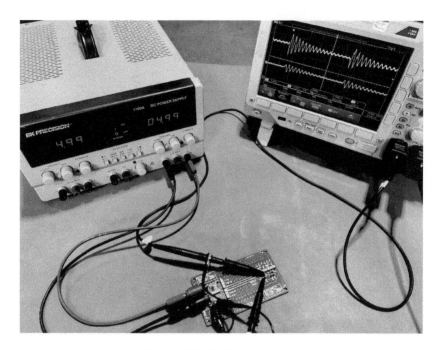

Figure 4.71 Laboratory setup.

4.4.2 Embedded Capacitance and Decoupling Capacitors

Objective: This laboratory exercise examines the impact of decoupling capacitors and embedded capacitance on the PCB impedance.

Note: This laboratory requires custom PCBs [Aldridge, 2022], for which the Altium files and BOM are provided.

4.4.2.1 Laboratory Equipment and Supplies

Network analyzer: KEYSIGHT ENA Network Analyzer E5080A (or equivalent), shown in Figure 4.72.

Adapter Coaxial Connector N Plug, Male Pin To SMA Jack, Female Socket 50 Ohm: Digikey PN: ACX1341-ND (or equivalent), shown in Figure 4.73.

SMA Male to SMA Male Cable Assembly, 4 Sets: Digikey PN: 3648-P1CA-STMSTM-ST085-36-ND (or equivalent), shown in Figure 4.74.

PCB: Custom PCBs – shown in Figure 4.75. These PCBs are similar to the ones discussed in Sections 4.3.1 and 4.3.2, with one distinct difference: the distance between the ground and power planes is 8 mils and 47 mils, respectively.

4.4.2.2 Laboratory Procedure

Create the following PCB configurations (described in Sections 4.2.1 and 4.3.2): 1A through 4A, 1B through 4B, 1C through 6C.

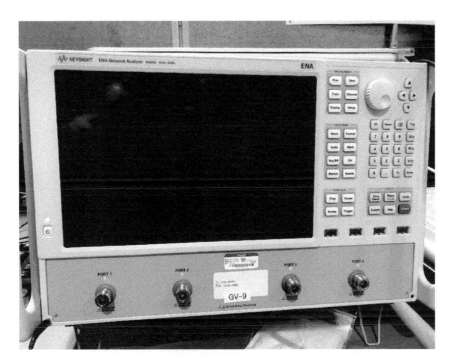

Figure 4.72 Laboratory equipment: network analyzer.

Figure 4.73 Laboratory equipment: N-type to SMA connector.

Figure 4.74   Laboratory equipment: SMA to SMA cable.

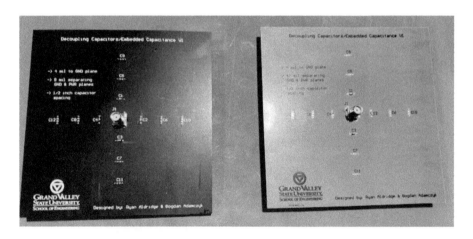

Figure 4.75   Laboratory equipment: custom PCBs.

Connect PCBs to the network analyzer as shown in Figure 4.76.

Note: Compare the impedance measurements for the Cases described in each step. Use markers as appropriate to identify relevant frequencies which define the specific regions where one design outperforms the other (if applicable). Comment on the results.

Step 1: Compare Case 1A with Case 1B.
Step 2: Compare Case 1A with Case 2A.

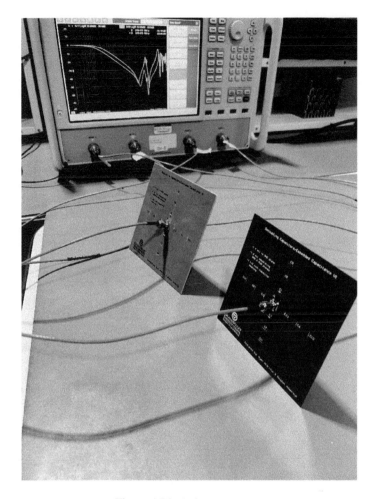

Figure 4.76 Laboratory setup.

Step 3: Compare Case 1B with Case 2B.
Step 4: Compare Case 2A with Case 3A.
Step 5: Compare Case 2B with Case 3B.
Step 6: Compare Case 2A with Case 4A.
Step 7: Compare Case 2B with Case 4B.
Step 8: Compare Case 1B with Case 1C.
Step 9: Compare Case 1C with Case 2C.
Step 10: Compare Case 2C with Case 3C.
Step 11: Compare Case 2C with Case 4C.
Step 12: Compare Case 2C with Case 5C.
Step 13: Compare Case 4C with Case 5C.
Step 14: Compare Case 3C with Case 6C.

References

Bogdan Adamczyk. *Foundations of Electromagnetic Compatibility*. John Wiley & Sons, Ltd, Chichester, UK, 2017. ISBN 9781119120810. doi: https://doi.org/10.1002/9781119120810.

Bogdan Adamczyk. Impact of a Decoupling Capacitor in a CMOS Inverter Circuit. *In Compliance Magazine*, September 2019. URL https://incompliancemag.com/article/impact-of-a-decoupling-capacitor-in-a-cmos-inverter-circuit/.

Bogdan Adamczyk. EMC Resonance Part I: Non-Ideal Passive Components. *In Compliance Magazine*, February 2021a. URL https://incompliancemag.com/article/emc-resonance-part-i-non-ideal-passive-components/.

Bogdan Adamczyk. EMC Resonance Part II: Decoupling Capacitors. *In Compliance Magazine*, March 2021b. URL https://incompliancemag.com/article/emc-resonance-part-ii-decoupling-capacitors/.

Bogdan Adamczyk and Jim Teune. Impact of Decoupling Capacitors and Embedded Capacitance on Impedance of Power and Ground Planes: Part I. *In Compliance Magazine*, March 2020a. URL https://incompliancemag.com/article/impact-of-decoupling-capacitors-and-embedded-capacitance/.

Bogdan Adamczyk and Jim Teune. Impact of Decoupling Capacitors and Embedded Capacitance on Impedance of Power and Ground Planes: Part II. *In Compliance Magazine*, April 2020b. URL https://incompliancemag.com/article/impact-of-decoupling-capacitors-and-embedded-capacitance-on-impedance-of-power-and-ground-planes-part-ii/.

Ryan Aldridge. PDN PCB Designer, 2022.

Matt French. Decoupling Capacitors PCB Designer, 2023.

T.H. Hubing, J.L. Drewniak, T.P. Van Doren, and D.M. Hockanson. Power bus decoupling on multilayer printed circuit boards. *IEEE Transactions on Electromagnetic Compatibility*, **37**:155–166, 1995. ISSN 00189375. doi: https://doi.org/10.1109/15.385878. URL http://ieeexplore.ieee.org/document/385878/.

Henry W. Ott. *Electromagnetic Compatibility Engineering*. John Wiley & Sons, Inc., Hoboken, New Jersey, 2009. ISBN 978-0-470-18930-6.

Clayton R. Paul. *Introduction to Electromagnetic Compatibility*. John Wiley & Sons, Inc., Hoboken, New Jersey, 2nd edition, 2006. ISBN 978-0-471-75500-5.

Chapter 5: EMC Filters

5.1 Insertion Loss Definition

EMC filters are described in terms of the Insertion Loss (IL), defined as [Paul, 2006]

$$IL_{dB} = 20 \log_{10} \frac{V_{L,without\,filter}}{V_{L,with\,filter}} \qquad (5.1)$$

where V_L is the magnitude of the complex voltage $\hat{V}_L$. Figure 5.1 illustrates this definition.

Since $V_{L,without\,filter} > V_{L,with\,filter}$, the insertion loss defined by Eq. (5.1) is a positive number in dB. The insertion loss could alternatively be defined as

$$IL_{dB} = 20 \log_{10} \frac{V_{L,with\,filter}}{V_{L,without\,filter}} \qquad (5.2)$$

In this case the insertion loss in dB is the negative of the loss defined by Eq. (5.1). We will use this definition when plotting the simulation results and comparing the simulation results to the Vector Network Analyzer (VNA) measurements.

5.2 Basic Filter Configurations

The most basic low-pass EMC filter configurations are shown in Figure 5.2.

A typical 2nd-order EMC low-pass filter consists of a series inductance and shunt capacitance [Adamczyk, 2017].

Figure 5.3 shows two different filter configurations: CL and LC filters.

The filters shown in Figure 5.3 can be cascaded to produce higher-order filters. This is shown in Figure 5.4.

Third-order π and T filters are shown in Figure 5.5.

The higher the order of the filter, the sharper the transition from the pass band to the rejection region. Note that for each order of the filter we have two different configurations. Which one will perform better, i.e. which one will result in the higher insertion loss (in the negative sense)? This question is answered in Section 5.3.

5.3 Source and Load Impedance Impact

The insertion loss of the filter depends (in most cases) on the load and source impedance. The general rule is that the inductor should be on the low-impedance side and the capacitor on the high-impedance side [Adamczyk and Haring, 2019].

Figure 5.6 shows the appropriate configurations when both the source and the load impedances are low.

Figure 5.7 shows the appropriate configurations when both the source and the load impedances are high.

Principles of Electromagnetic Compatibility: Laboratory Exercises and Lectures, First Edition. Bogdan Adamczyk.
© 2024 John Wiley & Sons Ltd. Published 2024 by John Wiley & Sons Ltd.
Companion website: www.wiley.com/go/principlesofelectromagneticcompatibility

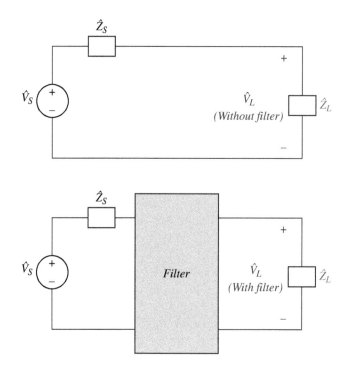

Figure 5.1 Illustration of the insertion loss of a filter.

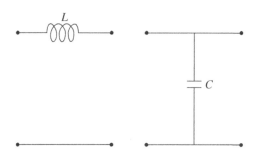

Figure 5.2 First-order low-pass filters.

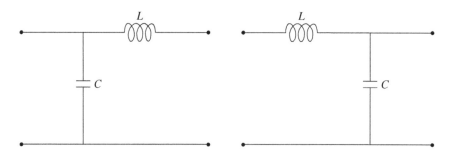

Figure 5.3 Second-order low-pass CL and LC filters.

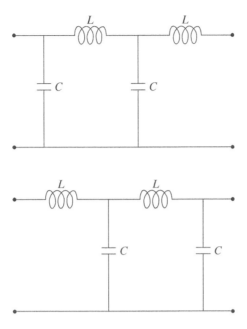

Figure 5.4 Cascaded CL and LC filters.

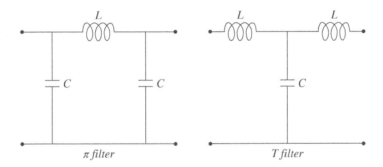

Figure 5.5 Third-order low-pass filters.

Figure 5.8 shows the appropriate configurations when the source impedance is low and the load impedance is high.

Figure 5.9 shows the appropriate configurations when the source impedance is high and the load impedance is low.

The natural question to ask is: *what do we mean by "low" or "high" impedance?* We will answer this question in Section 5.4.

5.4 What Do We Mean by Low or High Impedance?

Consider a typical EMC filter circuit shown in Figure 5.10.

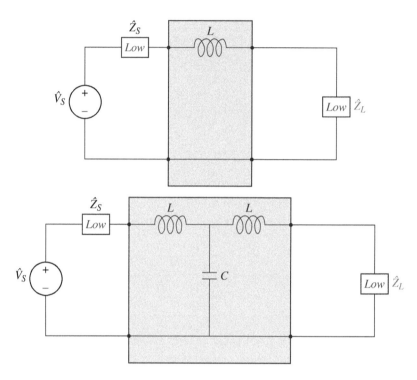

Figure 5.6 Filter configurations when both the source and load impedance are low.

In the following discussion, it is helpful to think of the passive EMC filter as a two-port network shown in Figure 5.11.

The input impedance $\hat{Z}_{IN}$ to the filter circuit in Figure 5.10 is obtained from the circuit shown in Figure 5.12

and calculated as

$$\hat{Z}_{IN} = \frac{\hat{V}_1}{\hat{I}_1} \tag{5.3}$$

The output impedance $\hat{Z}_{OUT}$ of the filter circuit in Figure 5.10 is obtained from the circuit shown in Figure 5.13, where $\hat{V}_G$ is an arbitrary voltage source with its impedance of $\hat{Z}_G$.

and calculated as

$$\hat{Z}_{OUT} = \frac{\hat{V}_2}{\hat{I}_2} \tag{5.4}$$

The effectiveness (insertion loss) of the filter is related to the four impedances shown in Figure 5.14. EMC filter is most effective when there is a large mismatch between $\hat{Z}_S$ and $\hat{Z}_{IN}$ as well as a large mismatch between $\hat{Z}_L$ and $\hat{Z}_{OUT}$ [Fan and Xia, 2012]. Therefore, the EMC filter is most effective when the conditions shown in Figure 5.14 are met.

The calculation of the input and output impedances is quite cumbersome if not impractical, even when the source and load impedances are known. An alternative approach is based on the fact that at high frequencies (where the filter should be most effective) the impedance of the inductor is high and the impedance of the capacitor is low.

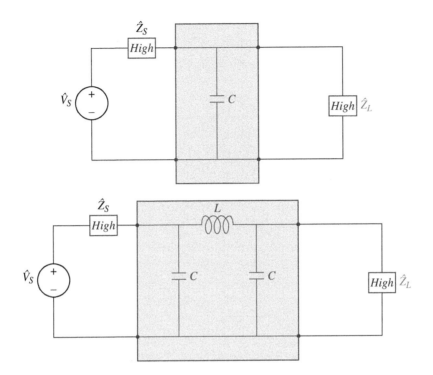

Figure 5.7   Filter configurations when both the source and load impedance are high.

This is the origin of the general rule that the inductor should be placed on the low-impedance side and the capacitor on the high-impedance side.

5.5 LC and CL Filters

5.5.1 LC Filter

Let's calculate the insertion loss of the LC filter shown in Figure 5.15.
With no filter, the load voltage, $\hat{V}_{L,NF}$, is

$$\hat{V}_{L,NF} = \frac{R_L}{R_S + R_L} \hat{V}_S \tag{5.5}$$

With the filter inserted the load voltage, $\hat{V}_{L,F}$, is

$$\hat{V}_{L,F} = \frac{R_L \parallel C}{R_S + sL + R_L \parallel C} \hat{V}_S \tag{5.6}$$

First, let's evaluate the impedance of the load resistor parallel with the capacitor.

$$R_L \parallel C = \frac{R_L \frac{1}{sC}}{R_L + \frac{1}{sC}} = \frac{R_L}{sR_L C + 1} \tag{5.7}$$

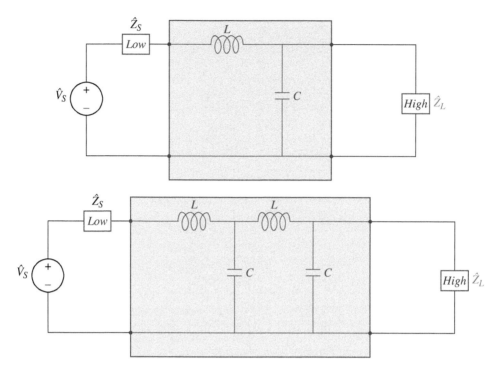

Figure 5.8   Filter configurations when the source is low and the load impedance is high.

Substituting Eq. (5.7) into Eq. (5.6) produces

$$\hat{V}_{L,F} = \frac{\frac{R_L}{sR_LC+1}}{R_S + sL + \frac{R_L}{sR_LC+1}}\hat{V}_S = \frac{R_L}{s^2R_LLC + s(L + R_SR_LC) + R_S + R_L}\hat{V}_S \tag{5.8}$$

Substituting $s = j\omega$ we get

$$\hat{V}_{L,F} = \frac{R_L}{(R_S + R_L - \omega^2R_LLC) + j\omega(L + R_SR_LC)}\hat{V}_S \tag{5.9}$$

The insertion loss is obtained from Eqs. (5.5) and (5.9) as

$$IL_{dB} = 20\log_{10}\frac{V_{L,F}}{V_{L,NF}} = 20\log_{10}\frac{R_S + R_L}{|(R_S + R_L - \omega^2R_LLC) + j\omega(L + R_SR_LC)|} \tag{5.10}$$

or

$$IL_{dB} = 20\log_{10}\frac{R_S + R_L}{\sqrt{(R_S + R_L - \omega^2R_LLC)^2 + \omega^2(L + R_SR_LC)^2}} \tag{5.11}$$

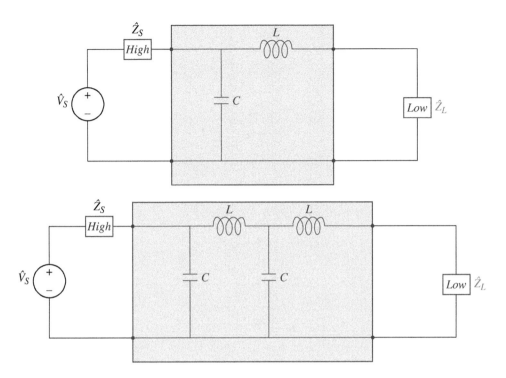

Figure 5.9 Filter configurations when the source is high and the load impedance is low.

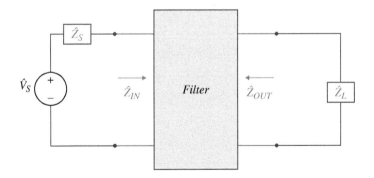

Figure 5.10 EMC filter circuit.

The input impedance, $\hat{Z}_{IN}$, to the filter is calculated from the circuit shown in Figure 5.16.

$$\hat{Z}_{IN}(s) = sL + \left(R_L \parallel \frac{1}{sC} \right) = sL + \frac{R_L \frac{1}{sC}}{R_L + \frac{1}{sC}} = sL + \frac{R_L}{sR_L C + 1} = \frac{s^2 R_L LC + sL + R_L}{sR_L C + 1} \qquad (5.12)$$

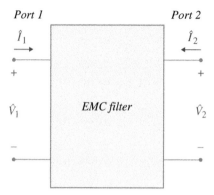

Figure 5.11 Two-port network.

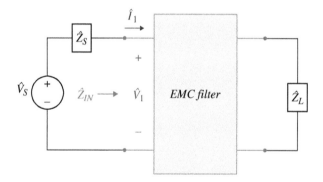

Figure 5.12 Circuit used to define the input impedance.

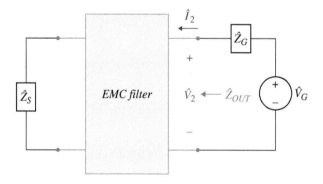

Figure 5.13 Circuit used to define the output impedance.

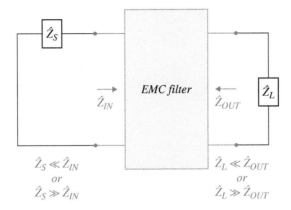

Figure 5.14　Mismatch conditions at Ports 1 and 2.

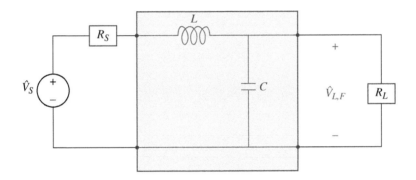

Figure 5.15　LC filter.

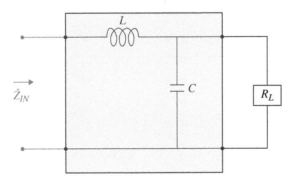

Figure 5.16　Input impedance of the LC filter.

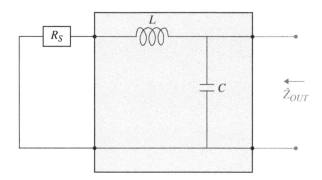

Figure 5.17   Output impedance of the LC filter.

or in terms of ω

$$\hat{Z}_{IN}(j\omega) = \frac{(R_L - \omega^2 R_L LC) + j\omega L}{1 + j\omega R_L C} \tag{5.13}$$

The magnitude of the input impedance is

$$Z_{IN} = \frac{\sqrt{(R_L - \omega^2 R_L LC)^2 + (\omega L)^2}}{\sqrt{1 + (\omega R_L C)^2}} \tag{5.14}$$

The output impedance, $\hat{Z}_{OUT}$, to the filter is calculated from the circuit shown in Figure 5.17.

$$\hat{Z}_{OUT}(s) = (sL + R_S) \parallel \frac{1}{sC} = \frac{\frac{sL+R_S}{sC}}{sL + R_S + \frac{1}{sC}} = \frac{sL + R_S}{s^2LC + sR_SC + 1} \tag{5.15}$$

or in terms of ω

$$\hat{Z}_{OUT}(j\omega) = \frac{R_S + j\omega L}{1 - \omega^2 LC + jR_S C} \tag{5.16}$$

The magnitude of the output impedance is

$$Z_{OUT} = \frac{\sqrt{R_S^2 + (\omega L)^2}}{\sqrt{(1 - \omega^2 LC)^2 + (\omega R_S C)^2}} \tag{5.17}$$

5.5.2 CL Filter

Let's calculate the insertion loss of the CL filter shown in Figure 5.18.
   With no filter, the load voltage, $\hat{V}_{L,NF}$, is

$$\hat{V}_{L,NF} = \frac{R_L}{R_S + R_L} \hat{V}_S \tag{5.18}$$

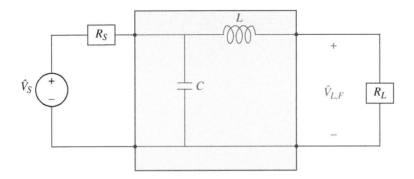

Figure 5.18   CL filter.

To obtain load voltage with the CL filter inserted, let's first calculate the impedance of the parallel combination of

$$(R_L + L) \parallel C = \frac{(R_L + sL)\frac{1}{sC}}{R_L + sL + \frac{1}{sC}} = \frac{R_L + sL}{s^2LC + sR_LC + 1} \tag{5.19}$$

Then the load voltage with the filter inserted is obtained from the voltage divider as

$$\hat{V}_{L,F} = \frac{\frac{R_L + sL}{s^2LC + sR_LC + 1}}{R_S + \frac{R_L + sL}{s^2LC + sR_LC + 1}}\hat{V}_S \tag{5.20}$$

or

$$\hat{V}_{L,F} = \frac{R_L + sL}{s^2 R_S LC + s(R_S R_L C + L) + R_S + R_L}\hat{V}_S \tag{5.21}$$

Substituting $s = j\omega$ we get

$$\hat{V}_{L,F} = \frac{R_L + j\omega L}{R_S + R_L - \omega^2 R_S LC + j\omega(R_S R_L C + L)}\hat{V}_S \tag{5.22}$$

The insertion loss is, therefore,

$$IL_{dB} = 20\log_{10}\frac{V_{L,F}}{V_{L,NF}} = 20\log_{10}\frac{\frac{R_L}{R_S + R_L}}{\frac{R_L + j\omega L}{R_S + R_L - \omega^2 R_S LC + j\omega(R_S R_L C + L)}} \tag{5.23}$$

or

$$IL_{dB} = 20\log_{10}\frac{R_L|R_S + R_L - \omega^2 R_S LC + j\omega(R_S R_L C + L)|}{(R_S + R_L)|R_L + j\omega L|} \tag{5.24}$$

leading to

$$IL_{dB} = 20\log_{10}\frac{R_L}{R_S + R_L}\frac{\sqrt{(R_S + R_L - \omega^2 R_S LC)^2 + [\omega(R_S R_L C + L)]^2}}{\sqrt{R_L^2 + (\omega L)^2}} \tag{5.25}$$

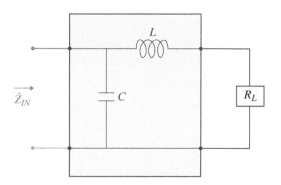

Figure 5.19   Input impedance of the CL filter.

The input impedance, $\hat{Z}_{IN}$, to the CL filter is calculated from the circuit shown in Figure 5.19.

$$\hat{Z}_{IN}(s) = (sL + R_L) \parallel \frac{1}{sC} \tag{5.26}$$

or, using the result of Eq. (5.19)

$$\hat{Z}_{IN}(s) = \frac{R_L + sL}{s^2LC + sR_LC + 1} \tag{5.27}$$

or in terms of ω

$$\hat{Z}_{IN}(j\omega) = \frac{R_L + j\omega L}{1 - \omega^2 LC + j\omega R_L C} \tag{5.28}$$

The magnitude of the input impedance is

$$Z_{IN} = \frac{\sqrt{R_L^2 + (\omega L)^2}}{\sqrt{(1 - \omega^2 LC)^2 + (\omega R_L C)^2}} \tag{5.29}$$

The output impedance, $\hat{Z}_{OUT}$, to the filter is calculated from the circuit shown in Figure 5.20.

$$\hat{Z}_{IN}(s) = sL + \left(R_S \parallel \frac{1}{sC}\right) = sL + \frac{R_S \frac{1}{sC}}{R_S + \frac{1}{sC}} = sL + \frac{R_S}{sR_SC + 1} = \frac{s^2 R_S LC + sL + R_S}{sR_SC + 1} \tag{5.30}$$

or in terms of ω

$$\hat{Z}_{IN}(j\omega) = \frac{(R_S - \omega^2 R_S LC) + j\omega L}{1 + j\omega R_S C} \tag{5.31}$$

The magnitude of the output impedance is

$$Z_{IN} = \frac{\sqrt{(R_S - \omega^2 R_S LC)^2 + (\omega L)^2}}{\sqrt{1 + (\omega R_S C)^2}} \tag{5.32}$$

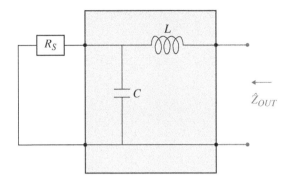

Figure 5.20 Output impedance of the CL filter.

5.5.3 LC Filter vs. CL Filter

Let's evaluate the LC and CL filter performance in terms of the input impedance and insertion loss.

5.5.3.1 Input Impedance – Simulations and Calculations

First, let's look at the input impedance of the two filters. The simulation circuit for this comparison is shown in Figure 5.21.

The input impedance of the two filter configurations is shown in Figure 5.22.

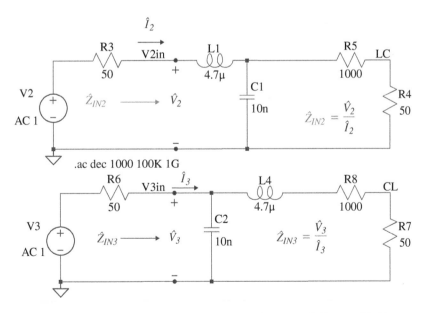

Figure 5.21 Circuit to determine the input impedance – LC filter vs. CL filter.

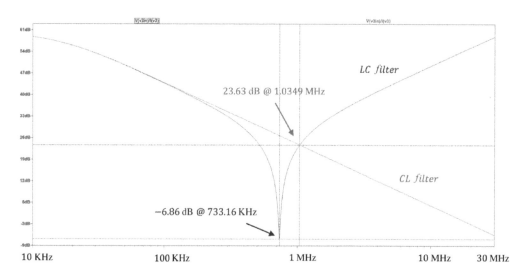

Figure 5.22   Input impedance – LC filter vs. CL filter.

The resonant frequency of the LC filter is at

$$f_r = \frac{1}{2\pi\sqrt{LC}} = \frac{1}{2\pi\sqrt{10 \times 10^{-9} \times 4.7 \times 10^{-6}}} = 734.127\,\text{kHz} \qquad (5.33)$$

This value is confirmed by the simulation result in Figure 5.22.

Let's determine the frequency at which the input impedances (and insertion losses) of both filters are equal. The magnitude of the input impedance to the LC filter is given by Eq. (5.14), repeated here

$$Z_{IN} = \frac{\sqrt{(R_L - \omega^2 R_L LC)^2 + (\omega L)^2}}{\sqrt{1 + (\omega R_L C)^2}} \qquad (5.34)$$

while the magnitude of the input impedance to the CL filter is given by Eq. (5.29), repeated here

$$Z_{IN} = \frac{\sqrt{R_L^2 + (\omega L)^2}}{\sqrt{(1 - \omega^2 LC)^2 + (\omega R_L C)^2}} \qquad (5.35)$$

When the two magnitudes are equal we have

$$\frac{\sqrt{(R_L - \omega^2 R_L LC)^2 + (\omega L)^2}}{\sqrt{1 + (\omega R_L C)^2}} = \frac{\sqrt{R_L^2 + (\omega L)^2}}{\sqrt{(1 - \omega^2 LC)^2 + (\omega R_L C)^2}} \qquad (5.36)$$

Comparing the numerators in Eq. (5.36) we get

$$(R_L - \omega^2 R_L LC)^2 + (\omega L)^2 = R_L^2 + (\omega L)^2 \qquad (5.37)$$

or
$$(R_L - \omega^2 R_L LC)^2 = R_L^2 \tag{5.38}$$

thus
$$R_L^2 - 2\omega^2 R_L^2 LC + \omega^4 R_L^2 L^2 C^2 = R_L^2 \tag{5.39}$$

which simplifies to
$$\omega^2 L^2 C^2 = 2LC \tag{5.40}$$

resulting in
$$\omega = \sqrt{\frac{2}{LC}} \tag{5.41}$$

or
$$f = \frac{1}{2\pi} \sqrt{\frac{2}{LC}} \tag{5.42}$$

Next, let's compare the denominators in Eq. (5.36):
$$1 + (\omega R_L C)^2 = (1 - \omega^2 LC)^2 + (\omega R_L C)^2 \tag{5.43}$$

which simplifies to
$$1 = 1 - 2\omega^2 LC + \omega^4 L^2 C^2 \tag{5.44}$$

resulting in
$$\omega = \sqrt{\frac{2}{LC}} \tag{5.45}$$

which agrees with the result obtained from the comparison of the numerators.
Substituting the component values we get

$$f = \frac{1}{2\pi} \sqrt{\frac{2}{4.7 \times 10^{-6} \times 10 \times 10^{-9}}} = 1.038\,\text{MHz} \tag{5.46}$$

which is consistent with the value obtained from the simulation in Figure 5.22.

5.5.3.2 Insertion Loss – Simulations and Measurements

Next, let's simulate and measure the insertion loss of the two filters. The circuits used in the simulations are shown in Figure 5.23 [Adamczyk and Haring, 2019].

The $50\,\Omega$ source impedance is provided by the network analyzer at Port 1. The measurement made by the network analyzer at Port 2 is across its internal $50\,\Omega$ impedance. To vary the impedance at the source or the load, an in-line resistance could be inserted at either or both sides. In the circuits shown, $1000\,\Omega$ impedance is inserted on the load side.

The insertion loss of the two filter configurations is shown in Figure 5.24.

At 1.04 MHz the insertion losses of the two filters are equal. This is the same frequency at which the input impedances of the two filters were equal!

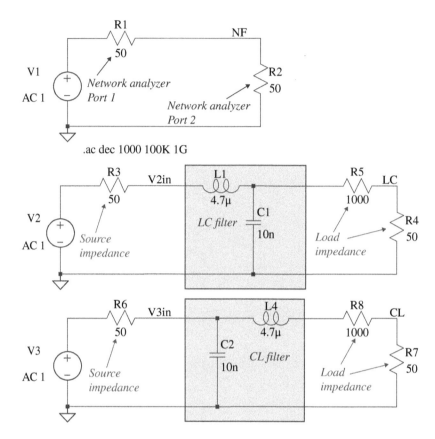

Figure 5.23 Circuit to determine the insertion loss – LC filter vs. CL filter.

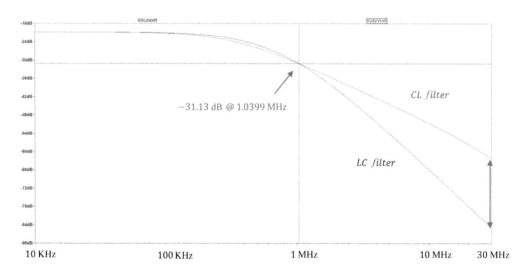

Figure 5.24 Insertion loss – LC filter vs. CL filter.

Note that up to the frequency of 1.04 MHz the insertion loss of the CL filter is larger (in the absolute sense) than that of the LC filter. Beyond that frequency the insertion loss of the LC filter is larger. This means that the LC filter is more effective than the CL filter beyond the frequency of 1.04 MHz.

We have arrived at a very important observation: once the filter components values L and C are chosen, we can determine the frequency at which the insertion losses of LC and CL filters are equal. This is the frequency at which the input impedances are equal, given by Eq. (5.42), repeated here

$$f = \frac{1}{2\pi}\sqrt{\frac{2}{LC}} \tag{5.47}$$

To verify the simulations results of the insertion loss the measurement setup shown in Figure 5.25 was used [Adamczyk and Haring, 2019].

The details of the filter PCBs are shown in Figure 5.26.

Figure 5.27 shows the measurement results for the two configurations simulated in Figure 5.24.

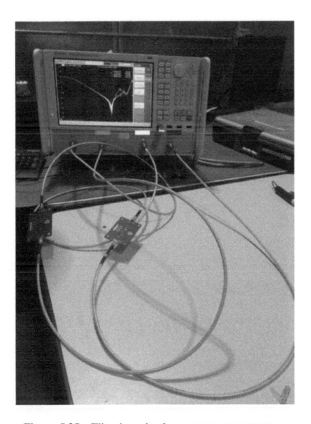

Figure 5.25   Filter insertion loss – measurement setup.

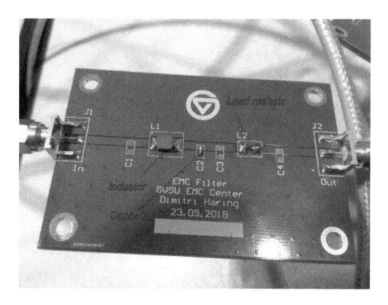

Figure 5.26 PCBs used for insertion loss measurements.

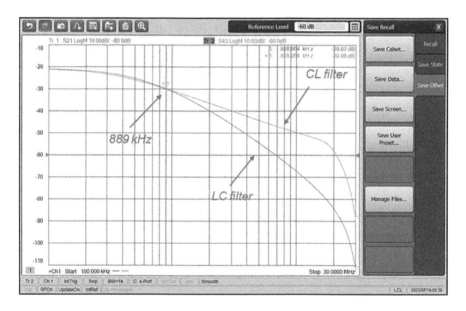

Figure 5.27 LC and CL filters – insertion loss measurement.

Beyond the frequency of 1.04 MHz, the LC filter outperforms the CL filter, which is consistent with the simulation results. In the frequency range 100 kHz – 10 MHz the simulated and measured results are remarkably close, as summarized in Figure 5.28.

At 10 MHz the difference between the simulated insertion losses of the two filters is 15.1 dB which is close to the measured difference of 16.7 dB. The measured results show

LC filter	$f = 100$ kHz	$f = 1$ MHz	$f = 10$ MHz
Simulated insertion loss	21.2 dB	30.8 dB	65.8 dB
Measured insertion loss	21.3 dB	30 dB	65.7 dB

CL Filter	$f = 100$ kHz	$f = 1$ MHz	$f = 10$ MHz
Simulated insertion loss	21.2 dB	30.8 dB	50.7 dB
Measured insertion loss	21.3 dB	30 dB	49 dB

Figure 5.28   LC and CL filters – measurement vs. simulation.

the self-resonant frequency at 30 MHz, with the insertion loss of 84.9 dB for the CL filter and 115 dB for the LC filter. A second resonance occurs at 60 MHz with the insertion loss of 83 dB for the CL filter and 95.5 dB for the LC filter. These resonances were not predicted by the simulation models, as those models assumed ideal components and did not account for the board parasitics.

The measured results clearly show that, beyond the frequency of 1.04 MHz, the LC filter (inductor on the low impedance side and capacitor on the high impedance side) has a higher insertion loss than the CL filter (capacitor on the low impedance side and inductor on the high impedance side).

5.6 Pi and T Filters

5.6.1 Pi Filter

The input impedance, $\hat{Z}_{IN}$, to the Pi filter is calculated from the circuit shown in Figure 5.29.

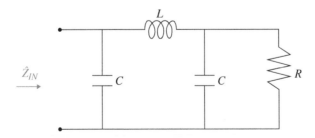

Figure 5.29   Input impedance of the Pi filter.

The equivalent impedance of the parallel RC configuration is

$$R \parallel \frac{1}{sC} = \frac{R \times \frac{1}{sC}}{R + \frac{1}{sC}} = \frac{R}{sRC + 1} \tag{5.48}$$

This impedance is in series with the impedance of the inductor

$$sL + \frac{R}{sRC + 1} = \frac{sL(sRC + 1) + R}{sRC + 1} = \frac{s^2LRC + sL + R}{sRC + 1} \tag{5.49}$$

which, in turn, is parallel with the impedance of the capacitor. Thus, the input impedance to the filter is

$$\hat{Z}_{IN}(s) = \left(\frac{1}{sC}\right) \parallel \left(\frac{s^2LRC + sL + R}{sRC + 1}\right) \tag{5.50}$$

or

$$\hat{Z}_{IN}(s) = \frac{\frac{1}{sC} \times \frac{s^2LRC+sL+R}{sRC+1}}{\frac{1}{sC} + \frac{s^2LRC+sL+R}{sRC+1}} \tag{5.51}$$

or

$$\hat{Z}_{IN}(s) = \frac{\frac{s^2LRC+sL+R}{sRC+1}}{1 + \frac{sC(s^2LRC+sL+R)}{sRC+1}} \tag{5.52}$$

resulting in

$$\hat{Z}_{IN}(s) = \frac{s^2LRC + sL + R}{s^3LRC^2 + s^2LC + s2RC + 1} \tag{5.53}$$

Replacing s by $j\omega$ we obtain

$$\hat{Z}_{IN}(j\omega) = \frac{R - \omega^2LRC + j\omega L}{j\omega(-\omega)^2LRC^2 - \omega^2LC + j\omega2RC + 1} \tag{5.54}$$

and thus

$$\hat{Z}_{IN}(j\omega) = \frac{(R - \omega^2LRC) + j\omega L}{(1 - \omega^2LC) + j(\omega2RC - \omega^3LRC^2)} \tag{5.55}$$

The magnitude of the input impedance is

$$|\hat{Z}_{IN}(j\omega)| = \frac{\sqrt{(R - \omega^2LRC)^2 + (\omega L)^2}}{\sqrt{(1 - \omega^2LC)^2 + (\omega2RC - \omega^3LRC^2)^2}} \tag{5.56}$$

5.6.2 T Filter

The input impedance, $\hat{Z}_{IN}$, to the filter is calculated from the circuit shown in Figure 5.30. The equivalent impedance of resistor/inductor in parallel with the capacitor is

$$(sL + R) \parallel \frac{1}{sC} = \frac{(sL + R) \times \frac{1}{sC}}{(sL + R) + \frac{1}{sC}} = \frac{sL + R}{s^2LC + sRC + 1} \tag{5.57}$$

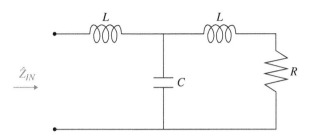

Figure 5.30 Input impedance of the T filter.

Thus, the input impedance to the filter is

$$\hat{Z}_{IN}(s) = sL + \frac{sL + R}{s^2LC + sRC + 1} = \frac{sL(s^2LC + sRC + 1) + sL + R}{s^2LC + sRC + 1} \tag{5.58}$$

or

$$\hat{Z}_{IN}(s) = \frac{s^3L^2C + s^2LRC + s2L + R}{s^2LC + sRC + 1} \tag{5.59}$$

Replacing s by $j\omega$ we obtain

$$\hat{Z}_{IN}(j\omega) = \frac{j\omega(-\omega)^2L^2C - \omega^2LRC + j\omega2L + R}{1 - \omega^2LC + j\omega RC} \tag{5.60}$$

or

$$\hat{Z}_{IN}(j\omega) = \frac{(R - \omega^2LRC) + j(\omega2L - \omega^3L^2C)}{(1 - \omega^2LC) + j\omega RC} \tag{5.61}$$

The magnitude of the input impedance is

$$|\hat{Z}_{IN}(j\omega)| = \frac{\sqrt{(R - \omega^2LRC)^2 + (\omega2L - \omega^3L^2C)^2}}{\sqrt{(1 - \omega^2LC)^2 + (\omega RC)^2}} \tag{5.62}$$

5.6.3 Pi Filter vs. T Filter

5.6.3.1 Input Impedance – Simulations and Calculations

Let's look at the input impedance of the two filters. The simulation circuit for this comparison is shown in Figure 5.31.

The input impedance of the two filter configurations is shown in Figure 5.32.

The magnitude of the input impedance to the Pi filter is given by Eq. (5.56), repeated here

$$|\hat{Z}_{IN}(j\omega)| = \frac{\sqrt{(R - \omega^2LRC)^2 + (\omega L)^2}}{\sqrt{(1 - \omega^2LC)^2 + (\omega2RC - \omega^3LRC^2)^2}} \tag{5.63}$$

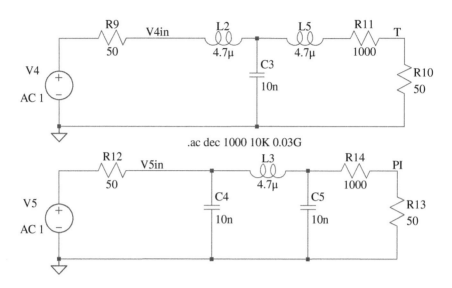

Figure 5.31 Circuit to determine the input impedance – Pi filter vs. T filter.

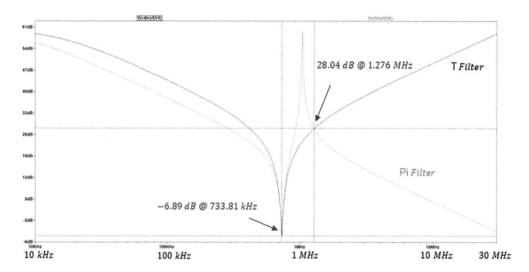

Figure 5.32 Input impedance – Pi filter vs. T filter.

while the magnitude of the input impedance to the T filter is given by Eq. (5.62). When the two magnitudes are equal we have

$$\frac{\sqrt{(R - \omega^2 LRC)^2 + (\omega L)^2}}{\sqrt{(1 - \omega^2 LC)^2 + (\omega 2RC - \omega^3 LRC^2)^2}} = \frac{\sqrt{(R - \omega^2 LRC)^2 + (\omega 2L - \omega^3 L^2 C)^2}}{\sqrt{(1 - \omega^2 LC)^2 + (\omega RC)^2}} \qquad (5.64)$$

To solve for the frequency or frequencies at which the two magnitudes are equal, we will compare the numerators and the denominators in Eq. (5.64).

Let's begin by comparing the numerators, which gives

$$(R - \omega^2 LRC)^2 + (\omega L)^2 = (R - \omega^2 LRC)^2 + (\omega 2L - \omega^3 L^2 C)^2 \tag{5.65}$$

or

$$(\omega L)^2 = (\omega 2L - \omega^3 L^2 C)^2 \tag{5.66}$$

resulting in

$$\omega^4 L^2 C^2 - 4\omega^2 LC + 3 = 0 \tag{5.67}$$

Let $\omega^2 = x$ in Eq. (5.67). Then,

$$L^2 C^2 x^2 - 4LCx + 3 = 0 \tag{5.68}$$

The roots of the quadratic Eq. (5.68) are

$$x_1 = \frac{3}{LC} \tag{5.69}$$

and

$$x_2 = \frac{1}{LC} \tag{5.70}$$

Thus, the two frequencies (in radians), at which the input impedance to the Pi filter equals the input impedance to the T filter, are

$$\omega_1 = \sqrt{\frac{3}{LC}} \tag{5.71}$$

and

$$\omega_2 = \sqrt{\frac{1}{LC}} \tag{5.72}$$

The corresponding frequencies in Hertz are

$$f_1 = \frac{1}{2\pi}\sqrt{\frac{3}{LC}} \tag{5.73}$$

and

$$f_2 = \frac{1}{2\pi}\sqrt{\frac{1}{LC}} \tag{5.74}$$

Next, let's compare the denominators in Eq. (5.64).

$$(1 - \omega^2 LC)^2 + (\omega 2RC - \omega^3 LRC^2)^2 = (1 - \omega^2 LC)^2 + (\omega RC)^2 \tag{5.75}$$

leading to

$$(\omega 2RC - \omega^3 LRC^2)^2 = (\omega RC)^2 \tag{5.76}$$

or

$$\omega^4 L^2 C^2 - 4\omega^2 LC + 3 = 0 \tag{5.77}$$

This equation is identical to Eq. (5.67) and therefore has the solutions given by Eqs. (5.73) and (5.74).

Substituting the component values we get

$$f_1 = \frac{1}{2\pi} \sqrt{\frac{3}{4.7 \times 10^{-6} \times 10 \times 10^{-9}C}} = 1.271\,\text{MHz} \tag{5.78}$$

$$f_2 = \frac{1}{2\pi} \sqrt{\frac{1}{4.7 \times 10^{-6} \times 10 \times 10^{-9}C}} = 733.98\,\text{kHz} \tag{5.79}$$

which is consistent with the values obtained from the simulation in Figure 5.32.

5.6.3.2 Insertion Loss – Simulations and Measurements

Next, let's determine the insertion loss of the two filters. The circuits used in the simulations are shown in Figure 5.33 [Adamczyk and Gilbert, 2020].

The insertion loss of the two filter configurations is shown in Figure 5.34.

At 734 kHz and 1.27 MHz the insertion losses of the two filters are equal. These are the same frequencies at which the input impedances of the two filters were equal!

Note that up to the frequency of 734 kHz the insertion loss of the Pi filter is larger than that of the T filter. Between the frequencies of 734 kHz and 1.27 MHz the insertion loss of the T filter is larger. Beyond the frequency of 1.27 MHz the insertion loss of the Pi filter is again larger.

Again, we have arrived at a very important observation: once the filter components values L and C are chosen, we can determine the frequencies at which the insertion losses of Pi and T filters are equal. These are the frequencies at which the input impedances are equal, given by Eqs. (5.73) and (5.74), repeated here

$$f_1 = \frac{1}{2\pi} \sqrt{\frac{3}{LC}} \tag{5.80}$$

and

$$f_2 = \frac{1}{2\pi} \sqrt{\frac{1}{LC}} \tag{5.81}$$

Once these frequencies are calculated, they provide us with the frequency regions where one filter outperforms the other.

To verify the simulations results of the insertion loss the measurement setup shown in Figure 5.35 was used [Adamczyk and Gilbert, 2020].

The details of the filter PCBs are shown in Figure 5.36.

Figure 5.37 shows the measurement results for the two configurations simulated in Figure 5.36.

The measurement results are consistent with the simulation results. In the frequency range 100 kHz – 10 MHz the simulated and measured results are remarkably close, as summarized in Figure 5.38.

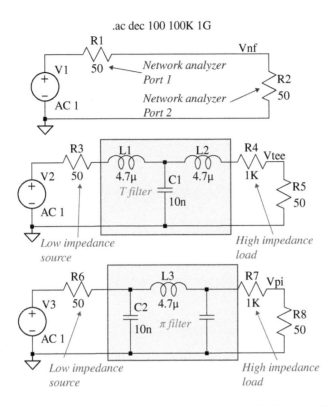

Figure 5.33 Circuit to determine the insertion loss – Pi filter vs. *T* filter.

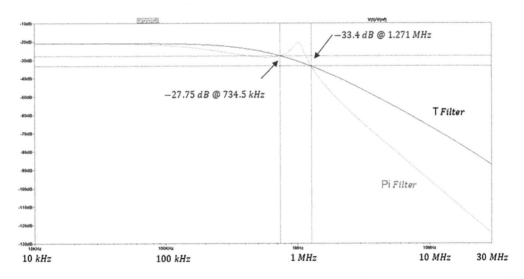

Figure 5.34 Insertion loss – Pi filter vs. *T* filter.

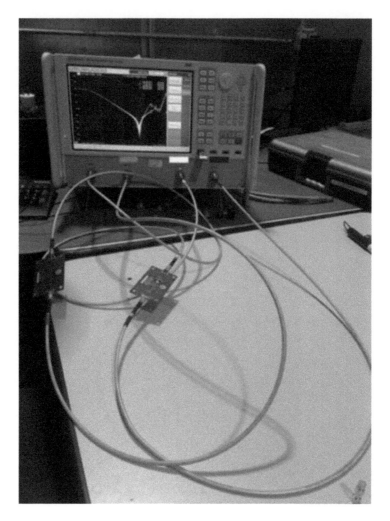

Figure 5.35   Filter insertion loss – measurement setup.

5.7 LCLC and CLCL Filters

5.7.1 LCLC Filter

The input impedance, $\hat{Z}_{IN}$, to the filter is calculated from the circuit shown in Figure 5.39.
   Note that the input impedance of this filter can be obtained by using the input impedance of the Pi filter in Eq. (5.53) and combining it in series with the impedance of an inductor.

$$\hat{Z}_{IN}(s) = sL + \frac{s^2LRC + sL + R}{s^3LRC^2 + s^2LC + s2RC + 1} \tag{5.82}$$

thus

$$\hat{Z}_{IN}(s) = \frac{sL(s^3LRC^2 + s^2LC + s2RC + 1) + s^2LRC + sL + R}{s^3LRC^2 + s^2LC + s2RC + 1} \tag{5.83}$$

Figure 5.36   PCBs used for insertion loss measurements.

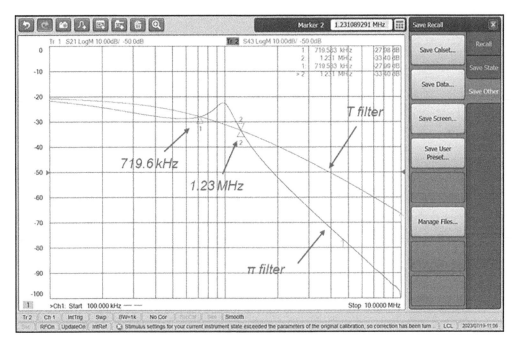

Figure 5.37   Pi and *T* filters – insertion loss measurement.

π filter	$f = 100\ \text{kHz}$	$f = 1\ \text{MHz}$	$f = 10\ \text{MHz}$
Simulated insertion loss	22 dB	20.6 dB	95.65 dB
Measured insertion loss	22 dB	23.8 dB	93.1 dB

T filter	$f = 100\ \text{kHz}$	$f = 1\ \text{MHz}$	$f = 10\ \text{MHz}$
Simulated insertion loss	21.1 dB	30.7 dB	66.2 dB
Measured insertion loss	21.2 dB	29.7 dB	65.7 dB

Figure 5.38   Pi and T filters – measurement vs. simulation.

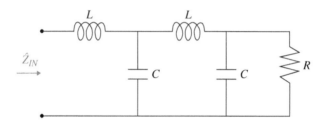

Figure 5.39   Input impedance of the LCLC filter.

resulting in

$$\hat{Z}_{IN}(s) = \frac{s^4 L^2 RC^2 + s^3 L^2 C + s^2 3LRC + s2L + R}{s^3 LRC^2 + s^2 LC + s2RC + 1} \tag{5.84}$$

Replacing s by $j\omega$ we obtain

$$\hat{Z}_{IN}(j\omega) = \frac{\omega^4 L^2 RC^2 + j\omega(-\omega^2)L^2 C - \omega^2 3LRC + j\omega 2L + R}{j\omega(-\omega^2)LRC^2 - \omega^2 LC + j\omega 2RC + 1} \tag{5.85}$$

or

$$\hat{Z}_{IN}(j\omega) = \frac{(\omega^4 L^2 RC^2 + R - \omega^2 3LRC) + j(\omega 2L - \omega^3 L^2 C)}{(1 - \omega^2 LC) + j(\omega 2RC - \omega^3 LRC^2)} \tag{5.86}$$

The magnitude of the input impedance is

$$|\hat{Z}_{IN}(j\omega)| = \frac{\sqrt{(\omega^4 L^2 RC^2 + R - \omega^2 3LRC)^2 + (\omega 2L - \omega^3 L^2 C)^2}}{\sqrt{(1 - \omega^2 LC)^2 + (\omega 2RC - \omega^3 LRC^2)^2}} \tag{5.87}$$

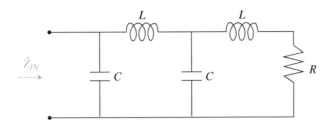

Figure 5.40   Input impedance of the CLCL filter.

5.7.2 CLCL Filter

The input impedance, $\hat{Z}_{IN}$, to the filter is calculated from the circuit shown in Figure 5.40.

Note that the input impedance of this filter can be obtained by using the input impedance of the T filter in Eq. (5.59) and combining it in parallel with the impedance of a capacitor.

$$\hat{Z}_{IN}(s) = \left(\frac{s^3 L^2 C + s^2 LRC + s2L + R}{s^2 LC + sRC + 1} \right) \parallel \left(\frac{1}{sC} \right) \tag{5.88}$$

or

$$\hat{Z}_{IN}(s) = \frac{\frac{1}{sC} \times \frac{s^3 L^2 C + s^2 LRC + s2L + R}{s^2 LC + sRC + 1}}{\frac{1}{sC} + \frac{s^3 L^2 C + s^2 LRC + s2L + R}{s^2 LC + sRC + 1}} \tag{5.89}$$

or

$$\hat{Z}_{IN}(s) = \frac{\frac{s^3 L^2 C + s^2 LRC + s2L + R}{s^2 LC + sRC + 1}}{1 + \frac{sC(s^3 L^2 C + s^2 LRC + s2L + R)}{s^2 LC + sRC + 1}} \tag{5.90}$$

Resulting in

$$\hat{Z}_{IN}(s) = \frac{s^3 L^2 C + s^2 LRC + s2L + R}{s^4 L^2 C^2 + s^3 LRC^2 + s^2 3LC + s2RC + 1} \tag{5.91}$$

Replacing s by $j\omega$ we obtain

$$\hat{Z}_{IN}(j\omega) = \frac{j\omega(-\omega^2)L^2 C - \omega^2 LRC + j\omega 2L + R}{\omega^4 L^2 C^2 + j\omega(-\omega^2)LRC^2 - \omega^2 3LC + j\omega 2RC + 1} \tag{5.92}$$

or

$$\hat{Z}_{IN}(j\omega) = \frac{(R - \omega^2 LRC) + j(\omega 2L - \omega^3 L^2 C)}{(\omega^4 L^2 C^2 - \omega^2 3LC + 1) + j(\omega 2RC - \omega^3 LRC^2)} \tag{5.93}$$

The magnitude of the input impedance is

$$| \hat{Z}_{IN}(j\omega) | = \frac{\sqrt{(R - \omega^2 LRC)^2 + (\omega 2L - \omega^3 L^2 C)^2}}{\sqrt{(\omega^4 L^2 C^2 - \omega^2 3LC + 1)^2 + (\omega 2RC - \omega^3 LRC^2)^2}} \tag{5.94}$$

5.7.3 LCLC Filter vs. CLCL Filter

5.7.3.1 Input Impedance – Simulations and Calculations

Let's look at the input impedance of the two filters. The simulation circuit for this comparison is shown in Figure 5.41.

The input impedance of the two filter configurations is shown in Figure 5.42.

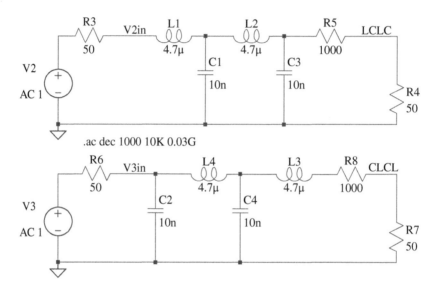

Figure 5.41 Circuit to determine the input impedance – LCLC filter vs. CLCL filter.

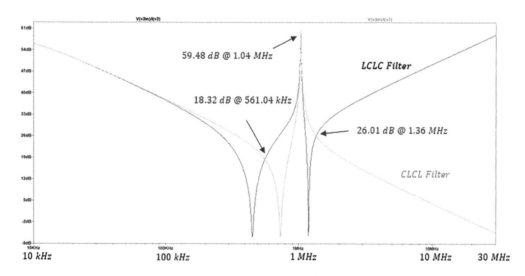

Figure 5.42 Input impedance – LCLC filter vs.CLCL filter.

The magnitude of the input impedance to the LCLC filter is given by Eq. (5.87), repeated here

$$| \hat{Z}_{IN}(j\omega) | = \frac{\sqrt{(\omega^4 L^2 RC^2 + R - \omega^2 3LRC)^2 + (\omega 2L - \omega^3 L^2 C)^2}}{\sqrt{(1 - \omega^2 LC)^2 + (\omega 2RC - \omega^3 LRC^2)^2}} \tag{5.95}$$

while the magnitude of the input impedance to the CLCL filter is given by Eq. (5.94). When the two magnitudes are equal we have

$$\frac{\sqrt{(\omega^4 L^2 RC^2 + R - \omega^2 3LRC)^2 + (\omega 2L - \omega^3 L^2 C)^2}}{\sqrt{(1 - \omega^2 LC)^2 + (\omega 2RC - \omega^3 LRC^2)^2}} = \frac{\sqrt{(\omega^4 L^2 RC^2 + R - \omega^2 3LRC)^2 + (\omega 2L - \omega^3 L^2 C)^2}}{\sqrt{(1 - \omega^2 LC)^2 + (\omega 2RC - \omega^3 LRC^2)^2}} \tag{5.96}$$

or

$$\frac{\sqrt{(R - \omega^2 LRC)^2 + (\omega 2L - \omega^3 L^2 C)^2}}{\sqrt{(\omega^4 L^2 C^2 - \omega^2 3LC + 1)^2 + (\omega 2RC - \omega^3 LRC^2)^2}} = \frac{(\omega^4 L^2 RC^2 + R - \omega^2 3LRC)^2 + (\omega 2L - \omega^3 L^2 C)^2}{(1 - \omega^2 LC)^2 + (\omega 2RC - \omega^3 LRC^2)^2} \tag{5.97}$$

Comparing the numerators in Eq. (5.97) we get

$$(R - \omega^2 LRC)^2 + (\omega 2L - \omega^3 L^2 C)^2 = (\omega^4 L^2 RC^2 + R - \omega^2 3LRC)^2 + (\omega 2L - \omega^3 L^2 C)^2 \tag{5.98}$$

or

$$(R - \omega^2 LRC)^2 = (\omega^4 L^2 RC^2 + R - \omega^2 3LRC)^2 \tag{5.99}$$

Equation (5.99) is of the form $a^2 = b^2$. It is satisfied when one of the three cases happens. Case 1: $a = b$, Case 2: $a = -b$, and Case 3: $-a = b$.

Case 1: Results in

$$R - \omega^2 LRC = \omega^4 L^2 RC^2 + R - \omega^2 3LRC \tag{5.100}$$

or

$$\omega^4 L^2 RC^2 - 2\omega^2 LRC = 0 \tag{5.101}$$

resulting in

$$\omega^2 = \frac{2}{LC} \tag{5.102}$$

and thus

$$\omega = \sqrt{\frac{2}{LC}} \tag{5.103}$$

Case 2: Results in

$$R - \omega^2 LRC = -(\omega^4 L^2 RC^2 + R - \omega^2 3LRC) \tag{5.104}$$

or

$$R - \omega^2 LRC = -\omega^4 L^2 RC^2 - R + \omega^2 3LRC \tag{5.105}$$

resulting in

$$L^2 C^2 \omega^4 - 4LC\omega^2 + 2 = 0 \tag{5.106}$$

Let $\omega^2 = x$ in Eq. (5.106). Then,

$$L^2C^2x^2 - 4LCx + 2 = 0 \tag{5.107}$$

The roots of the quadratic Eq. (5.107) are

$$x_1 = \frac{2 - \sqrt{2}}{LC} \tag{5.108}$$

and

$$x_2 = \frac{2 + \sqrt{2}}{LC} \tag{5.109}$$

thus, the solutions to Eq. (5.106) are

$$\omega_1 = \sqrt{\frac{2 - \sqrt{2}}{LC}} \tag{5.110}$$

and

$$\omega_2 = \sqrt{\frac{2 + \sqrt{2}}{LC}} \tag{5.111}$$

Case 3: Results in

$$-(R - \omega^2 LRC) = (\omega^4 L^2 RC^2 + R - \omega^2 3LRC) \tag{5.112}$$

or

$$L^2C^2\omega^4 - 4LC\omega^2 + 2 = 0 \tag{5.113}$$

Equation (5.113) has the same form as Eq. (5.106), and thus the solutions to Eq. (5.113) are given by Eqs. (5.110) and (5.111).

Next, let's compare the denominators in Eq. (5.97). Then

$$(\omega^4 L^2 C^2 - \omega^2 3LC + 1)^2 + (\omega 2RC - \omega^3 LRC^2)^2 = (1 - \omega^2 LC)^2 + (\omega 2RC - \omega^3 LRC^2)^2 \tag{5.114}$$

or

$$(\omega^4 L^2 C^2 - \omega^2 3LC + 1)^2 = (1 - \omega^2 LC)^2 \tag{5.115}$$

Equation (5.115) is of the form $a^2 = b^2$. It is satisfied when one of the three cases happens. Case 1: $a = b$, Case 2: $a = -b$, and Case 3: $-a = b$.

Case 1: Results in

$$\omega^4 L^2 C^2 - \omega^2 3LC + 1 = 1 - \omega^2 LC \tag{5.116}$$

or

$$\omega^4 L^2 C^2 = 2\omega^2 LC \tag{5.117}$$

resulting in

$$\omega^2 = \frac{2}{LC} \tag{5.118}$$

and thus

$$\omega = \sqrt{\frac{2}{LC}} \tag{5.119}$$

This is the same solution as the one in Eq. (5.103) which validates the Case 1 solution of the denominators comparison.

Case 2: Results in

$$L^2C^2\omega^4 - 4LC\omega^2 + 1 = -(1 - \omega^2 LC) \tag{5.120}$$

or

$$L^2C^2\omega^4 - \omega^2 4LC + 2 = 0 \tag{5.121}$$

Equation (5.121) has the same form as Eq. (5.106) and thus the solutions to Eq. (5.121) are given by Eqs. (5.110) and (5.111).

Case 3: Results in

$$-(L^2C^2\omega^4 - 4LC\omega^2 + 1) = 1 - \omega^2 LC \tag{5.122}$$

or

$$L^2C^2\omega^4 - 4LC\omega^2 + 2 = 0 \tag{5.123}$$

Equation (5.123) has the same form as Eq. (5.106) and thus the solutions to Eq. (5.113) are given by Eqs. (5.110) and (5.111).

In summary, the input impedances of the LCLC and CLCL filters are equal at the frequencies

$$\omega_1 = \sqrt{\frac{2}{LC}} \tag{5.124}$$

$$\omega_2 = \sqrt{\frac{2 - \sqrt{2}}{LC}} \tag{5.125}$$

and

$$\omega_3 = \sqrt{\frac{2 + \sqrt{2}}{LC}} \tag{5.126}$$

The corresponding frequencies in Hertz are

$$f_1 = \frac{1}{2\pi}\sqrt{\frac{2}{LC}} \tag{5.127}$$

$$f_2 = \frac{1}{2\pi}\sqrt{\frac{2 - \sqrt{2}}{LC}} \tag{5.128}$$

and

$$f_3 = \frac{1}{2\pi}\sqrt{\frac{2 + \sqrt{2}}{LC}} \tag{5.129}$$

Substituting the component values we get

$$f_1 = \frac{1}{2\pi}\sqrt{\frac{2}{4.7 \times 10^{-6} \times 10 \times 10^{-9}}} = 1.038\,\text{MHz} \qquad (5.130)$$

$$f_2 = \frac{1}{2\pi}\sqrt{\frac{2 - \sqrt{2}}{4.7 \times 10^{-6} \times 10 \times 10^{-9}}} = 561.8\,\text{kHz} \qquad (5.131)$$

$$f_3 = \frac{1}{2\pi}\sqrt{\frac{2 + \sqrt{2}}{4.7 \times 10^{-6} \times 10 \times 10^{-9}}} = 1.356\,\text{MHz} \qquad (5.132)$$

These results are consistent with the values obtained from the simulation in Figure 5.42.

5.7.3.2 Insertion Loss – Simulations

Next, let's determine the insertion loss of the two filters. The circuits used in the simulations are shown in Figure 5.43.

The insertion loss of the two filter configurations is shown in Figure 5.44.

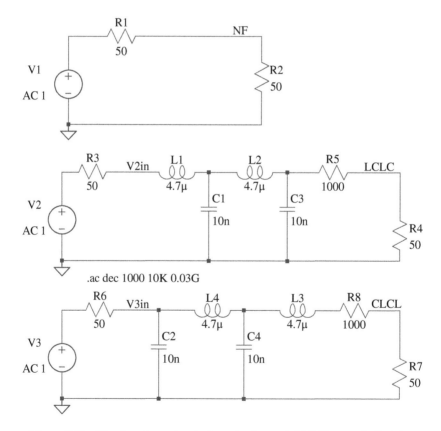

Figure 5.43 Circuit to determine the insertion loss – LCLC filter vs. CLCL filter.

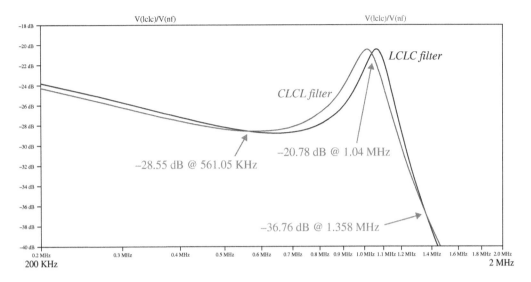

Figure 5.44 Insertion loss – LCLC filter vs. CLCL filter.

At 561 kHz, 1.04 MHz, and 1.36 MHz the insertion losses of the two filters are equal. These are the same frequencies at which the input impedances of the two filters were equal!

Note that up to the frequency of 561 kHz the insertion loss of the CLCL filter is larger than that of the LCLC filter. Between the frequencies of 561 kHz and 1.04 MHz the insertion loss of the LCLC filter is larger. Between the frequencies of 1.04 MHz kHz and 1.36 MHz the insertion loss of the CLCL filter is larger. Beyond the frequency of 1.36 MHz the insertion loss of the LCLC filter is again larger.

Once again, we have arrived at a very important observation: once the filter components values L and C are chosen, we can determine the frequencies at which the insertion losses of LCLC and CLCL filters are equal. These are the frequencies at which the input impedances are equal, given by Eqs.(5.127), (5.128), and (5.129), repeated here

$$f_1 = \frac{1}{2\pi}\sqrt{\frac{2}{LC}} \tag{5.133}$$

$$f_2 = \frac{1}{2\pi}\sqrt{\frac{2-\sqrt{2}}{LC}} \tag{5.134}$$

and

$$f_3 = \frac{1}{2\pi}\sqrt{\frac{2+\sqrt{2}}{LC}} \tag{5.135}$$

Once these frequencies are calculated, they provide us with the frequency regions where one filter outperforms the other.

5.8 Laboratory Exercises

Note: This laboratory requires a custom PCB [Haring, 2020], for which Altium files and BOM are provided.

5.8.1 Input Impedance and Insertion Loss of EMC Filters

Objective: This laboratory exercise investigates the input impedance and insertion loss of several filter configurations through simulations and measurements.

5.8.2 Laboratory Equipment and Supplies

Network Analyzer: KEYSIGHT ENA Network Analyzer E5080A (or equivalent), shown in Figure 5.45.
   Adapter Coaxial Connector N Plug, Male Pin To SMA Jack, Female Socket 50 Ohm: Digikey PN: ACX1341-ND (or equivalent), shown in Figure 5.46.
   SMA Male to SMA Male, 2 or 4 cable Assemblies: Digikey PN: 3648-P1CA-STMSTM-ST085-36-ND (or equivalent), shown in Figure 5.47.
   PCB: Custom PCBs.

Figure 5.45   Laboratory equipment: network analyzer.

Figure 5.46   Laboratory equipment: N-type to SMA connector.

Figure 5.47   Laboratory equipment: SMA to SMA cable.

5.8.3 Laboratory Procedure

5.8.3.1 LC Filter vs. CL Filter

Part I – Calculations and Simulations

Step 1: Using the chosen capacitor and inductor values calculate the frequency at which the input impedances of the two filters are equal.

$$f = \frac{1}{2\pi}\sqrt{\frac{2}{LC}} \tag{5.136}$$

Step 2: Use LTspice software to plot the input impedance of the LC and CL filters. Determine the frequency at which the two impedances are equal. Verify that this frequency is equal to the one obtained in Step 1.

Step 3: Use LTspice software to plot the insertion loss of the two filters. Determine the regions where the insertion loss of one filter is greater than that of the other. Verify that the border frequency defining these regions is the same as obtained in Step 2.

Part II – Measurements

This part utilizes two PCB assemblies – one for the LC filter and the other for the CL filter.

Step 1: Solder components onto each PCB to create an LC and CL filter, shown in Figure 5.48.

Step 2: Connect the LC filter PCB to channels 1 and 2 of the network analyzer, as shown in Figure 5.49. If using the 4-channel network analyzer, connect the CL filter PCB between channels 3 and 4. If using 2-channel network analyzer, repeat the entire Step 2 for the CL filter.

Measure the insertion loss and compare it to the simulation results. Use markers at the border frequency, and below and above that frequency. Tabulate the measured and simulated values. Comment on the results.

Figure 5.48 PCBs used for the LC and CL filters insertion loss measurements.

Figure 5.49 Measurement setup - LC and CL filters.

5.8.3.2 Pi Filter vs. *T* Filter

Part I – Calculations and Simulations

Step 1: Using the chosen capacitor and inductor values calculate the frequencies at which the input impedances of the two filters are equal.

$$f_1 = \frac{1}{2\pi}\sqrt{\frac{1}{LC}} \tag{5.137}$$

and

$$f_2 = \frac{1}{2\pi}\sqrt{\frac{3}{LC}} \tag{5.138}$$

Step 2: Use LTspice software to plot the input impedance of the Pi and *T* filters. Determine the frequencies at which the two impedances are equal. Verify that these frequency are equal to the ones obtained in Step 1.

Step 3: Use LTspice software to plot the insertion loss of the two filters. Determine the regions where the insertion loss of one filter is greater than that of the other. Verify that the border frequencies defining these regions are the same as obtained in Step 2.

Part II – Measurements

This part utilizes two PCB assemblies – one for the Pi filter and the other for the T filter.

Step 1: Solder components onto each PCB to create Pi and T filters, like the ones shown in Figure 5.50.

Step 2: Connect the Pi filter PCB to channels 1 and 2 of the network analyzer, as shown in Figure 5.51. If using the four-channel network analyzer, connect the T filter PCB

Figure 5.50   PCBs used for the Pi and T filters insertion loss measurements.

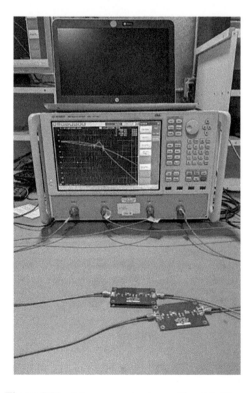

Figure 5.51   Measurement setup – Pi and T filters.

between channels 3 and 4. If using 2-channel network analyzer, repeat the entire Step 2 for the *T* filter.

Measure the insertion loss and compare it to the simulation results. Use markers at the border frequencies, and below and above these frequencies. Tabulate the measured and simulated values. Comment on the results.

References

Bogdan Adamczyk. *Foundations of Electromagnetic Compatibility*. John Wiley & Sons, Ltd, Chichester, UK, 2017. ISBN 9781119120810. doi: https://doi.org/10.1002/9781119120810.

Bogdan Adamczyk and Brian Gilbert. EMC Filters Comparison Part II: Pi and T Filters. *In Compliance Magazine*, January 2020. URL https://incompliancemag.com/article/emc-filters-comparison-part-ii--%cf-%80-and-t-filters/.

Bogdan Adamczyk and Dimitri Haring. EMC Filters Comparison Part I: CL and LC Filters. *In Compliance Magazine*, December 2019. URL https://incompliancemag.com/article/emc-filters-comparison-part-i-cl-and-lc-filters/.

Li Fan and Fei Xia. The Impedance Mismatching on the EMI Power Filter Design, 2012. URL https://link.springer.com/content/pdf/10.1007/978-3-642-27326-1_90.pdf.

Dimitri Haring. EMC Filter PCB Designer, 2020.

Clayton R. Paul. *Introduction to Electromagnetic Compatibility*. John Wiley & Sons, Inc., Hoboken, New Jersey, 2nd edition, 2006. ISBN 978-0-471-75500-5.

Chapter 6: Transmission Lines – Time Domain

6.1 Introduction

In a transmission line structure the signal travels from the source to load along one conductor (forward path) and returns to the source along the other conductor (parallel return path).

The simplest circuit model of a wire-type transmission line is shown in Figure 6.1.

Wire ground-plane transmission line is shown in Figure 6.2.

Coaxial transmission line is shown in Figure 6.3.

Co-planar transmission line is shown in Figure 6.4.

Microstrip transmission line is shown in Figure 6.5.

Finally, the stripline transmission line is shown in Figure 6.6.

6.1.1 Transmission Line Effects

There are two basic transmission line effects: propagation delay and reflections. The propagation delay effect is rather obvious – as the signal travels from the source to the load the connecting line introduces a time delay.

The reflection effect is not so obvious. As we shall see, one of the parameters characterizing a transmission line is *characteristic impedance*, Z_C. When the load, R_L (or the source R_S) is not matched to this impedance, reflections take place. These reflections cause a shift in the signal level when it arrives at the load or the source. We refer to this phenomenon as the *signal integrity*.

Consider a transmission line shown in Figure 6.7 (we will recreate this experiment in the laboratory section).

Figure 6.8 shows the propagation delay and the reflections at the load and the source.

6.1.2 When a Line Is not a Transmission Line

A very important question to ask is: *when the line is not a transmission line?*, i.e. when can we ignore the time delay and the reflections?

To answer this question consider a circuit shown in Figure 6.9 with the time delay (one-way travel time) of $T_D = 9$ ns (we will recreate this experiment in the laboratory section) [Adamczyk and Gilbert, 2020].

Let the source signal be a pulse with a variable rise time. Figure 6.10 shows the voltages at the source and the load with the rise time $t_r = 2.5$ ns

Principles of Electromagnetic Compatibility: Laboratory Exercises and Lectures, First Edition. Bogdan Adamczyk.
© 2024 John Wiley & Sons Ltd. Published 2024 by John Wiley & Sons Ltd.
Companion website: www.wiley.com/go/principlesofelectromagneticcompatibility

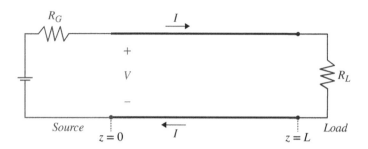

Figure 6.1 Two-wire transmission line.

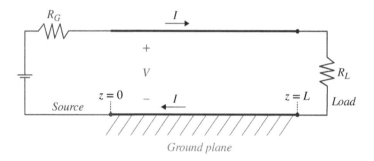

Figure 6.2 Wire ground-plane transmission line.

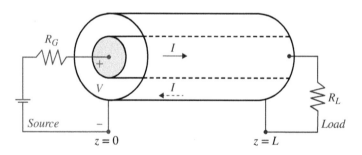

Figure 6.3 Coaxial transmission line.

We observe a noticeable time delay and reflections. Figure 6.11 shows the voltages at the source and the load with the rise time $t_r = 10\,\text{ns}$

Note that the time delay and the reflections become smaller. Let's increase the rise time to 50 ns. The result is shown in Figure 6.12.

Notice the further decrease in the time delay and reflections. Finally, let's increase the rise time to 100 ns. The result is shown in Figure 6.13.

The results shown in Figure 6.13 lead to the conclusion that the transmission line effects are negligible.

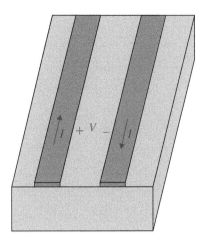

Figure 6.4   Co-planar transmission line.

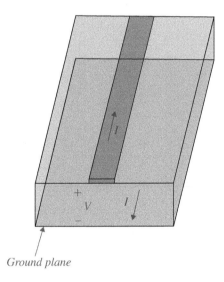

Figure 6.5   Microstrip transmission line.

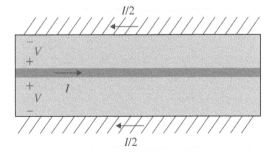

Figure 6.6   Stripline transmission line.

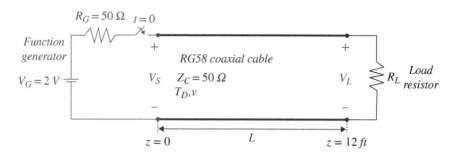

Figure 6.7   Wire-type transmission line.

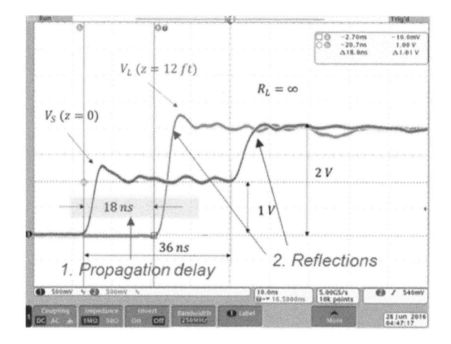

Figure 6.8   Transmission line effects.

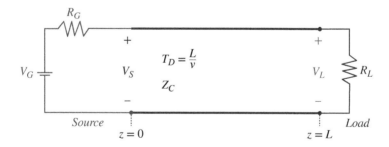

Figure 6.9   Circuit used to determine the transmission line criterion.

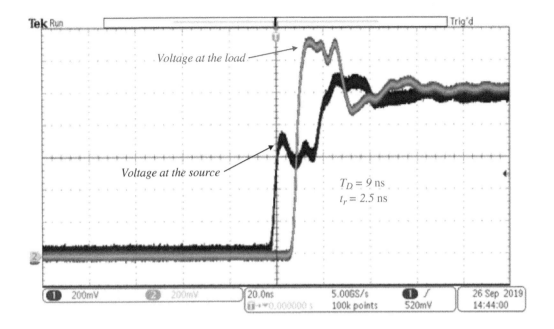

Figure 6.10   Transmission line effects with $t_r = 2.5$ ns.

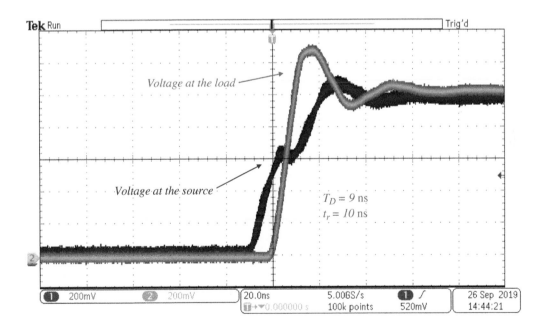

Figure 6.11   Transmission line effects with $t_r = 10$ ns.

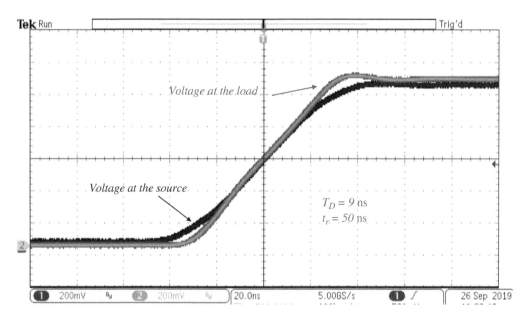

Figure 6.12   Transmission line effects with $t_r = 50$ ns.

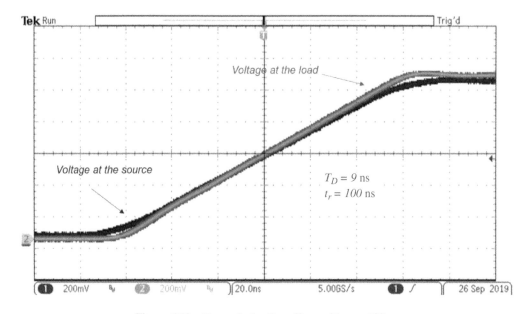

Figure 6.13   Transmission line effects with $t_r = 100$ ns.

This leads us to the (time domain) transmission line criterion:

The line does not need to be treated as a transmission line when the pulse rise time t_r is greater than, or equal to, 10 one-way time delay T_D of the line

$$t_r \geq 10T_D \tag{6.1}$$

Next, we will derive an equivalent frequency domain criterion. From Eq. (6.1) we get

$$T_D \leq \frac{1}{10}t_r \tag{6.2}$$

Recall from Eq. (1.52) that the highest significant frequency present in a pulse signal is

$$f = \frac{1}{t_r} \tag{6.3}$$

Using Eq. (6.3) in Eq. (6.2) we get

$$T_D \leq \frac{1}{10f} \tag{6.4}$$

The one-way time delay T_D is related to the line length, L, and the velocity of propagation, v, by

$$T_D = \frac{L}{v} \tag{6.5}$$

Using Eq. (6.5) in Eq. (6.4) we get

$$\frac{L}{v} \leq \frac{1}{10f} \tag{6.6}$$

or

$$L \leq \frac{v}{10f} \tag{6.7}$$

The wavelength λ is related to the velocity of propagation, v, and the frequency f by

$$\lambda = \frac{v}{f} \tag{6.8}$$

Using Eq. (6.8) in Eq. (6.7) we get

$$L \leq \frac{\lambda}{10} \tag{6.9}$$

We have arrived at the (frequency domain) transmission line criterion:

The line does not need to be treated as a transmission line when its length L is smaller than, or equal to, $\frac{\lambda}{10}$, where λ is the wavelength at the highest frequency of interest.

The condition in Eq. (6.9) is also the criterion for the *electrically short* structures.

A physical structure, or a physical distance, is considered electrically short when its physical length L is smaller than, or equal to, $\frac{\lambda}{10}$, where λ is the wavelength at the highest frequency of interest.

6.1.3 Transmission Line Equations

Transmission line can be modeled as a distributed-parameter circuit consisting of a series of small segments of length Δz as shown in Figure 6.14.

The distributed parameters (per unit length) describing the transmission line are: r resistance in $\frac{\Omega}{m}$, l inductance in $\frac{H}{m}$, g conductance in $\frac{S}{m}$, and c capacitance in $\frac{F}{m}$.

The transmission line model in Figure 6.14 describes a *lossy* transmission line. To gain an insight into the transmission line theory it is very helpful to consider a *lossless* transmission line first. Such a transmission line is shown in Figure 6.15.

To obtain the transmission line equations, let us consider a single segment of a lossless transmission line shown in Figure 6.16.

The distributed, per-unit-length parameters describing the transmission line are the inductance l in H/m and the capacitance c in F/m.

Writing Kirchhoff's voltage law (KVL), around the outside loop results in

$$-V(z,t) + l\Delta z\frac{\partial I(z + \Delta z, t)}{\partial t} + V(z + \Delta z, t) = 0 \tag{6.10}$$

or

$$V(z + \Delta z, t) - V(z, t) = -l\Delta z\frac{\partial I(z + \Delta z, t)}{\partial t} \tag{6.11}$$

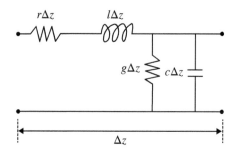

Figure 6.14   Circuit model of a lossy transmission line.

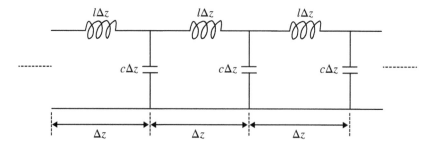

Figure 6.15   Circuit model of a lossless transmission line.

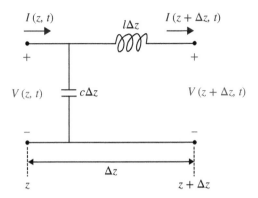

Figure 6.16 Single segment of a lossless transmission line.

Dividing both sides by Δz and taking the limit gives

$$\lim_{\Delta z \to 0} \frac{V(z + \Delta z, t) - V(z, t)}{\Delta z} = -\lim_{\Delta z \to 0} l\frac{\partial I(z + \Delta z, t)}{\partial t} \tag{6.12}$$

or

$$\frac{\partial V(z, t)}{\partial z} = -l\frac{\partial I(z, t)}{\partial t} \tag{6.13}$$

Writing Kirchhoff's current law (KCL), at the upper node of the capacitor results in

$$I(z, t) = c\Delta z\frac{\partial V(z, t)}{\partial t} + I(z + \Delta z, t) \tag{6.14}$$

or

$$I(z + \Delta z, t) - I(z, t) = -c\Delta z\frac{\partial V(z, t)}{\partial t} \tag{6.15}$$

Dividing both sides by Δz and taking the limit gives

$$\lim_{\Delta z \to 0} \frac{I(z + \Delta z, t) - I(z, t)}{\Delta z} = -\lim_{\Delta z \to 0} c\frac{\partial V(z, t)}{\partial t} \tag{6.16}$$

or

$$\frac{\partial I(z, t)}{\partial z} = -c\frac{\partial V(z, t)}{\partial t} \tag{6.17}$$

Equations (6.13) and (6.17) constitute a set of first-order, coupled transmission line equations. These equations can be decoupled [Adamczyk, 2017b], resulting in

$$\frac{\partial^2 V(z, t)}{\partial z^2} = lc\frac{\partial^2 V(z, t)}{\partial t^2} \tag{6.18}$$

$$\frac{\partial^2 I(z, t)}{\partial z^2} = lc\frac{\partial^2 I(z, t)}{\partial t^2} \tag{6.19}$$

6.1.3.1 Time-Domain Solution

The general time-domain solutions to these transmission-line equations are

$$V(z,t) = V^+ \left(t - \frac{z}{v} \right) + V^- \left(t + \frac{z}{v} \right) \tag{6.20}$$

$$I(z,t) = I^+ \left(t - \frac{z}{v} \right) + I^- \left(t + \frac{z}{v} \right) \tag{6.21}$$

where

$$I^+ \left(t - \frac{z}{v} \right) = \frac{1}{Z_C} V^+ \left(t - \frac{z}{v} \right) \tag{6.22}$$

$$I^- \left(t + \frac{z}{v} \right) = -\frac{1}{Z_C} V^- \left(t + \frac{z}{v} \right) \tag{6.23}$$

Z_C is the characteristic impedance of the line equal to

$$Z_C = \sqrt{\frac{l}{c}} \tag{6.24}$$

The function $V^+ \left(t - \frac{z}{v} \right)$ represents a forward-traveling (in the $+z$ direction) voltage wave, while the function $V^- \left(t + \frac{z}{v} \right)$ represents a backward-traveling (in the $-z$ direction) voltage wave [Adamczyk, 2021a].

Similar statements are valid for the current waves. The total solution consists of the sum of the forward-traveling and backward-traveling waves.

The velocity of the wave propagation along the line is given by

$$v = \frac{1}{\sqrt{lc}} \tag{6.25}$$

To simplify the notation in the following discussion let's rewrite the solutions in Eqs. (6.20) and (6.21) in a concise form [Rao, 2004],

$$V = V^+ + V^- \tag{6.26}$$

$$I = I^+ + I^- \tag{6.27}$$

Then, from Eqs. (6.22) and (6.23) we have

$$I^+ = \frac{V^+}{Z_C} \tag{6.28}$$

$$I^- = -\frac{V^-}{Z_C} \tag{6.29}$$

Also, from Eqs. (6.28) and (6.29) we obtain

$$Z_C = \frac{V^+}{I^+} \tag{6.30}$$

$$Z_C = -\frac{V^-}{I^-} \tag{6.31}$$

6.2 Transient Analysis

6.2.1 Reflections at a Resistive Load

Consider a transmission line of length L driven by a constant voltage source V_G with a source resistance R_G, as shown in Figure 6.17 [Adamczyk, 2017c]

When the switch closes at $t = 0$ a forward voltage wave, V^+, originates at $z = 0$ and travels toward the load. This is shown in Figure 6.18.

The value of this voltage is [Adamczyk, 2017b]

$$V^+ = \frac{Z_C}{R_G + Z_C} V_G \tag{6.32}$$

As this wave travels along the transmission line, the voltage along the line changes from 0 to V^+ and remains at that value (for now). At the time T the voltage wave reaches the load and sets up a reflection, V^-. This is shown in Figure 6.19.

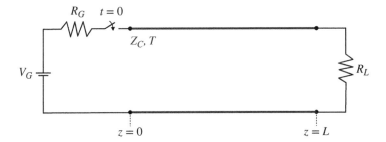

Figure 6.17   Transmission line driven by a constant voltage source and terminated by a resistive load.

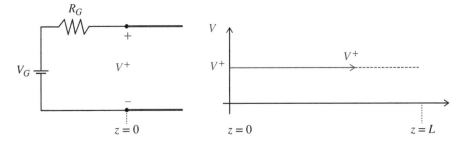

Figure 6.18   Forward voltage wave originates at the source and travels toward the load.

The reflected voltage, V^-, is related to the incident voltage, V^+, by

$$V^- = \Gamma_L V^+ \tag{6.33}$$

where Γ_L is the *(voltage) load reflection coefficient* given by

$$\Gamma_L = \frac{V^-}{V^+} = \frac{R_L - Z_C}{R_L + Z_C} \tag{6.34}$$

The total voltage at the load is sum of the incident and reflected voltages.

$$V = V^+ + V^- \tag{6.35}$$

The reflected voltage, V^-, now travels back to the source, as shown in Figure 6.19. As this wave travels back to the source, voltage along the line changes from V^+ to $V^+ + V^-$. This wave reaches the source at time $2T$ and sets up another reflection, V^{-+}. This is shown in Figure 6.20.

Voltage reflected at the source, V^{-+}, is related to the incident voltage, V^-, by

$$V^{-+} = \Gamma_S V^- \tag{6.36}$$

where Γ_S is the *(voltage) source reflection coefficient* given by

$$\Gamma_S = \frac{V^{-+}}{V^-} = \frac{R_G - Z_C}{R_G + Z_C} \tag{6.37}$$

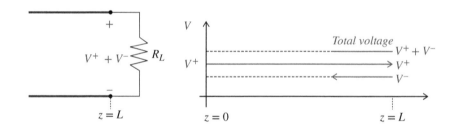

Figure 6.19   Incident wave is reflected at the load and travels back to the source.

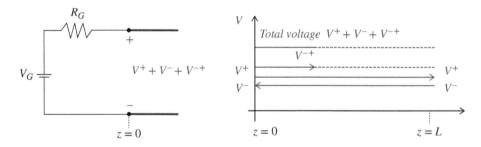

Figure 6.20   Reflected wave arrives at the source and travels back to the load.

The total voltage at the source is now

$$V = V^+ + V^- + V^{-+} \tag{6.38}$$

6.2.1.1 Reflection Coefficient – Special Cases

There are three important special cases of the load reflection coefficient: short-circuit load, open-circuit load, and matched load.

Case 1: Short-circuit load $R_L = 0$
  In this case the load reflection coefficient is

$$\Gamma_L = \frac{R_L - Z_C}{R_L + Z_C} = \frac{0 - Z_C}{0 + Z_C} = -1 \tag{6.39}$$

The reflected voltage is

$$V^- = \Gamma_L V^+ = -V^+ \tag{6.40}$$

The total voltage at the load is

$$V_{total} = V^+ + V^- = V^+ - V^+ = 0 \tag{6.41}$$

Case 2: Open-circuit load $R_L = \infty$
  In this case the load reflection coefficient is

$$\Gamma_L = \frac{R_L - Z_C}{R_L + Z_C} = \left(\frac{1 - \frac{Z_C}{R_L}}{1 + \frac{Z_C}{R_L}} \right)_{R_L \to \infty} = 1 \tag{6.42}$$

The reflected voltage is

$$V^- = \Gamma_L V^+ = V^+ \tag{6.43}$$

The total voltage at the load is

$$V_{total} = V^+ + V^- = V^+ + V^+ = 2V^+ \tag{6.44}$$

Case 3: Matched load $R_L = Z_C$
  In this case the load reflection coefficient is

$$\Gamma_L = \frac{R_L - Z_C}{R_L + Z_C} = \frac{Z_C - Z_C}{Z_C + Z_C} = 0 \tag{6.45}$$

The reflected voltage is

$$V^- = \Gamma_L V^+ = 0 \tag{6.46}$$

The total voltage at the load is

$$V_{total} = V^+ + V^- = V^+ + 0 = V^+ \tag{6.47}$$

There is no reflection at the load; the total voltage at the load is equal to the incident voltage only.

6.2.1.2 Bounce Diagram

Consider the circuit shown in Figure 6.21 [Adamczyk, 2018a].

When the switch closes the forward voltage wave travels towards the load and reaches it at $t = T$ ($T =$ one-way travel time). Since the line and the load are mismatched a reflection is created and travels back to the source, reaching it at $t = 2T$ (assuming zero-rise time). Since the line and the source are mismatched, another reflection is created which travels forward to the load reaching it at $t = 3T$.

This process theoretically continues indefinitely; practically, it continues until the steady-state voltages are reached at the source and at the load. A bounce diagram is a plot of the voltage (or current) at the source or the load (or any other location) after each reflection. Let's create a plot of the voltages at the source and the load for the circuit shown in Figure 6.21.

The initial voltage at the location $z = 0$ is

$$V_S(z = 0, t = 0) = V^+ = V_G \frac{Z_C}{R_G + Z_C} = (10)\frac{75}{50 + 75} = 6 \text{ V} \tag{6.48}$$

This is shown in Figure 6.22.

The reflection coefficient at the load is

$$\Gamma_L = \frac{R_L - Z_C}{R_L + Z_C} = \frac{216 - 75}{216 + 75} = 0.4845 \tag{6.49}$$

The initial voltage wave of 6 V travels to the load and reaches it at $t = T$ creating a reflection

$$V^- = \Gamma_L V^+ = (0.4845)(6) = 2.907 \text{ V} \tag{6.50}$$

The total voltage at the load (at $t = T$) is

$$V_L(z = L, t = T) = V^+ + V^- = 6 + 2.907 = 8.907 \text{ V} \tag{6.51}$$

This is shown in Figure 6.23.

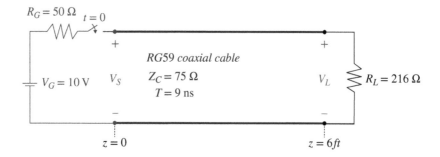

Figure 6.21   Circuit used to create bounce diagram.

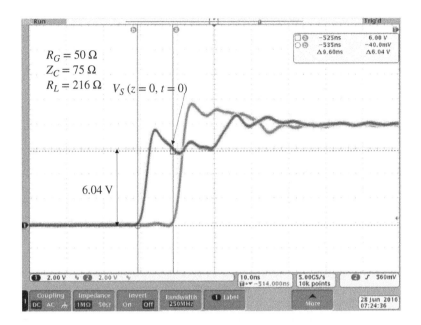

Figure 6.22 Initial voltage wave at $z = 0, t = 0$.

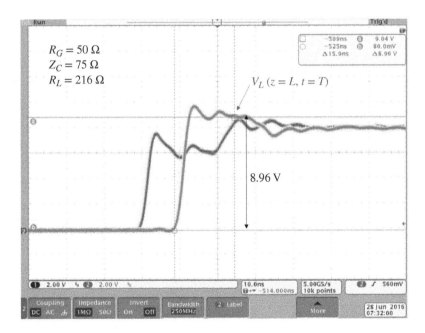

Figure 6.23 Voltage at the load $z = L, t = T$.

Voltage reflected at the load ($V^- = 2.907$ V) travels back to the source. The reflection coefficient at the source is

$$\Gamma_S = \frac{R_S - Z_C}{R_S + Z_C} = \frac{50 - 75}{50 + 75} = -0.2 \tag{6.52}$$

The re-reflected voltage at the source is

$$V^{-+} = \Gamma_S V^- = (-0.2)(2.907) = -0.5814 \text{ V} \tag{6.53}$$

The total voltage at the source (at $t = 2T$) is

$$V_S(z = 0, t = 2T) = V^+ + V^- + V^{-+} = 6 + 2.907 - 0.5814 = 8.3256 \text{ V} \tag{6.54}$$

This is shown in Figure 6.24.

The voltage reflected at the source ($V^{(-+)} = -0.5814$ V) travels toward the load where it will create another reflection which will travel toward the source. This process will continue until the steady-state is reached. The bounce diagram showing the voltages at the source and the load after each reflection is shown in Figure 6.25.

Figure 6.26 shows the voltage at the source ($z = 0$) while Figure 6.27 shows the voltage at the load ($z = L$) during the period $0 \leq t \leq 8T$.

It is apparent the source and load voltages eventually reach the steady state. Recall that a transmission line can be modeled as a sequence of in-line inductors and shunt capacitors. Under the dc conditions (steady state when driven by a dc source) inductors act as short circuits and capacitors act as open circuits. Thus, in steady state the circuit

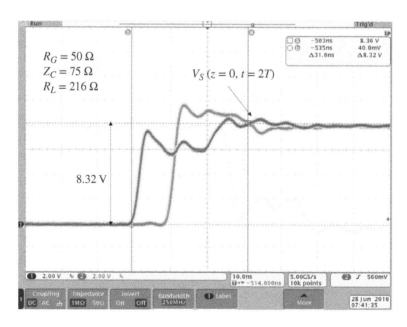

Figure 6.24 Voltage at the source $z = 0, t = 2T$.

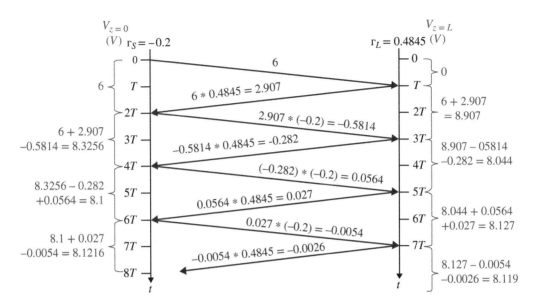

Figure 6.25 Bounce diagram: voltages at the source and the load.

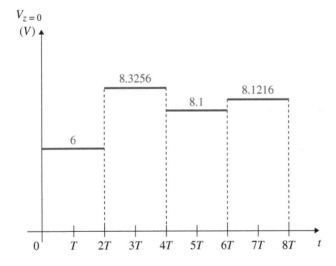

Figure 6.26 Voltage at the source during $0 \leq t \leq 8T$.

in Figure 6.21 is equivalent to the circuit in Figure 6.28 where the transmission line is modeled as an ideal conductor.

The steady-state value of the voltage at $z = 0$ is the same as the value at $z = L$ and can be obtained from the voltage divider as

$$V_{SS} = \frac{216}{50 + 216}(10) = 8.1203 \text{ V} \qquad (6.55)$$

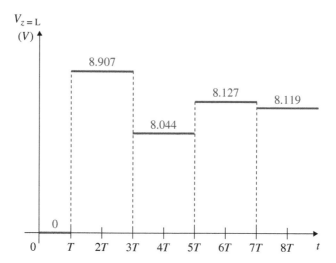

Figure 6.27   Voltage at the load during $0 \leq t \leq 8T$.

Figure 6.28   Voltage at the load during $0 \leq t \leq 8T$.

6.2.2 Reflections at a Resistive Discontinuity

In this section we consider the effects of discontinuity along the transmission line which occurs when the transmission line's characteristic impedance changes from Z_{C1} to Z_{C2} as shown in Figure 6.29 [Adamczyk, 2017a].

Let's consider a voltage and current waves v_{i1} and i_{i1} traveling on transmission line 1 and incident onto the junction. Upon their arrival at the junction the reflected waves v_{r1} and $i_{r1}1$, and the transmitted waves v_{t2} and i_{t2} are created.

KVL at the junction produces

$$v_{i1} + v_{r1} = v_{t2} \tag{6.56}$$

while the KCL gives

$$i_{i1} + i_{r1} = i_{t2} \tag{6.57}$$

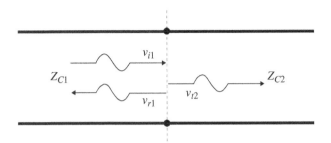

Figure 6.29 Discontinuity along a transmission line.

The current waves are related to the voltage waves by

$$i_{i1} = \frac{v_{i1}}{Z_{C1}} \tag{6.58}$$

$$i_{r1} = -\frac{v_{r1}}{Z_{C1}} \tag{6.59}$$

$$i_{t2} = \frac{v_{t2}}{Z_{C2}} \tag{6.60}$$

Substituting Eqs. (6.58), (6.59), and (6.60) into Eq. (6.57) produces

$$\frac{v_{i1}}{Z_{C1}} - \frac{v_{r1}}{Z_{C1}} = \frac{v_{t2}}{Z_{C2}} \tag{6.61}$$

Thus

$$\frac{v_{t2}}{Z_{C2}} = \frac{Z_{C2}}{Z_{C1}}(v_{i1} - v_{r1}) \tag{6.62}$$

Using Eq. (6.62) in Eq. (6.56) results in

$$v_{i1} + v_{r1} = \frac{Z_{C2}}{Z_{C1}}(v_{i1} - v_{r1}) \tag{6.63}$$

or

$$v_{i1} + v_{r1} = \frac{Z_{C2}}{Z_{C1}}v_{i1} - \frac{Z_{C2}}{Z_{C1}}v_{r1} \tag{6.64}$$

or

$$v_{r1} + \frac{Z_{C2}}{Z_{C1}}v_{r1} = \frac{Z_{C2}}{Z_{C1}}v_{i1} - v_{i1} \tag{6.65}$$

leading to

$$\left(1 + \frac{Z_{C2}}{Z_{C1}}\right)v_{r1} = \left(\frac{Z_{C2}}{Z_{C1}} - 1\right)v_{i1} \tag{6.66}$$

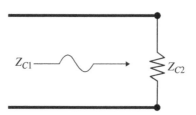

Figure 6.30   Incoming wave sees a termination impedance Z_{C2}.

or

$$\left(\frac{Z_{C1} + Z_{C2}}{Z_{C1}}\right) v_{r1} = \left(\frac{Z_{C2} - Z_{C1}}{Z_{C1}}\right) v_{i1} \tag{6.67}$$

and thus

$$\frac{v_{r1}}{v_{i1}} = \frac{Z_{C2} - Z_{C1}}{Z_{C2} + Z_{C1}} = \Gamma_{12} \tag{6.68}$$

where Γ_{12} is the voltage reflection coefficient for the wave incident from the left onto the boundary. In terms of this reflection coefficient, the reflected voltage can be expressed as

$$v_{r1} = \Gamma_{12} v_{i1} \tag{6.69}$$

Thus, to the incident wave, the transmission line to the right looks like its characteristic impedance Z_{C2}, as shown in Figure 6.30.

We also define the voltage *transmission coefficient* as the ratio of the transmitted voltage v_{t2} to the incident voltage v_{i1}

$$T_{12} = \frac{v_{t2}}{v_{i1}} \tag{6.70}$$

Since

$$v_{i1} + v_{r1} = v_{t2} \tag{6.71}$$

we have

$$T_{12} = \frac{v_{t2}}{v_{i1}} = \frac{v_{i1} + v_{r1}}{v_{i1}} = 1 + \frac{v_{r1}}{v_{i1}} \tag{6.72}$$

or

$$T_{12} = 1 + \Gamma_{12} \tag{6.73}$$

Utilizing Eq. (6.68) in Eq. (6.73) results in

$$T_{12} = 1 + \frac{Z_{C2} - Z_{C1}}{Z_{C2} + Z_{C1}} = \frac{Z_{C2} + Z_{C1} + Z_{C2} - Z_{C1}}{Z_{C2} + Z_{C1}} \tag{6.74}$$

or

$$T_{12} = \frac{2 Z_{C2}}{Z_{C2} + Z_{C1}} \tag{6.75}$$

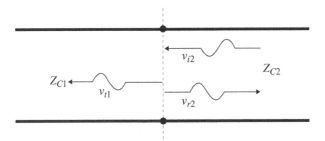

Figure 6.31   Wave incident from the right.

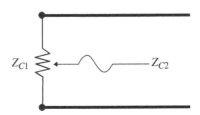

Figure 6.32   Incoming wave sees a termination impedance Z_{C1}.

A similar derivation can be performed for the wave incident on the boundary from the right, as shown in Figure 6.31.

In this case the voltage reflection coefficient is given by

$$\Gamma_{21} = \frac{v_{r2}}{v_{i2}} = \frac{Z_{C1} - Z_{C2}}{Z_{C1} + Z_{C2}} \tag{6.76}$$

Again, to the incident wave, the transmission line to the left looks like its characteristic impedance Z_{C1}, as shown in Figure 6.32.

The voltage transmission coefficient for the wave incident from the right is

$$T_{21} = \frac{2Z_{C1}}{Z_{C1} + Z_{C2}} \tag{6.77}$$

and the reflection and transmission coefficients for the wave incident from the right are related by

$$T_{21} = 1 + \Gamma_{21} \tag{6.78}$$

6.2.3 Reflections at a Shunt Resistive Discontinuity

Consider the circuit shown in Figure 6.33, where the transmission line of length l has a shunt resistive discontinuity in the middle of line, at a location $z = d$ [Adamczyk, 2022a].

Note that the transmission line is matched at the source, and the resistive discontinuity and the load resistor values are equal to the characteristic impedance of the line.

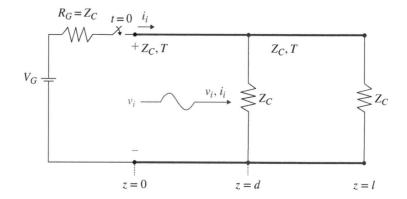

Figure 6.33 Shunt resistive discontinuity along a transmission line.

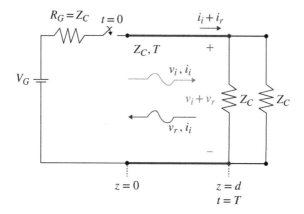

Figure 6.34 Reflection at the resistive discontinuity.

When the switch closes at $t = 0$, a wave originates at $z = 0$, and travels toward the discontinuity. At the time $t = T$ this wave arrives at the discontinuity. The transmission line immediately to the right of the discontinuity looks to the circuit on the left of the discontinuity like a shunt resistance equal to the characteristic impedance of the right line.

When this wave arrives at the discontinuity (at the time $t = T$), the reflected wave, v_r and i_r, is created, and we have a situation depicted in Figure 6.34.

The circuit in Figure 6.34 is equivalent to the one shown in Figure 6.35.

The reflection coefficient at the discontinuity is

$$\Gamma = \frac{\frac{Z_C}{2} - Z_C}{\frac{Z_C}{2} + Z_C} = \frac{-\frac{1}{2}}{\frac{3}{2}} = -\frac{1}{3} \tag{6.79}$$

The reflected voltage is related to the incident voltage by the reflection coefficient as

$$v_r = \Gamma v_i = -\frac{1}{3} v_i \tag{6.80}$$

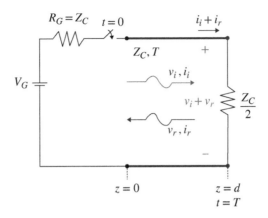

Figure 6.35   Equivalent circuit.

The total voltage at the discontinuity is

$$v_d = v_i + v_r = v_i - \frac{1}{3}v_i = \frac{2}{3}v_i \tag{6.81}$$

Since [Adamczyk, 2017c],

$$v_i = \frac{V_G}{2} \tag{6.82}$$

the total voltage at the discontinuity is

$$v_d = \frac{2}{3}\frac{V_G}{2} = \frac{V_G}{3} \tag{6.83}$$

The reflected voltage travels back to the source, arriving there at $t = 2T$. Since the source is matched to the transmission line, there is no reflection, and the total voltage at the source becomes

$$v_s = v_i + v_r = v_i - \frac{1}{3}v_i = \frac{2}{3}v_i \tag{6.84}$$

6.2.4 Reflections with Transmission Lines in Parallel

Consider the circuit shown in Figure 6.36, where the transmission line of length d is connected to two other transmission lines [Adamczyk, 2022b].

Note that load resistors are matched to the transmission lines and all transmission lines have the same characteristic impedances. This makes the simulations and measurements easier to follow, since there are no reflections at the loads.

When the switch closes at $t = 0$, a wave originates at $z = 0$ and travels toward the discontinuity. The transmission lines immediately to the right of discontinuity look to the circuit on the left of discontinuity like two resistances in parallel. Their resistance values are equal to the corresponding values of their characteristic impedances. When the incident wave arrives at the discontinuity (at the time $t = T$), the reflected wave, v_r and i_r, is created, and we have a situation depicted in Figure 6.37.

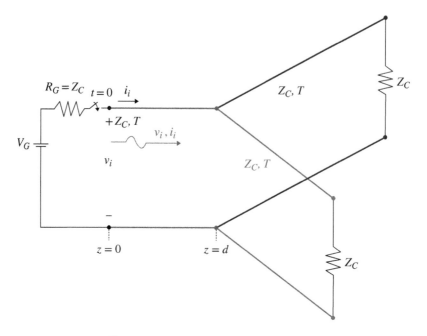

Figure 6.36   Transmission lines in parallel.

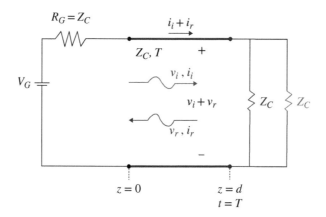

Figure 6.37   Incident and reflected waves at the discontinuity.

This part of the circuit is identical to the one discussed in Section 6.2.3, where the transmission line had a shunt resistive discontinuity. Thus, the total voltage across at the discontinuity is

$$v(z = d, t) = \frac{V_G}{3}, \ t > T \tag{6.85}$$

and the total voltage at the source (after $t = 2T$) is

$$v_S = v_i + v_r = \frac{V_G}{3}, \ t > 2T \tag{6.86}$$

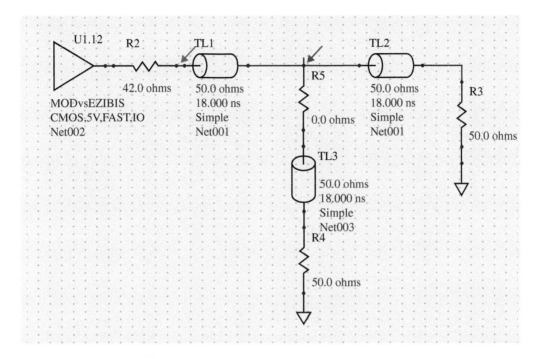

Figure 6.38 Transmission lines in parallel – HyperLynx schematic.

Figure 6.38 shows HyperLynx schematic of the transmission line in parallel.
The simulation results are shown in Figure 6.39.
Note that the simulation results agree with the analytical results.
Figure 6.40 shows the measurement setup with transmission lines in parallel.
The measurement results are shown in Figure 6.41.
Note that the measurement results agree with the analytical and simulation results.

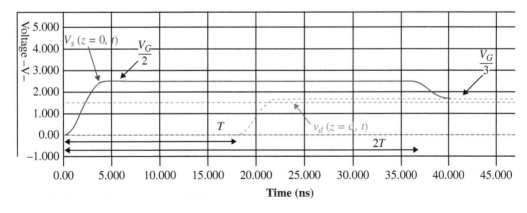

Figure 6.39 Transmission lines in parallel – voltages at the source ($z = 0$) and the discontinuity ($z = d$).

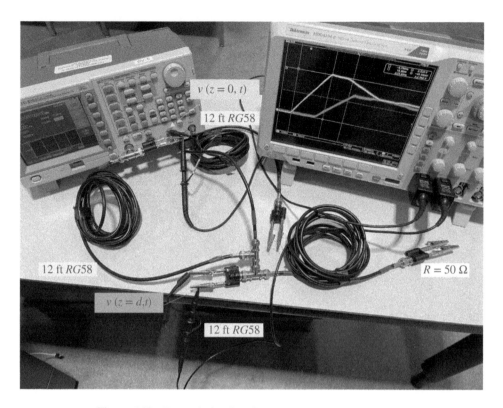

Figure 6.40 Transmission lines in parallel – measurement setup.

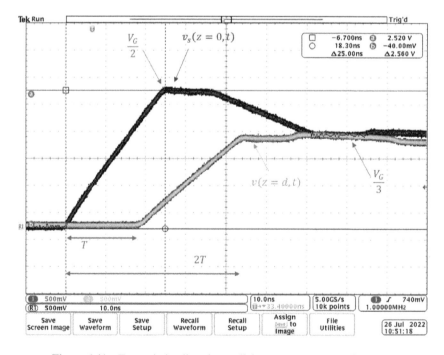

Figure 6.41 Transmission lines in parallel – measurement results.

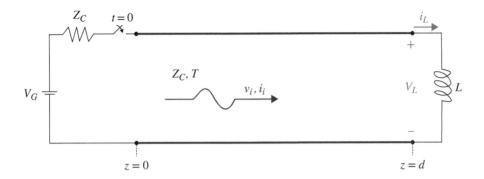

Figure 6.42   Inductive termination of a transmission line.

6.2.5 Reflections at a Reactive Load

6.2.5.1 Reflections at an Inductive Load

Consider the circuit shown in Figure 6.42 where the transmission line of length d is terminated by an inductor L [Adamczyk, 2018b].

The source impedance is matched to the characteristic impedance of the line; it is also assumed that the initial current in the inductor is zero.

$$i_L(0^-) = 0 \tag{6.87}$$

When the switch closes at $t = 0$, a wave originates at $z = 0$, with

$$v_i = \frac{V_G}{2} \tag{6.88}$$

$$i_i = \frac{V_G}{2Z_C} \tag{6.89}$$

and travels toward the load. When this wave arrives at the load (at the time $t = T$), the reflected waves v_r and i_r are created. This is shown in Figure 6.43.

The reflected current wave is related to the reflected voltage wave by

$$i_r(t) = -\frac{v_r(t)}{Z_C} \tag{6.90}$$

KVL at the load produces

$$v_i + v_r(t) = v_L(t) \tag{6.91}$$

while the KCL gives

$$i_i + i_r(t) = i_L(t) \tag{6.92}$$

Our initial goal is to determine the reflected voltage $v_r(t)$ at the location $z = d$, i.e. $v_r(d, t)$. The ultimate goal is to determine the total voltage at the load, $v_L(d, t)$.

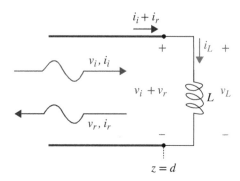

Figure 6.43 Incident and reflected waves at the inductive load.

The differential $v - i$ relationship for the load produces

$$v_L(t) = L\frac{di_L(t)}{dt} \tag{6.93}$$

Utilizing Eqs. (6.91) and (6.92) in Eq. (6.93) we get

$$v_i + v_r(t) = L\frac{d}{dt}[i_i + i_r(t)] \tag{6.94}$$

Using Eqs. (6.88), (6.89), and (6.90) in Eq. (6.94) we obtain

$$\frac{V_G}{2} + v_r(t) = L\frac{d}{dt}\left[\frac{V_G}{2Z_C} - \frac{v_r(t)}{Z_C}\right] \tag{6.95}$$

Since V_G and Z_C are constant Eq. (6.95) reduces to

$$\frac{V_G}{2} + v_r(t) = -\frac{L}{Z_C}\frac{dv_r(t)}{dt} \tag{6.96}$$

or

$$\frac{L}{Z_C}\frac{dv_r(t)}{dt} + v_r(t) = -\frac{V_G}{2} \tag{6.97}$$

This differential equation needs to be solved for $v_r(t)$, for $t > T$, subject to the initial condition $v_r(t = T)$. Let's determine this initial condition.

KCL at the inductor node produces

$$i_i + i_r(t) = i_L(t) \tag{6.98}$$

or

$$\frac{V_G}{2Z_C} - \frac{v_r(t)}{Z_C} = i_L(t) \tag{6.99}$$

Evaluating it at $t = T$, we get

$$\frac{V_G}{2Z_C} - \frac{v_r(T)}{Z_C} = i_L(T) \qquad (6.100)$$

Since the current through the inductor cannot change instantaneously, we have $i_L(T) = 0$, and thus

$$\frac{V_G}{2Z_C} - \frac{v_r(T)}{Z_C} = 0 \qquad (6.101)$$

or

$$v_r(T) = \frac{V_G}{2} \qquad (6.102)$$

Now we are ready to solve Eq. (6.97), subject to the initial condition in Eq. (6.102):

$$\frac{L}{Z_C}\frac{dv_r(t)}{dt} + v_r(t) = -\frac{V_G}{2}, \quad v_r(T) = \frac{V_G}{2}, \quad t > T \qquad (6.103)$$

First, let's rewrite this equation in a standard form:

$$\frac{dv_r(t)}{dt} + \frac{v_r(t)}{L/Z_C} = -\frac{V_G Z_C}{2L} \qquad (6.104)$$

or

$$\frac{dv_r(t)}{dt} + \frac{v_r(t)}{\tau} = K \qquad (6.105)$$

where

$$\tau = \frac{L}{Z_C} \qquad (6.106)$$

$$K = -\frac{V_G Z_C}{2L} \qquad (6.107)$$

We will solve Eq. (6.105) using the separation of variables method. Rearranging Eq. (6.105) we get

$$\frac{dv_r(t)}{dt} = -\frac{v_r}{\tau} + K \qquad (6.108)$$

or

$$\frac{dv_r(t)}{dt} = -\frac{(v_r - K\tau)}{\tau} \qquad (6.109)$$

Separating the variables, we get

$$\frac{dv_r(t)}{v_r - K\tau} = -\frac{1}{\tau}dt \qquad (6.110)$$

Now, integrating Eq. (6.110) we obtain

$$\int_{v_r(t_0 = T)}^{v_r(T)} \frac{dv_r(t)}{v_r - K\tau} = -\int_{t_0 = T}^{t} \frac{1}{\tau}dt \qquad (6.111)$$

resulting in

$$\ln(v_r - K\tau)\Big|_{v_r(T)}^{v_r(t)} = -\frac{1}{\tau}\Big|_T^t \tag{6.112}$$

or

$$\ln\frac{v_r(t) - K\tau}{v_r(T) - K\tau} = -\frac{1}{\tau}(t - T) \tag{6.113}$$

and thus

$$\frac{v_r(t) - K\tau}{v_r(T) - K\tau} = e^{-\frac{1}{\tau}(t-T)} \tag{6.114}$$

leading to

$$v_r(t) = K\tau + [v_r(T) - K\tau]e^{-\frac{1}{\tau}(t-T)} \tag{6.115}$$

Utilizing Eqs. (6.102), (6.106), and (6.107) in Eq. (6.115) we get

$$v_r(t) = \left(-\frac{V_G Z_C}{2L}\right)\left(\frac{L}{Z_C}\right) + \left[\frac{V_G}{2} - \left(-\frac{V_G Z_C}{2L}\right)\left(\frac{L}{Z_C}\right)\right]e^{-\frac{Z_C}{L}(t-T)} \tag{6.116}$$

or

$$v_r(d, t) = -\frac{V_G}{2} + V_G e^{-\frac{Z_C}{L}(t-T)}, \quad t \geq T \tag{6.117}$$

The total voltage across the inductor is

$$v_L(d, t) = v_i + v_r(d, t) = \frac{V_G}{2} + v_r(d, t) \tag{6.118}$$

or

$$v_L(d, t) = V_G e^{-\frac{Z_C}{L}(t-T)}, \quad t \geq T \tag{6.119}$$

6.2.5.2 Reflections at an RL Load

Consider the circuit shown in Figure 6.44 where the transmission line of length d is terminated by an RL load [Adamczyk, 2021b].

Note that the load resistor value is equal to the characteristic impedance of the transmission line; it is also assumed that the initial current through the inductor is zero.

$$i_L(0^-) = 0 \tag{6.120}$$

When the switch closes at $t = 0$, a wave originates at $z = 0$, with

$$v_i = \frac{V_G}{2} \tag{6.121}$$

$$i_i = \frac{V_G}{2Z_C} \tag{6.122}$$

and travels toward the load. When this wave arrives at the load (at the time $t = T$), the reflected waves v_r and i_r are created. This is shown in Figure 6.45.

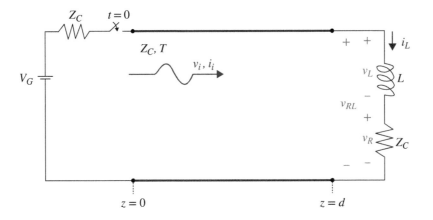

Figure 6.44   RL termination of a transmission line.

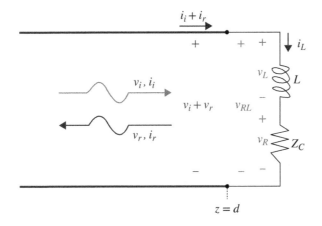

Figure 6.45   Incident and reflected waves at the RL load.

The reflected current wave is related to the reflected voltage wave by

$$i_r(t) = -\frac{v_r(t)}{Z_C} \tag{6.123}$$

KVL at the load produces

$$v_i + v_r(t) = v_L(t) + v_R(t) \tag{6.124}$$

while the KCL gives

$$i_i + i_r(t) = i_L(t) \tag{6.125}$$

Our initial is to determine the reflected voltage $v_r(t)$ at the location $z = d$, i.e. $v_r(d, t)$. The ultimate goal is to determine the total voltage at the load, $v_{RL}(d, t)$.

From Eq. (6.124) we obtain the inductor voltage as

$$v_L(t) = v_i + v_r(t) - v_R(t) \tag{6.126}$$

The load resistor voltage can be obtained from

$$v_R(t) = Z_C i_L(t) \tag{6.127}$$

Using Eqs. (6.121) and (6.127) in Eq. (6.126) produces

$$v_L(t) = \frac{V_G}{2} + v_r(t) - Z_C i_L(t) \tag{6.128}$$

The differential $v - i$ relationship for the inductor is

$$v_L(t) = L\frac{di_L(t)}{dt} \tag{6.129}$$

Utilizing Eqs. (6.125) and (6.128) in Eq. (6.129) we get

$$\frac{V_G}{2} + v_r(t) - Z_C[i_i + i_r(t)] = L\frac{d}{dt}[i_i + i_r(t)] \tag{6.130}$$

Using Eqs. (6.122) and (6.123) in Eq. (6.130) we have

$$\frac{V_G}{2} + v_r(t) - Z_C\left[\frac{V_G}{2Z_C} - \frac{v_r(t)}{Z_C}\right] = L\frac{d}{dt}\left[\frac{V_G}{2Z_C} - \frac{v_r(t)}{Z_C}\right] \tag{6.131}$$

Since V_G and Z_C are constant, Eq. (6.131) reduces to

$$\frac{V_G}{2} + v_r(t) - \left[\frac{V_G}{2} - v_r(t)\right] = \frac{L}{Z_C}\frac{dv_r(t)}{dt} \tag{6.132}$$

or

$$\frac{L}{2Z_C}\frac{dv_r(t)}{dt} + v_r(t) = 0 \tag{6.133}$$

This differential equation needs to be solved for $v_r(t)$, for $t > T$, subject to the initial condition $v_r(t = T)$. Let's determine this initial condition.

KCL at the inductor node produces

$$i_i + i_r(t) = i_L(t) \tag{6.134}$$

or

$$\frac{V_G}{2Z_C} - \frac{v_r(t)}{Z_C} = i_L(t) \tag{6.135}$$

Evaluating it at $t = T$, we get

$$\frac{V_G}{2Z_C} - \frac{v_r(T)}{Z_C} = i_L(T) \tag{6.136}$$

Since the current through the inductor cannot change instantaneously, we have $i_L(T) = 0$, and thus

$$\frac{V_G}{2Z_C} - \frac{v_r(T)}{Z_C} = 0 \tag{6.137}$$

or

$$v_r(T) = \frac{V_G}{2} \tag{6.138}$$

Now we are ready to solve Eq. (6.133), subject to the initial condition in Eq. (6.138):

$$\frac{L}{2Z_C} \frac{dv_r(t)}{dt} + v_r(t) = 0, \quad v_r(T) = \frac{V_G}{2}, \quad t > T \tag{6.139}$$

First, let's rewrite this equation in a standard form:

$$\frac{dv_r(t)}{dt} + \frac{v_r(t)}{L/2Z_C} = 0 \tag{6.140}$$

or

$$\frac{dv_r(t)}{dt} + \frac{v_r(t)}{\tau} = K \tag{6.141}$$

where

$$\tau = \frac{L}{2Z_C} \tag{6.142}$$

$$K = 0 \tag{6.143}$$

The solution of Eq. (6.141) was derived in Section 6.2.5.1 as

$$v_r(t) = K\tau + [v_r(T) - K\tau]e^{-\frac{1}{\tau}(t-T)} \tag{6.144}$$

Utilizing Eqs. (6.138), (6.142), and (6.143) in Eq. (6.144) we obtain

$$v_r(t) = \frac{V_G}{2}e^{-\frac{2Z_C}{L}(t-T)} \tag{6.145}$$

or

$$v_r(d,t) = \frac{V_G}{2}e^{-\frac{2Z_C}{L}(t-T)}, \quad t \geq T \tag{6.146}$$

The total voltage across the RL load is

$$v_{RL}(d,t) = v_i + v_r(d,t) = \frac{V_G}{2} + v_r(d,t) \tag{6.147}$$

or

$$v_{RL}(d,t) = \frac{V_G}{2} + \frac{V_G}{2}e^{-\frac{2Z_C}{L}(t-T)}, \quad t \geq T \tag{6.148}$$

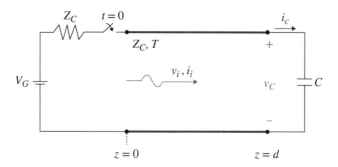

Figure 6.46 Capacitive termination of a transmission line.

6.2.5.3 Reflections at a Capacitive Load

Consider the circuit shown in Figure 6.46 where the transmission line of length d is terminated by a capacitor C [Adamczyk, 2018b].

The source impedance is matched to the characteristic impedance of the line; it is also assumed that the initial voltage across the capacitor is zero

$$v_C(0^-) = 0 \tag{6.149}$$

When the switch closes at $t = 0$, a wave originates at $z = 0$, with

$$v_i = \frac{V_G}{2} \tag{6.150}$$

$$i_i = \frac{V_G}{2Z_C} \tag{6.151}$$

and travels toward the load. When this wave arrives at the load (at the time $t = T$), the reflected waves v_r and i_r are created. This is shown in Figure 6.47.

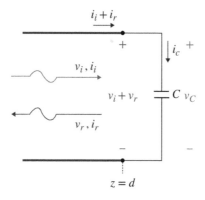

Figure 6.47 Incident and reflected waves at the capacitive load.

The reflected current wave is related to the reflected voltage wave by

$$i_r(t) = -\frac{v_r(t)}{Z_C} \tag{6.152}$$

KVL at the load produces

$$v_i + v_r(t) = v_C(t) \tag{6.153}$$

while the KCL gives

$$i_i + i_r(t) = i_C(t) \tag{6.154}$$

Our initial goal is to determine the reflected voltage $v_r(t)$ at the location $z = d$, i.e. $v_r(d, t)$. The ultimate goal is to determine the total voltage at the load, $v_C(d, t)$.

The differential $v - i$ relationship for the load produces

$$i_C(t) = C\frac{dv_C(t)}{dt} \tag{6.155}$$

Utilizing Eqs. (6.153) and (6.154) in Eq. (6.155) we get

$$i_i + i_r(t) = C\frac{d}{dt}[v_i + v_r(t)] \tag{6.156}$$

Using Eqs. (6.150), (6.151), and (6.152) in Eq. (6.156) we obtain

$$\frac{V_G}{2Z_C}\frac{v_r(t)}{Z_C} = C\frac{d}{dt}\left[\frac{V_G}{2} - v_r(t)\right] \tag{6.157}$$

Since V_G is constant Eq. (6.157) reduces to

$$\frac{V_G}{2} + v_r(t) = -\frac{L}{Z_C}\frac{dv_r(t)}{dt} \tag{6.158}$$

or

$$Z_C\frac{dv_r(t)}{dt} + v_r(t) = \frac{V_G}{2} \tag{6.159}$$

This differential equation needs to be solved for $v_r(t)$, for $t > T$, subject to the initial condition $v_r(t = T)$. Let's determine this initial condition.

KVL at the load produces

$$v_i + v_r(t) = v_C(t) \tag{6.160}$$

or

$$\frac{V_G}{Z_C} + v_r(t) = v_C(t) \tag{6.161}$$

Evaluating it at $t = T$, we get

$$\frac{V_G}{Z_C} + v_r(T) = v_C(T) \tag{6.162}$$

Since the voltage across the capacitor cannot change instantaneously, we have $v_C(T) = 0$, and thus

$$\frac{V_G}{Z_C} + v_r(T) = 0 \tag{6.163}$$

or

$$v_r(T) = -\frac{V_G}{2} \tag{6.164}$$

Now we are ready to solve Eq. (6.159), subject to the initial condition in Eq. (6.164):

$$Z_C\frac{dv_r(t)}{dt} + v_r(t) = \frac{V_G}{2}, \quad v_r(T) = -\frac{V_G}{2}, \quad t > T \tag{6.165}$$

First, let's rewrite this equation in a standard form:

$$\frac{dv_r(t)}{dt} + \frac{v_r(t)}{CZ_C} = \frac{V_G}{2CZ_C} \tag{6.166}$$

or

$$\frac{dv_r(t)}{dt} + \frac{v_r(t)}{\tau} = K \tag{6.167}$$

where

$$\tau = CZ_C \tag{6.168}$$

$$K = -\frac{V_G}{2CZ_C} \tag{6.169}$$

The solution of Eq. (6.167) was derived in Section 6.2.5.1 as

$$v_r(t) = K\tau + [v_r(T) - K\tau]e^{-\frac{1}{\tau}(t-T)} \tag{6.170}$$

Utilizing Eqs. (6.164), (6.168), and (6.169) in Eq. (6.170) we get

$$v_r(t) = \left(\frac{V_G}{2CZ_C}\right)(CZ_C) + \left[-\frac{V_G}{2} - \left(-\frac{V_G}{2CZ_C}\right)(CZ_C)\right]e^{-\frac{1}{CZ_C}(t-T)} \tag{6.171}$$

or

$$v_r(d,t) = \frac{V_G}{2} - V_Ge^{-\frac{1}{CZ_C}(t-T)}, \quad t \geq T \tag{6.172}$$

The total voltage across the capacitor is

$$v_C(d,t) = v_i + v_r(d,t) = \frac{V_G}{2} + v_r(d,t) \tag{6.173}$$

or

$$v_C(d,t) = V_G - V_Ge^{-\frac{1}{CZ_C}(t-T)}, \quad t \geq T \tag{6.174}$$

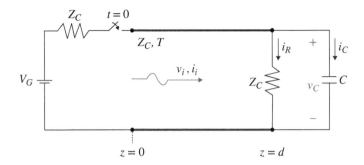

Figure 6.48   RC termination of a transmission line.

6.2.5.4 Reflections at an RC Load

Consider the circuit shown in Figure 6.48 where the transmission line of length d is terminated by an RC load [Adamczyk, 2021b].

Note that the load resistor value is equal to the characteristic impedance of the transmission line; it is also assumed that the voltage across the capacitor is zero.

$$v_C(0^-) = 0 \qquad (6.175)$$

When the switch closes at $t = 0$, a wave originates at $z = 0$, with

$$v_i = \frac{V_G}{2} \qquad (6.176)$$

$$i_i = \frac{V_G}{2Z_C} \qquad (6.177)$$

and travels toward the load. When this wave arrives at the load (at the time $t = T$), the reflected waves v_r and i_r are created. This is shown in Figure 6.49.

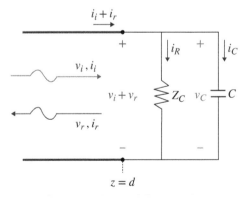

Figure 6.49   Incident and reflected waves at the RC load.

The reflected current wave is related to the reflected voltage wave by

$$i_r(t) = -\frac{v_r(t)}{Z_C} \tag{6.178}$$

KVL at the load produces

$$v_i + v_r(t) = v_C(t) \tag{6.179}$$

while the KCL gives

$$i_i + i_r(t) = i_R(t) + i_C(t) \tag{6.180}$$

Our initial goal is to determine the reflected voltage $v_r(t)$ at the location $z = d$, i.e. $v_r(d, t)$. The ultimate goal is to determine the total voltage at the load, $v_C(d, t)$.

From Eq. (6.180) we obtain the capacitor current as

$$i_C(t) = i_i + i_r(t) - i_R(t) \tag{6.181}$$

The load resistor current can be obtained from

$$i_R(t) = \frac{v_i + v_r(t)}{Z_C} \tag{6.182}$$

Using Eqs. (6.177) and (6.178) and (6.182) in Eq. (6.181) produces

$$i_C(t) = \frac{V_G}{2Z_C} + \frac{v_r(t)}{Z_C} - \frac{v_i + v_r(t)}{Z_C} \tag{6.183}$$

The differential $v - i$ relationship for the capacitor is

$$i_C(t) = C\frac{dv_C(t)}{dt} \tag{6.184}$$

Utilizing Eqs. (6.179) and (6.183) in Eq. (6.184) we get

$$\frac{V_G}{2Z_C} - \frac{v_r(t)}{Z_C} - \frac{v_i + v_r(t)}{Z_C} = C\frac{d}{dt}[v_i + v_r(t)] \tag{6.185}$$

Using Eq. (6.176) in Eq. (6.185) we have

$$\frac{V_G}{2Z_C} - \frac{v_r(t)}{Z_C} - \frac{V_G}{2Z_C} - \frac{v_r(t)}{Z_C} = C\frac{d}{dt}\left[\frac{V_G}{2} + v_r(t)\right] \tag{6.186}$$

Since V_G is constant, Eq. (6.186) reduces to

$$-\frac{2}{Z_C}v_r(t) = C\frac{dv_r(t)}{dt} \tag{6.187}$$

or

$$\frac{CZ_C}{2}\frac{dv_r(t)}{dt} + v_r(t) = 0 \tag{6.188}$$

This differential equation needs to be solved for $v_r(t)$, for $t > T$, subject to the initial condition $v_r(t = T)$. Let's determine this initial condition.

KVL at the load produces

$$vi_i + v_r(t) = v_C(t) \tag{6.189}$$

or

$$\frac{V_G}{2} + v_r(t) = v_C(t) \tag{6.190}$$

Evaluating it at $t = T$, we get

$$\frac{V_G}{2} + v_r(T) = v_C(T) \tag{6.191}$$

Since the voltage across the capacitor cannot change instantaneously, we have $v_C(T) = 0$, and thus

$$v_r(T) = -\frac{V_G}{2} \tag{6.192}$$

Now we are ready to solve Eq. (6.188), subject to the initial condition in Eq. (6.192):

$$\frac{CZ_C}{2}\frac{dv_r(t)}{dt} + v_r(t) = 0, \quad v_r(T) = -\frac{V_G}{2}, \quad t > T \tag{6.193}$$

First, let's rewrite this equation in a standard form:

$$\frac{dv_r(t)}{dt} + \frac{v_r(t)}{CZ_C/2} = 0 \tag{6.194}$$

or

$$\frac{dv_r(t)}{dt} + \frac{v_r(t)}{\tau} = K \tag{6.195}$$

where

$$\tau = \frac{CZ_C}{2} \tag{6.196}$$

$$K = 0 \tag{6.197}$$

The solution of Eq. (6.195) was derived in Section 6.2.5.1 as

$$v_r(t) = K\tau + [v_r(T) - K\tau]e^{-\frac{1}{\tau}(t-T)} \tag{6.198}$$

Utilizing Eqs. (6.192), (6.196), and (6.197) in Eq. (6.198) we obtain

$$v_r(t) = -\frac{V_G}{2}e^{-\frac{2}{cz_C}(t-T)} \tag{6.199}$$

or

$$v_r(d,t) = -\frac{V_G}{2}e^{-\frac{2}{cz_C}(t-T)}, \quad t \geq T \tag{6.200}$$

The total voltage across the RC load is

$$v_C(d,t) = v_i + v_r(d,t) = \frac{V_G}{2} + v_r(d,t) \tag{6.201}$$

or

$$v_C(d,t) = \frac{V_G}{2} - \frac{V_G}{2}e^{-\frac{2}{cz_C}(t-T)}, \quad t \geq T \tag{6.202}$$

6.2.6 Reflections at a Shunt Reactive Discontinuity

6.2.6.1 Reflections at a Shunt Capacitive Discontinuity

Consider the circuit shown in Figure 6.50, where the transmission line of length l has a shunt capacitive discontinuity in the middle of line, at a location $z = d$ [Adamczyk, 2022a].

Note that the load resistor value is equal to the characteristic impedance of the transmission line; it is also assumed that the initial voltage across the capacitor is zero, $v_C(0^-) = 0$.

When the switch closes at $t = 0$, a wave originates at $z = 0$ and travels toward the discontinuity.

The transmission line immediately to the right of discontinuity looks to the circuit on the left of discontinuity like a shunt resistance equal to the characteristic impedance of

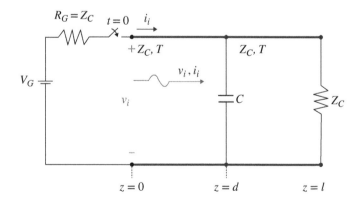

Figure 6.50   Shunt capacitive discontinuity along a transmission line.

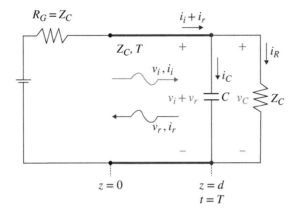

Figure 6.51   Incident and reflected waves at the capacitive discontinuity.

the right line. When the incident wave arrives at the discontinuity (at the time $t = T$), the reflected wave, v_r and i_r, is created, and we have a situation depicted in Figure 6.51.

This part of the circuit is identical to the one discussed in Section 6.2.5.4, where the transmission line was terminated with an RC load. Thus, the total voltage across the capacitive discontinuity is

$$v_C(d, t) = \frac{V_G}{2} - \frac{V_G}{2} e^{-\frac{2}{CZ_C}(t-T)}, \quad t \geq T \tag{6.203}$$

Equation (6.203) predicts that at $t = T$, the voltage at the discontinuity is zero and increases exponentially to $\frac{V_G}{2}$. Let's verify these observations through simulations and measurements.

Figure 6.52 shows HyperLynx schematic of the transmission line with a capacitive discontinuity.

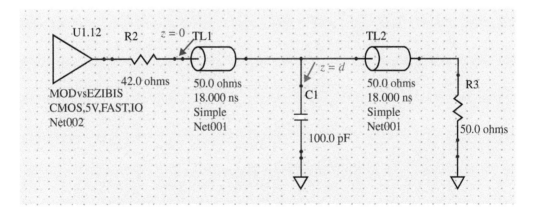

Figure 6.52   Capacitive discontinuity – HyperLynx schematic.

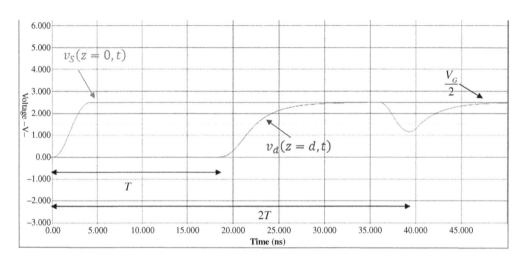

Figure 6.53 Capacitive discontinuity – voltages at the source ($z = 0$) and the discontinuity ($z = d$).

The simulation results are shown in Figure 6.53.
Note that the simulation results agree with the analytical results.
Figure 6.54 shows the measurement setup.
The measurement results are shown in Figure 6.55.
Note that the measurement results agree with the analytical and simulation results.

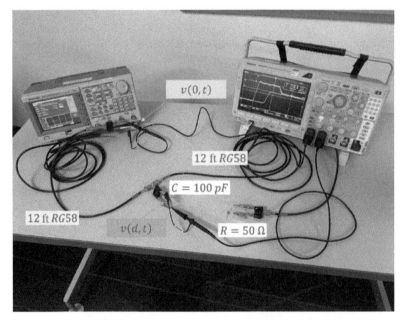

Figure 6.54 Capacitive discontinuity measurement setup.

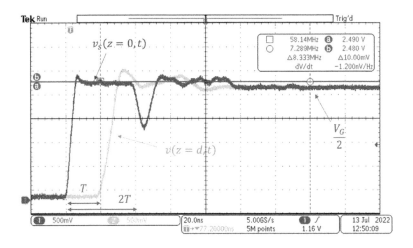

Figure 6.55 Capacitive discontinuity – measurement results.

6.2.6.2 Reflections at a Shunt Inductive Discontinuity

Consider the circuit shown in Figure 6.56 where the transmission line of length l has a shunt inductive discontinuity in the middle of the line, at a location $z = d$ [Adamczyk, 2022b].

Note that the transmission line is matched at the source and at the load; it is also assumed that the initial current through the inductor is zero, $i_L(0^-) = 0$.

When the switch closes at $t = 0$, a wave originates at $z = 0$, with

$$v_i = \frac{V_G}{2} \tag{6.204}$$

$$i_i = \frac{V_G}{2Z_C} \tag{6.205}$$

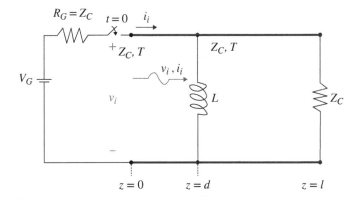

Figure 6.56 Shunt inductive discontinuity along a transmission line.

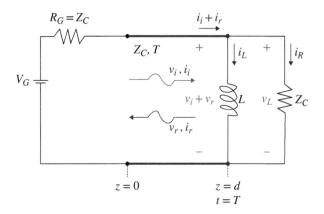

Figure 6.57   Incident and reflected waves at the inductive discontinuity.

and travels toward the discontinuity. When this wave arrives at the discontinuity (at the time $t = T$), the reflected waves v_r and i_r are created. The transmission line immediately to the right of discontinuity looks to the circuit on the left of discontinuity like a shunt resistance equal to the characteristic impedance of the right line. Figure 6.57 illustrates this.

The reflected current wave is related to the reflected voltage wave by

$$i_r(t) = -\frac{v_r(t)}{Z_C} \tag{6.206}$$

KVL and KCL at the discontinuity produce

$$v_i + v_r(t) = v_L(t) \tag{6.207}$$

$$i_i + i_r(t) = i_L(t) + i_R(t) \tag{6.208}$$

Our initial is to determine the reflected voltage $v_r(t)$ at the location $z = d$, i.e. $v_r(d, t)$. The ultimate goal is to determine the total voltage at the discontinuity, $v_L(d, t)$.

The current $i_R(t)$ can be expressed as

$$i_R(t) = \frac{v_i + v_r(t)}{Z_C} \tag{6.209}$$

From Eq. (6.208) we obtain the inductor current as

$$i_L(t) = i_i + i_r(t) - i_R(t) \tag{6.210}$$

Utilizing Eqs. (6.205), (6.206), and (6.209) in Eq. (6.210) we get

$$i_L(t) = \frac{V_G}{2Z_C} - \frac{v_r(t)}{Z_C} - \frac{V_G}{2Z_C} - \frac{v_r(t)}{Z_C} \tag{6.211}$$

or

$$i_L(t) = -2\frac{v_r(t)}{Z_C} \tag{6.212}$$

The differential $v - i$ relationship for the inductor is

$$v_L(t) = L\frac{di_L(t)}{dt} \tag{6.213}$$

Using Eqs. (1.1a), (1.2a), and (1.5) in Eq. (1.6) we obtain
Using Eqs. (6.204), (6.207), and (6.212) in Eq. (6.213) we obtain

$$\frac{V_G}{2} + v_r(t) = L\frac{d}{dt}\left[-2\frac{v_r(t)}{Z_C}\right] \tag{6.214}$$

or

$$\frac{dv_r(t)}{dt} + \frac{v_r(t)}{\frac{2L}{Z_C}} = \frac{V_GZ_C}{4L} \tag{6.215}$$

This differential equation needs to be solved for $v_r(t)$, for $t > T$, subject to the initial condition $v_r(t = T)$. Let's determine this initial condition. Initially, the current through an inductor $i_L(T^-)$ is zero. Thus, the inductor acts as an open circuit, and the incident wave sees the discontinuity as a matched load. Therefore, there is no reflection and

$$v_r(T) = 0 \tag{6.216}$$

Equation (6.215) can be written as

$$\frac{dv_r(t)}{dt} + \frac{v_r(t)}{\tau} = K \tag{6.217}$$

where

$$\tau = \frac{2L}{Z_C} \tag{6.218}$$

$$K = -\frac{V_GZ_C}{4L} \tag{6.219}$$

The solution of Eq. (6.217) was presented earlier in Eq. (6.115), repeated here, as

$$v_r(t) = K\tau + [v_r(T) - K\tau]e^{-\frac{1}{\tau}(t-T)} \tag{6.220}$$

Utilizing Eqs. (6.216), (6.218), and (6.219) in Eq. (6.220) we obtain

$$v_r(d,t) = -\frac{V_G}{2} + \frac{V_G}{2}e^{-\frac{Z_C}{2L}(t-T)}, \quad t \geq T \tag{6.221}$$

The total voltage across the discontinuity is

$$v_L(d,t) = v_i + v_r(d,t) = \frac{V_G}{2} + v_r(d,t) \tag{6.222}$$

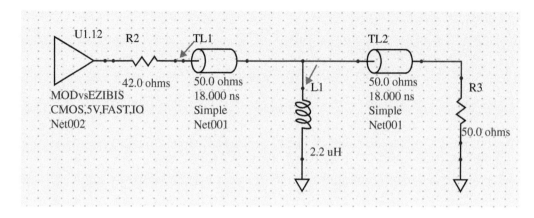

Figure 6.58 Inductive discontinuity – HyperLynx schematic.

or

$$v_L(d, t) = \frac{V_G}{2} e^{-\frac{z_C}{2L}(t-T)}, \quad t \geq T \tag{6.223}$$

Equation (6.223) predicts that at $t = T$, the voltage at the load rises from zero to $\frac{V_G}{2}$, and then decays exponentially to zero. Let's verify these observations through simulations and measurements.

Figure 6.58 shows the HyperLynx schematic of the transmission line with an inductive discontinuity.

The simulation results are shown in Figure 6.59.

Note that the simulation results agree with the analytical results.

The measurement setup is shown in Figure 6.60.

The measurement results are shown in Figure 6.61.

Note that the measurement results agree with the analytical and simulation results.

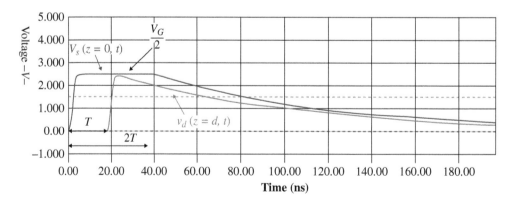

Figure 6.59 Inductive discontinuity – voltages at the source ($z = 0$) and the discontinuity ($z = d$).

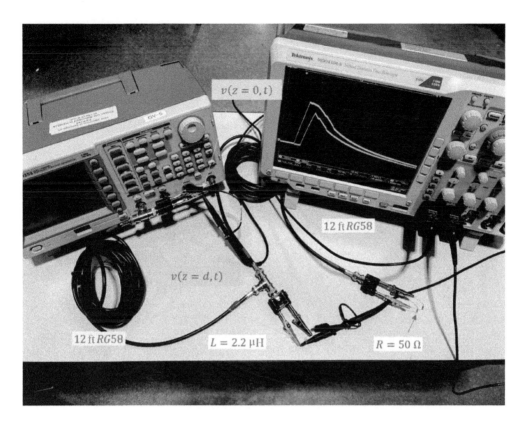

Figure 6.60   Inductive discontinuity – measurement setup.

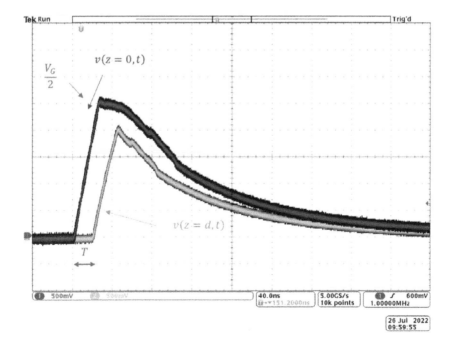

Figure 6.61   Inductive discontinuity – measurement results.

6.3 Eye Diagram

This section devoted to an Eye Diagram [Adamczyk et al., 2022a]. First, the fundamental definitions and concepts are presented. Then, we show the impact of driver, receiver, and interconnect properties on signal quality using data eye and data eye mask concepts, while evaluating several different HDMI cables.

6.3.1 Fundamental Concepts

Consider a digital signal as it travels from a transmitter to a receiver. The quality of the signal arriving at the receiver can be affected by many factors including the transmitter, cables or PCB traces, and connectors. The signal quality is also referred to as signal integrity.

An eye-diagram is a graphical tool used to quickly evaluate the quality of a digital signal. The name eye-diagram has been coined due to the fact that it has the appearance of a human eye. Eye diagrams are commonly used for testing at both receivers and transmitters.

An eye-diagram is basically an infinite, persisted overlay of all bits captured by an oscilloscope to show when bits are valid. This provides a composite picture of the overall quality of a system's physical layer characteristics. This picture covers all possible combinations of variations affecting the signal: amplitude, timing uncertainties, and infrequent signal anomalies.

The eye diagram is created by superimposing successive bit sequences of the data. Consider all possible 3-bit sequences shown in Figure 6.62a–h.

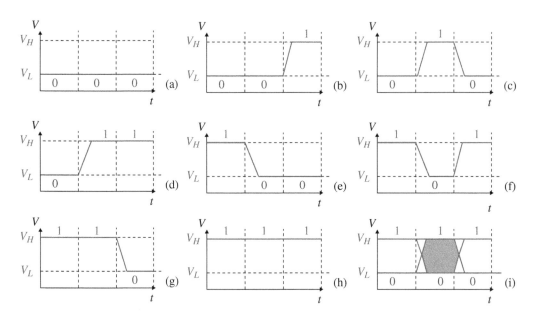

Figure 6.62   3-bit sequences (a)–(h), and eye diagram (i).

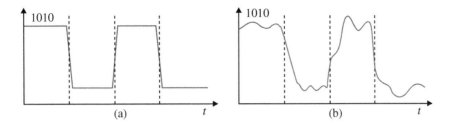

Figure 6.63   (a) Ideal bit sequences and (b) actual bit sequences.

It should be noted that the data sequences shown in Figure 6.62 and in Figure 6.63a, are shown as straight lines; the actual data stream looks like the one shown in Figure 6.63b.

6.3.1.1 Eye Diagram Parameters

Ideally, the eye diagram would consist of two parallel horizontal lines and two parallel vertical lines (assuming instantaneous rise and fall times), as shown in Figure 6.64a. Assuming a more realistic case with finite rise and fall times, the less "ideal" eye diagram would look like the one shown in Figure 6.64b.

An even more realistic signal would exhibit some degree of amplitude and rise/fall time variation. These amplitude and time variations give rise to several parameters associated with an eye diagram, as shown in Figure 6.65.

Note that the eye area has been reduced. The eye crossing in Figure 6.65 is often referred to as a zero crossing, since the data used for an eye diagram creation is usually transmitted as a differential pair signal.

The eye diagram shown in Figure 6.65 is still an "ideal" diagram, as it consists of perfectly straight lines. An actual (real data) eye diagram looks more like the one shown in Figure 6.66.

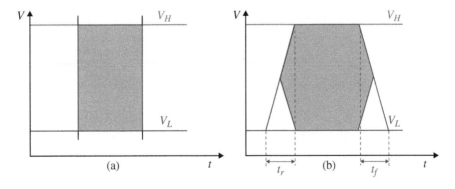

Figure 6.64   "Ideal" eye diagram - (a) instantaneous rise and fall times and (b) finite rise and fall times.

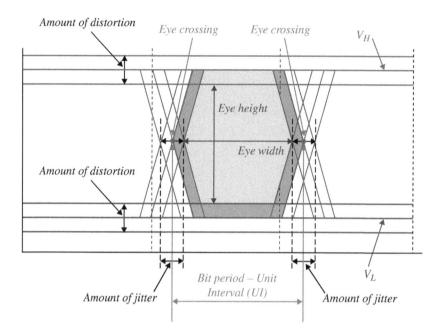

Figure 6.65   Eye diagram parameters.

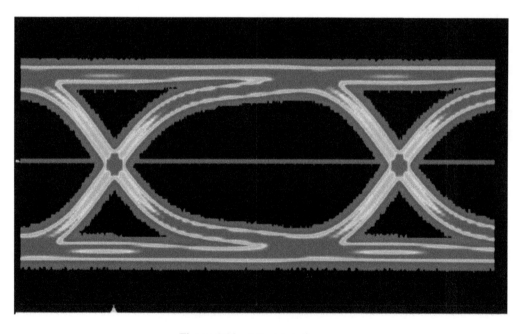

Figure 6.66   Actual eye diagram.

6.3.1.2 Data and Clock Dependencies

To achieve high reliability of data transfer, a synchronization signal is introduced. This signal is used to trigger data transfer operation. The data transfer occurs when the synchronization signal transitions its state (e.g. the rising edge of a clock signal), at which time the data signal state will be read as either low or high.

The high state will be read when the data signal is above a certain voltage threshold level (V_{IHmin}), and it will be read as low when it is below another voltage threshold (V_{ILmax}). This synchronization signal is typically referred to as a clock or strobe.

However, data signal voltage levels being below or above a predefined voltage threshold at the time of data transfer is an insufficient condition for reliable data transfer. It is also necessary to meet certain timing dependencies between data and synchronization signals.

To explain those dependencies let's look at a specific case, described in Figure 6.67.

The synchronization signal in this case is the clock, and the data is transferred (read) at the rising edge of the clock. We will assume that both the clock and data signals are transitioning very quickly.

To guarantee that the proper data will be read, valid data signal must be present for a certain time duration prior to the clock signal transition. This duration is referred to as a setup time (t_{SETUP}). Additionally, it is also required that the data signal remains valid for a certain time duration after the transition of the clock signal. This duration is referred to as a hold time (t_{HOLD}).

Setup and hold times are properties of devices receiving the data and are often referred to as their timing requirements. If the timing requirements are not met, incorrect data can be read by the receiver.

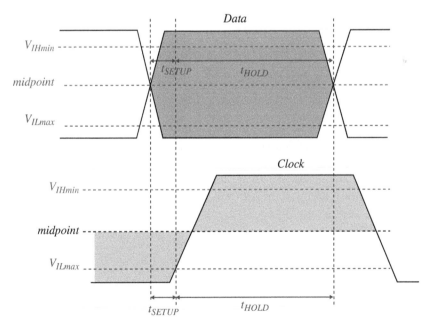

Figure 6.67 Data and clock synchronization for signals with fast transition times.

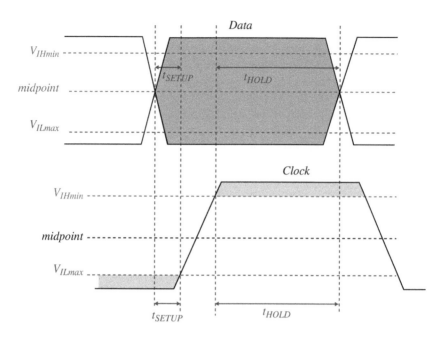

Figure 6.68 Data and clock synchronization for signals with slow transition times.

The assumption of very fast signal transitions allows us to measure timing dependencies between the data and clock (setup and hold time) using midpoint signal levels. In other words, this assumption means that we can neglect signal rise/fall time duration if those durations are much shorter than duration when the data bit is valid.

If clock period gets shorter and we can no longer neglect signal rise/fall time duration, the evaluation of timing dependencies between the data and clock (setup and hold time) has to account for slow signal transition. Figure 6.68 illustrates such a case.

The rising edge of the data is still very fast, but the clock transition is much slower. The duration of time between the clock transition from low and high level is now substantial when compared to the duration of the data bit. During this long clock transition time the clock state can be either high or low, so we no longer can measure setup and hold time using midpoint levels. This case would require setup and hold time to be measured when the signals are crossing the low or high voltage threshold levels, V_{ILmax}, V_{IHmin}.

Evaluation and visualization of valid signal timing using setup and hold time shown in Figure 6.67, is relatively easy, even with the clock and data jitter. It is, however, quite difficult, for the case shown in Figure 6.68, when taking jitter into account. That's where the eye diagram can help.

6.3.1.3 Data Eye Mask

The concept of a data eye diagram can be used to not only evaluate the quality of data signal, but also to evaluate whether the signal meets timing requirements. To accomplish this, receiver timing requirements are used to define horizontal dimension of a region.

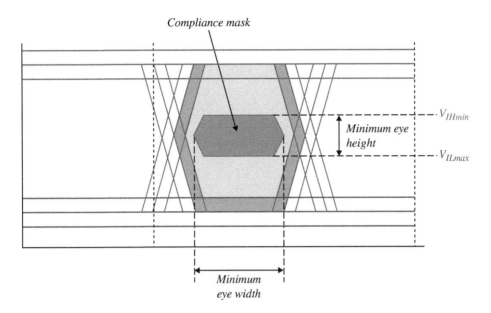

Compliance mask

Minimum eye height

V_{IHmin}

V_{ILmax}

Minimum eye width

Figure 6.69   HDMI data eye mask.

Voltage level thresholds V_{ILmax}, V_{IHmin} are used to define vertical boundaries of that region. The resulting region is referred to as the data eye mask. A sample of a data eye mask, representing requirements for a video HDMI standards receiver, is shown in Figure 6.69.

The data eye mask represents the "keep-out" region. Signals at the receiver must not cross the data eye mask region or a violation of receiver timing requirements occurs. The mask is defined based on receiver properties, and it can have various shapes (rectangular, triangular, etc.). The data eye mask can be of many different shapes, as shown in Figure 6.70.

6.3.2 Impact of Driver, HDMI Cable, and Receiver

This section addresses the impact of driver, HDMI cable, and receiver on signal quality using data eye, based on the following criteria: data eye opening, data mask violation, and data jitter [Adamczyk et al., 2022b].

6.3.2.1 Measurement Setup

The study includes three different HDMI signal sources, four different HDMI cables, and two different receivers. The block diagram of the measurement setup is shown in Figure 6.71.

The study focuses on evaluation of eye diagrams using the following criteria: eye opening, eye mask violations, and data jitter. The data jitter is presented in the form of a histogram.

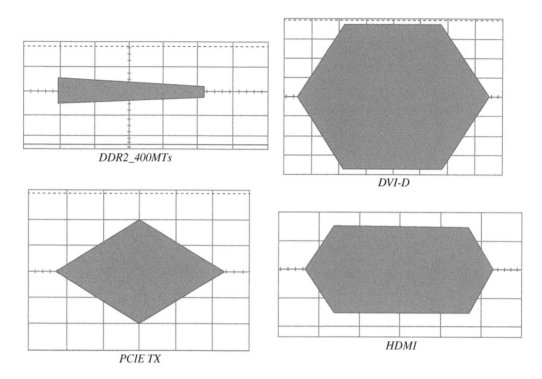

Figure 6.70   Examples of data eye masks.

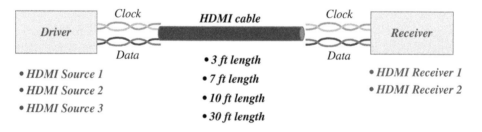

Figure 6.71   Block diagram of the measurement setup.

6.3.2.2 Impact of HDMI Sources

In this part of the study we compare three different HDMI Sources, while the cable length was the same (3-ft), and the same HDMI Receiver was used (Receiver 1). HDMI Sources used in the study had significant implementation differences. Differences consisted of: Driver IC and its configuration, differential trace routing, and HDMI connector style. Figure 6.72 shows the resulting eye diagrams.

Observations: HDMI Sources 1 and 3 passed the eye diagram test with a significant margin, while the HDMI Source 2 had failed (eye mask violation with large data jitter). Data jitter from HDMI Source 1 was smaller than that from HDMI Source 3.

Next, the impact of cable length was evaluated.

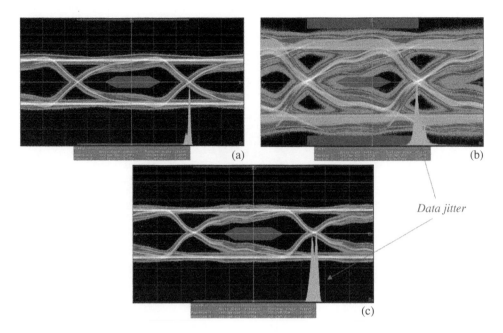

Figure 6.72 3-ft cable, same receiver driven by: (a) HDMI Source 1, (b) HDMI Source 2, and (c) HDMI Source 3.

6.3.2.3 Impact of HDMI Cable Length

In this part of the study we evaluate the impact of the cable length with HDMI Source 1 or 2, while keeping the HDMI Receiver unchanged. Figure 6.73 shows the eye diagram for HDMI Source 1 and four different cable lengths.

Observations: HDMI Source 1 passed the test for cable lengths: 3-ft, 7-ft, and 10-ft, but a failure was observed for the 30-ft cable. As the cable length increased, the eye opening became smaller, and the data jitter increased.

Figure 6.74 shows the eye diagram for HDMI Source 2 and four different cable lengths.

Observations: The HDMI Source 2 failed the test for all four cable lengths. Generally, as the cable length increased, the eye opening became smaller, and the data jitter became larger.

6.3.2.4 Impact of HDMI Receiver

In the final stage of the study, we use the same driver, same cable, and two different receivers. Figure 6.75 shows the corresponding eye diagrams.

Observations: Both receivers passed the test with a similar amount of jitter. The eye opening of the Receiver 1 was slightly larger than that of the Receiver 2.

6.3.2.5 Summary and Conclusions

This study addressed the impact of driver, HDMI cable, and receiver on signal quality using data eye, based on the following criteria: data eye opening, data mask violation,

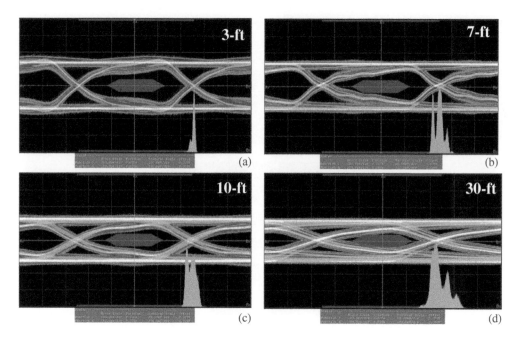

Figure 6.73 Impact of cable length driven by HDMI Source 1: (a) 3-ft cable, (b) 7-ft cable, (c) 10-ft cable, and (d) 30-ft cable.

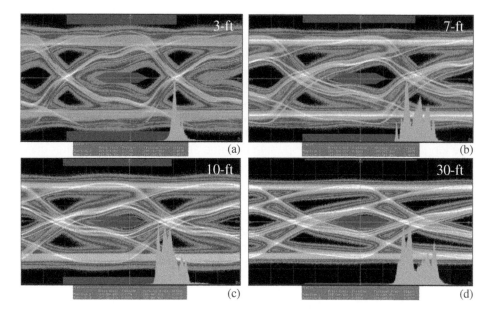

Figure 6.74 Impact of cable length driven by HDMI Source 2: (a) 3-ft cable, (b) 7-ft cable, (c) 10-ft cable, and (d) 30-ft cable.

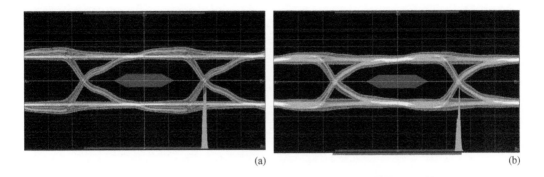

(a)   (b)

Figure 6.75   Eye diagram: (a) HDMI Receiver 1and (b) HDMI Receiver 2.

and data jitter. The study has shown that all three system components affect the eye diagram. Measurement results have shown a correlation between data jitter and data eye mask opening; as the data jitter increases, the data eye opening gets smaller. The impact of the receiver in our study was less pronounced than the impact of the driver. The most obvious observation was: the shorter the cable, the better the data quality.

6.4 Laboratory Exercises

6.4.1 Transmission Line Reflections

Objective: This laboratory exercise investigates the transmission line reflections at the resistive loads and at a resistive line discontinuity.

6.4.2 Laboratory Equipment and Supplies

Function Generator: Tektronix AFG3252 Function Generator (or equivalent), shown in Figure 6.76.

Oscilloscope: Tektronix MDO4104-6 Oscilloscope (or equivalent), shown in Figure 6.77.

Adapter Coaxial Connector BNC Plug, Male Pin To BNC Jack, Female Socket: Digi-Key Part Number ACX1064-ND, shown in Figure 6.78.

Adapter Connector BNC Female To Banana Plug, Double, Stackable Black: Digi-Key Part Number 314-1294-ND, shown in Figure 6.79.

Adapter Connector BNC Male To Banana Plug, Double, Stackable Black: Digi-Key Part Number 314-1295-ND, shown in Figure 6.80.

Alligator Test Clip (x4): Digi-Key Part Number 314-1033-ND, shown in Figure 6.81.

BNC Male to BNC Male (RG58) 50 Ohm Coaxial Cable Assembly, 12 ft: Amphenol Part Number CO-058BNCX200-012D, like the one shown in Figure 6.82.

BNC Male to BNC Male (RG59) 75 Ohm Coaxial Cable Assembly, 12 ft: Digi-Key Part Number ACX1803-ND.

Through Hole Resistors: Digi-Key Part Number A105936CT-ND - 330Ω, like the one shown in Figure 6.83.

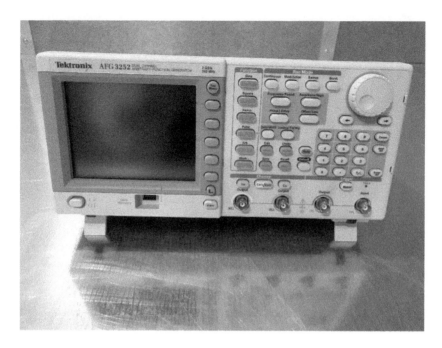

Figure 6.76   Laboratory equipment: function generator.

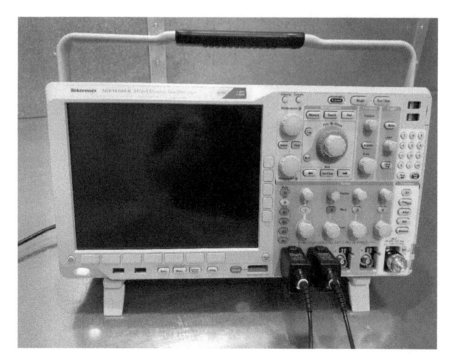

Figure 6.77   Laboratory equipment: oscilloscope.

Figure 6.78   Laboratory equipment: Adapter Coaxial Connector.

Figure 6.79   Laboratory equipment: Adapter Coaxial Connector.

Figure 6.80   Laboratory equipment: Adapter Coaxial Connector.

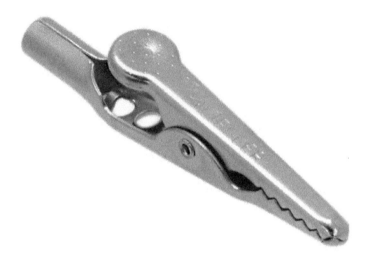

Figure 6.81   Laboratory equipment: Alligator test clip.

6.4.3 Reflections at a Resistive Load

6.4.3.1 Laboratory Procedure

Connect the equipment as shown in Figure 6.84.

Attach the $33\,\Omega$ resistor to the end of the coaxial cable as shown in Figure 6.85.

Function Generator Settings: Function = pulse, frequency = 1 MHz, amplitude = 1 Vpp ($load = 50\,\Omega$, offset = 500 mV, duty cycle = 50%, rise time = 20 ns, fall time = 20 ns).

Initial Oscilloscope Settings: Time scale = 10 ns, coupling = DC, impedance = $1\,M\Omega$

Figure 6.82 Laboratory equipment: RG58 Coaxial cable.

Step 1: Calculate the value of the initial voltage wave

$$V(z = 0, t = 0) = V^+ = \frac{Z_C}{R_G + Z_C} V_G \tag{6.224}$$

Step 2: Calculate the value of the load reflection coefficient

$$\Gamma_L = \frac{R_L - Z_C}{R_L + Z_C} \tag{6.225}$$

Step 3: Calculate the value of the reflected voltage

$$V^- = \Gamma_L V^+ \tag{6.226}$$

Step 4: Calculate the value of the total voltage at the load

$$V(z = L, t = T) = V^+ + V^- \tag{6.227}$$

Figure 6.83   Laboratory equipment: Through hole resistor.

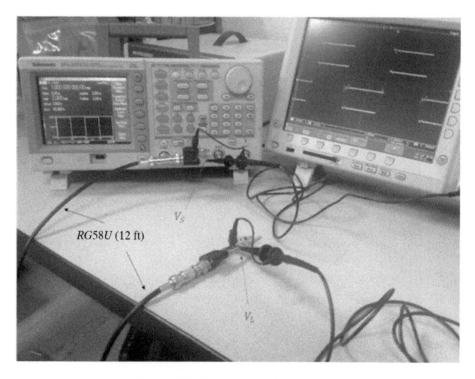

Figure 6.84   Laboratory equipment setup.

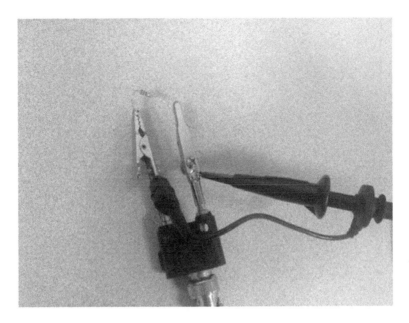

Figure 6.85 Resistive termination.

Step 5: Capture the oscilloscope plots to show the voltage values $V(z = 0, t = 0)$ and $V(z = L, t = T)$. Compare the measured values to the analytical results.

      Using the oscilloscope plots show the one-way travel time T. (It should be around 18 ns for a 12 ft cable).

Step 6: Replace the 33 Ω resistor by a 47 Ω resistor and repeat Steps 2 through 5.

Step 7: Replace the 47 Ω resistor by a 330 Ω resistor and repeat Steps 2 through 5.

6.4.4 Bounce Diagram

6.4.4.1 Laboratory Procedure

Connect the equipment as shown in Figure 6.86.

    Terminate the cable with a 330 Ω resistor.

Step 1: Calculate the value of the initial voltage wave

$$V_S(z = 0, t = 0) = V^+ = V_G \frac{Z_C}{R_G + Z_C} \tag{6.228}$$

Capture the oscilloscope plot and verify this result.

Step 2: Calculate the load reflection coefficient, the reflected voltage, and the total voltage at the load.

$$\Gamma_L = \frac{R_L - Z_C}{R_L + Z_C} \tag{6.229}$$

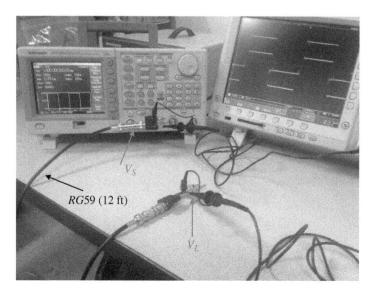

Figure 6.86   Equipment setup – bounce diagram.

$$V^- = \Gamma_L V^+ \tag{6.230}$$

$$V_L(z = L, t = T) = V^+ + V^- \tag{6.231}$$

Capture the oscilloscope plot and verify this result.

Step 3: Calculate the source reflection coefficient, the re-reflected voltage, and the total voltage at the source.

$$\Gamma_S = \frac{R_S - Z_C}{R_S + Z_C} \tag{6.232}$$

$$V^{-+} = \Gamma_S V^- \tag{6.233}$$

The total voltage at the source is

$$V_S(z = 0, t = 2T) = V^+ + V^- + V^{-+} \tag{6.234}$$

Capture the oscilloscope plot and verify this result.

Step 4: Create the bounce diagram plots like the ones shown in Figures 6.26–6.28.

6.4.5 Reflections at a Resistive Discontinuity

6.4.5.1 Laboratory Procedure

Connect the equipment as shown in Figure 6.87.
   The circuit diagram is shown in Figure 6.88.

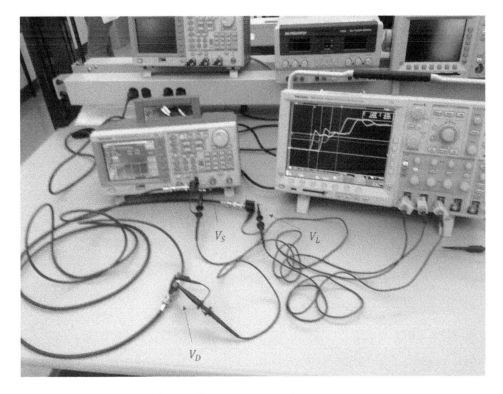

Figure 6.87 Laboratory equipment setup.

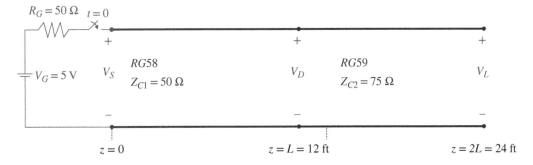

Figure 6.88 Reflections at a discontinuity – circuit diagram.

Step 1: Calculate the value of the initial voltage wave

$$V_{i1} = V_S(z = 0, t = 0) = \frac{Z_{C1}}{R_G + Z_{C1}} V_G \qquad (6.235)$$

Step 2: Calculate the reflection coefficient (from the left) at the discontinuity, the reflected voltage, and the total voltage.

$$\Gamma_{12} = \frac{Z_{C2} - Z_{C1}}{Z_{C2} + Z_{C1}} \qquad (6.236)$$

$$v_{r1} = \Gamma_{12} v_{i1} \qquad (6.237)$$

$$V_D(z = L, t = T) = v_{i1} + v_{r1} \qquad (6.238)$$

Step 3: Calculate the transmission coefficient at the discontinuity (from the left) and the transmitted voltage

$$T_{12} = \frac{2Z_{C2}}{Z_{C2} + Z_{C1}} \qquad (6.239)$$

$$V_D(z = L, t = T) = v_{t2} = T_{12} v_{i1} \qquad (6.240)$$

Step 4: Capture the oscilloscope plots to show the voltage values $V_S(z = 0, t = 0)$ and $V_D(z = L, t = T)$. Compare the measured values to the analytical results.

Step 5: Calculate the total voltage at the source at $t = 2T$.

$$V_S(z = 0, t = 2T) = v_{i1} + v_{r1} \qquad (6.241)$$

Compare this analytical result to the measured value (show the measured value on the captured oscilloscope plot).

Step 6: Calculate the total voltage at the load at $t = 2T$

$$V_L(z = L, t = 2T) = v_{t2} + \Gamma_L v_{t2} \qquad (6.242)$$

Compare this analytical result to the measured value (show the measured value on the captured oscilloscope plot).

Step 7: Calculate the reflection coefficient (from the right) at the discontinuity, the reflected voltage, and the total voltage.

$$\Gamma_{21} = \frac{Z_{C1} - Z_{C2}}{Z_{C1} + Z_{C2}} \qquad (6.243)$$

$$v_{r2} = \Gamma_{21} v_{i2} \qquad (6.244)$$

$$V_D(z = L, t = 3T) = V_D(z = L, t = T) + v_{r2} + \Gamma_{21} v_{r2} \qquad (6.245)$$

Compare this analytical result to the measured value (show the measured value on the captured oscilloscope plot).

Step 8: Calculate the transmission coefficient at the discontinuity (from the right) and the transmitted voltage

$$T_{21} = \frac{2Z_{C1}}{Z_{C1} + Z_{C2}} \tag{6.246}$$

$$V_D(z = L, t = 3T) = V_D(z = L, t = T) + T_{21}v_{i2} \tag{6.247}$$

Compare this result with the one obtained in Step 7.

Step 9: Calculate the total voltage at the source at $t = 4T$.

$$V_S(z = 0, t = 4T) = V_S(z = 0, t = 2T) + v_{t2} \tag{6.248}$$

Compare this analytical result to the measured value (show the measured value on the captured oscilloscope plot).

Step 10: Calculate the total voltage at the load at $t = 4T$.

$$V_L(z = 2L, t = 4T) = V_L(z = 2L, t = 2T) + v_{t2} + \Gamma_L v_{t2} \tag{6.249}$$

Compare this analytical result to the measured value (show the measured value on the captured oscilloscope plot).

References

Bogdan Adamczyk. Transmission Line Reflections at a Discontinuity. *In Compliance Magazine*, February 2017a. URL https://incompliancemag.com/article/transmission-line-reflections-at-a-discontinuity/.

Bogdan Adamczyk. *Foundations of Electromagnetic Compatibility*. John Wiley & Sons, Ltd, Chichester, UK, 2017b. ISBN 9781119120810. doi: https://doi.org/10.1002/9781119120810.

Bogdan Adamczyk. Transmission Line Reflections at a Resistive Load. *In Compliance Magazine*, January 2017c. URL https://incompliancemag.com/article/transmission-line-reflections-at-a-resistive-load/.

Bogdan Adamczyk. Transmission Line Reflections: Bounce Diagram. *In Compliance Magazine*, October 2018a. URL https://incompliancemag.com/article/transmission-line-reflections-bounce-diagram/.

Bogdan Adamczyk. Transmission Line Reflections at a Reactive Load. *In Compliance Magazine*, December 2018b. URL https://incompliancemag.com/article/transmission-line-reflections-at-a-reactive-load/.

Bogdan Adamczyk. EM Waves, Voltage, and Current Waves. *In Compliance Magazine*, April 2021a. URL https://incompliancemag.com/article/em-waves-voltage-and-current-waves/.

Bogdan Adamczyk. Transmission Line Reflections at the RL and RC Loads. *In Compliance Magazine*, January 2021b. URL https://incompliancemag.com/article/transmission-line-reflections-at-the-rl-and-rc-loads/.

Bogdan Adamczyk. Transmission Line Reflections at a Shunt Resistive and Reactive Discontinuity Along the Line. *In Compliance Magazine*, October 2022a. URL https://incompliancemag.com/article/transmission-line-reflections-at-a-shunt-resistive-and-reactive-discontinuity-along-the-line//.

Bogdan Adamczyk. Shunt Inductive Discontinuity along the Transmission Line and Transmission Lines in Parallel. *In Compliance Magazine*, November 2022b. URL https://incompliancemag.com/article/shunt-inductive-discontinuity-along-the-transmission-line-and-transmission-lines-in-parallel/.

Bogdan Adamczyk and Brian Gilbert. Basic EMC Rules. *In Compliance Magazine*, May 2020. URL https://incompliancemag.com/article/basic-emc-rules/.

Bogdan Adamczyk, Krzysztof Russa, and Nicholas Hare. Eye Diagram −Part 1: Fundamental Concepts. *In Compliance Magazine*, August 2022a. URL https://incompliancemag.com/article/eye-diagram-part1/.

Bogdan Adamczyk, Krzysztof Russa, and Nicholas Hare. Eye Diagram −Part 2: Impact of Driver, HDMI Cable and Receiver. *In Compliance Magazine*, September 2022b. URL https://incompliancemag.com/article/eye-diagram-part2/.

Nannapaneni N. Rao. *Elements of Engineering Electromagnetics*. Pearson Prentice-Hall, Upper Saddle River, New Jersey, 2004. ISBN 0-13-113961-4.

Chapter 7: Transmission Lines – Frequency Domain

7.1 Frequency-Domain Solution

Recall the lossless transmission line equations, (6.18) and (6.19), repeated here

$$\frac{\partial^2 V(z,t)}{\partial z^2} = lc\frac{\partial^2 V(z,t)}{\partial t^2} \tag{7.1}$$

$$\frac{\partial^2 I(z,t)}{\partial z^2} = lc\frac{\partial^2 I(z,t)}{\partial t^2} \tag{7.2}$$

In a sinusoidal steady state, the transmission line is driven by a sinusoid

$$v(t) = V\cos(\omega t + \theta) \tag{7.3}$$

which has a corresponding phasor [Adamczyk, 2022],

$$\hat{V} = Ve^{j\theta} \tag{7.4}$$

In phasor domain, Eqs. (7.1) and (7.2) become

$$\frac{\partial^2 \hat{V}(z)}{\partial z^2} = -\omega^2 lc\hat{V}(z) \tag{7.5}$$

$$\frac{\partial^2 \hat{I}(z)}{\partial z^2} = -\omega^2 lc\hat{I}(z) \tag{7.6}$$

Equations (7.5) and (7.6) describe a wave propagating with the velocity of

$$v = \frac{1}{\sqrt{lc}} \tag{7.7}$$

This velocity is related to frequency ω and phase constant β by

$$v = \frac{\omega}{\beta} \tag{7.8}$$

and thus

$$\beta = \omega\sqrt{lc} \tag{7.9}$$

or

$$\beta^2 = \omega^2 lc \tag{7.10}$$

Principles of Electromagnetic Compatibility: Laboratory Exercises and Lectures, First Edition. Bogdan Adamczyk.
© 2024 John Wiley & Sons Ltd. Published 2024 by John Wiley & Sons Ltd.
Companion website: www.wiley.com/go/principlesofelectromagneticcompatibility

Therefore, Eqs. (7.5) and (7.6) can be expressed as

$$\frac{\partial^2 \hat{V}(z)}{\partial z^2} = -\beta^2 \hat{V}(z) \qquad (7.11)$$

$$\frac{\partial^2 \hat{I}(z)}{\partial z^2} = -\beta^2 \hat{I}(z) \qquad (7.12)$$

The general solution of Eqs. (7.11) and (7.12) is of the form [Adamczyk, 2023a],

$$\hat{V}(z) = \hat{V}^+ e^{-j\beta z} + \hat{V}^- e^{j\beta z} \qquad (7.13)$$

$$\hat{I}(z) = \frac{\hat{V}^+}{Z_C} e^{-j\beta z} - \frac{\hat{V}^-}{Z_C} e^{j\beta z} \qquad (7.14)$$

where the *characteristic impedance* of the transmission line, Z_C, is given by

$$Z_C = \sqrt{\frac{l}{c}} \qquad (7.15)$$

and the constants $\hat{V}^+$ and $\hat{V}^-$ are obtained from the knowledge of a complete transmission line model (as we shall see in Section 7.1.1).

Before proceeding any further, let us verify that Eqs. (7.13) and (7.14) are indeed the solutions of Eqs. (7.11) and (7.12). First, let us demonstrate this for Eq. (7.13).

Differentiating Eq. (7.13) with respect to z gives

$$\frac{d\hat{V}(z)}{dz} = -j\beta \hat{V}^+ e^{-j\beta z} + j\beta \hat{V}^- e^{j\beta z} \qquad (7.16)$$

Differentiating once more produces

$$\frac{d^2 \hat{V}(z)}{dz^2} = (-j\beta)(-j\beta)\hat{V}^+ e^{-j\beta z} + (j\beta)(j\beta)\hat{V}^- e^{j\beta z} \qquad (7.17)$$

or

$$\frac{d^2 \hat{V}(z)}{dz^2} = -\beta^2 \hat{V}^+ e^{-j\beta z} - \beta^2 \hat{V}^- e^{j\beta z} \qquad (7.18)$$

Thus

$$\frac{d^2 \hat{V}(z)}{dz^2} = -\beta^2 (\hat{V}^+ e^{-j\beta z} + \hat{V}^- e^{j\beta z}) \qquad (7.19)$$

or

$$\frac{d^2\hat{V}(z)}{dz^2} = -\beta^2 \hat{V}(z) \tag{7.20}$$

proving that Eq. (7.13) is the solution of Eq. (7.11).

Next, let us look at Eq. (7.14). Differentiating Eq. (7.14) with respect to z gives

$$\frac{d\hat{I}(z)}{dz} = -j\beta\frac{\hat{V}^+}{Z_C}e^{-j\beta z} - j\beta\frac{\hat{V}^-}{Z_C}e^{j\beta z} \tag{7.21}$$

Differentiating once more produces

$$\frac{d^2\hat{I}(z)}{dz^2} = (-j\beta)(-j\beta)\frac{\hat{V}^+}{Z_C}e^{-j\beta z} - (j\beta)(j\beta)\frac{\hat{V}^-}{Z_C}e^{j\beta z} \tag{7.22}$$

or

$$\frac{d^2\hat{I}(z)}{dz^2} = -\beta^2\frac{\hat{V}^+}{Z_C}e^{-j\beta z} + \beta^2\frac{\hat{V}^-}{Z_C}e^{j\beta z} \tag{7.23}$$

Thus

$$\frac{d^2\hat{I}(z)}{dz^2} = -\beta^2\left(\frac{\hat{I}^+}{Z_C}e^{-j\beta z} - \frac{\hat{I}^-}{Z_C}e^{j\beta z}\right) \tag{7.24}$$

or

$$\frac{d^2\hat{I}(z)}{dz^2} = -\beta^2\hat{I}(z) \tag{7.25}$$

proving that Eq. (7.14) is the solution of Eq. (7.12).

The solutions in Eqs. (7.13) and (7.14) consist of the forward- and backward-traveling waves [Adamczyk, 2021],

$$\hat{V}(z) = \hat{V}_f(z) + \hat{V}_b(z) \tag{7.26}$$

$$\hat{I}(z) = \hat{I}_f(z) + \hat{I}_b(z) \tag{7.27}$$

The forward-traveling waves are described by

$$\hat{V}_f(z) = \hat{V}^+ e^{-j\beta z} \tag{7.28}$$

$$\hat{I}(z) = \frac{\hat{V}^+}{Z_C}e^{-j\beta z} \tag{7.29}$$

while the backward-traveling waves are given by

$$\hat{V}_b(z) = \hat{V}^- e^{j\beta z} \tag{7.30}$$

$$\hat{I}_b(z) = -\frac{\hat{V}^-}{Z_C}e^{j\beta z} \tag{7.31}$$

7.1.1 The Complete Circuit Model – Voltage, Current, and Input Impedance along the Transmission Line

7.1.1.1 The Complete Circuit Model of Transmission Line

To determine the voltages and currents along the transmission line we need to consider a complete circuit model consisting of the source, the transmission line, and the load, as shown in Figure 7.1.

In this model (Model 1), we are moving from the source located at $z = 0$, toward the load located at $z = L$. It is often convenient to use an alternate circuit, shown in Figure 7.2.

In this alternate model (Model 2), we are moving from the load located at $d = 0$, toward the source located at $d = L$.

Before analyzing Model 1 and Model 2 configurations, let us introduce a concept of the input impedance to the line.

7.1.1.2 Input Impedance to the Transmission Line

Consider a transmission line circuit shown in Figure 7.3.

The input impedance to the line at any location z, $Z_{in}(z)$, is always calculated looking toward the load, regardless whether we use Model 1 or Model 2.

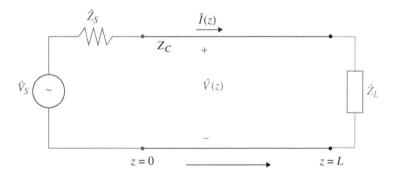

Figure 7.1   Model 1: Transmission line circuit with the source at $z = 0$ and the load at $z = L$.

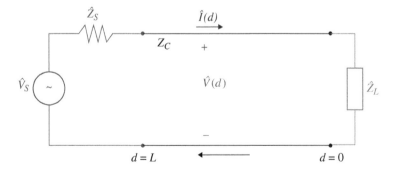

Figure 7.2   Model 2: Transmission line circuit with the load at $d = 0$ and the source at $d = L$.

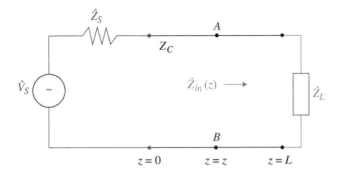

Figure 7.3 Input impedance to the line at any location z.

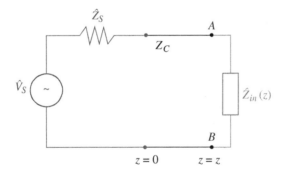

Figure 7.4 Equivalent circuit.

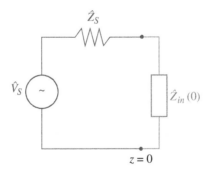

Figure 7.5 Input impedance to the line at the input to the line.

Figure 7.4 shows the equivalent circuit where the circuit to the right of nodes *AB* has been replaced with the input impedance to the line at that location [Ulaby, 1999].

The input impedance can be calculated at any location, including $z = 0$, as shown in Figure 7.5.

We refer to this impedance as the input impedance to the line at the input to the line.

7.1.1.3 Voltage, Current, and Input Impedance – Circuit Model 1

In this section we analyze Circuit Model 1, shown in Figure 7.6 [Adamczyk, 2023b].

Figure 7.6 shows a lossless transmission line with the characteristic impedance Z_C, driven by the source located at $z = 0$ and terminated by the load located at $z = L$.

Recall the frequency-domain solution of the transmission line equations, (7.13) and (7.14):

$$\hat{V}(z) = \hat{V}_z^+ e^{-j\beta z} + \hat{V}_z^- e^{j\beta z} \tag{7.32}$$

$$\hat{I}(z) = \frac{\hat{V}_z^+}{Z_C} e^{-j\beta z} - \frac{\hat{V}_z^-}{Z_C} e^{j\beta z} \tag{7.33}$$

Note: In equations, (7.13) and (7.14), the undetermined constants were denoted as $\hat{V}^+$ and $\hat{V}^-$.

In Eqs. (7.32) and (7.33) we use a different notation to distinguish between the constants for two different circuit models. Using Model 1, shown in Figure 7.6, we move from the source at $z = 0$ to the load at $z = L$, and use constants $\hat{V}_z^+$ and $\hat{V}_z^-$.

In Model 2, discussed in Section 7.1.1.4, we move from the load at $d = 0$ to the source at $d = L$, and use constants $\hat{V}_d^+$ and $\hat{V}_d^-$. These two sets of constants are different.

The solutions in equations, (7.32) and (7.33), consist of the forward- and backward-traveling waves. The forward-traveling voltage wave is described by

$$\hat{V}_f(z) = \hat{V}_z^+ e^{-j\beta z} \tag{7.34}$$

while the backward-traveling voltage wave is given by

$$\hat{V}_b(z) = \hat{V}_z^- e^{j\beta z} \tag{7.35}$$

Using these two waves, we define the voltage reflection coefficient *at any location z*, as the ratio of the backward-propagating wave to the forward-propagating wave

$$\hat{\Gamma}(z) = \frac{\hat{V}_b(z)}{\hat{V}_f(z)} = \frac{\hat{V}_z^- e^{j\beta z}}{\hat{V}_z^+ e^{-j\beta z}} \tag{7.36}$$

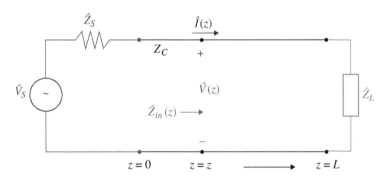

Figure 7.6   Circuit Model 1.

or

$$\hat{\Gamma}(z) = \frac{\hat{V}_z^-}{\hat{V}_z^+} e^{j2\beta z} \tag{7.37}$$

From Eq. (7.37) we obtain

$$\hat{V}_z^- = \hat{\Gamma}(z)\hat{V}_z^+ e^{-j2\beta z} \tag{7.38}$$

Utilizing Eq. (7.38) in Eq. (7.32) gives

$$\hat{V}(z) = \hat{V}_z^+ e^{-j\beta z} + \hat{\Gamma}(z)\hat{V}_z^+ e^{-j2\beta z} e^{j\beta z} \tag{7.39}$$

or

$$\hat{V}(z) = \hat{V}_z^+ e^{-j\beta z}[1 + \hat{\Gamma}(z)] \tag{7.40}$$

Utilizing Eq. (7.38) in Eq. (7.33) gives

$$\hat{I}(z) = \frac{\hat{V}_z^+}{Z_C} e^{-j\beta z} - \frac{\hat{\Gamma}(z)\hat{V}_z^+ e^{-j2\beta z}}{Z_C} e^{j\beta z} \tag{7.41}$$

or

$$\hat{I}(z) = \frac{\hat{V}_z^+}{Z_C} e^{-j\beta z}[1 - \hat{\Gamma}(z)] \tag{7.42}$$

Equations (7.40) and (7.42) express voltage and current at any location z, away from the source, in terms of the unknown constant $\hat{V}_z^+$ and the voltage reflection coefficient $\hat{\Gamma}(z)$, at any location z away from the source.

Let us return to this reflection coefficient, given by (7.37). Letting $z = L$, we obtain the voltage reflection coefficient at the *load*

$$\hat{\Gamma}(L) = \Gamma_L = \frac{\hat{V}_z^-}{\hat{V}_z^+} e^{j2\beta L} \tag{7.43}$$

Note that the load reflection coefficient can always be obtained directly from the knowledge of the load and the characteristic impedance of the line as

$$\Gamma_L = \frac{\hat{Z}_L - Z_C}{\hat{Z}_L + Z_C} \tag{7.44}$$

Let us return again to the reflection coefficient given by (7.37).

$$\hat{\Gamma}(z) = \frac{\hat{V}_z^-}{\hat{V}_z^+} e^{j2\beta z} = \frac{\hat{V}_z^-}{\hat{V}_z^+} e^{j2\beta(z+L-L)} = \frac{\hat{V}_z^-}{\hat{V}_z^+} e^{j2\beta L} e^{j2\beta(z-L)} \tag{7.45}$$

or

$$\hat{\Gamma}(z) = \hat{\Gamma}_L e^{j2\beta(z-L)} \tag{7.46}$$

Equation (7.46) expresses the voltage reflection coefficient at any location z, away from the source, in terms of the load reflection coefficient.

Equation (7.46) can be used to determine the voltage reflection coefficient at the input to the line, i.e. at $z = 0$ (we will need it shortly),

$$\hat{\Gamma}(0) = \hat{\Gamma}_L e^{-j2\beta L} \tag{7.47}$$

Utilizing Eq. (7.46) in Eqs. (7.40) and (7.42) gives

$$\hat{V}(z) = \hat{V}_z^+ e^{-j\beta z}[1 + \hat{\Gamma}_L e^{j2\beta(z-L)}] \tag{7.48}$$

$$\hat{I}(z) = \frac{\hat{V}_z^+}{Z_C} e^{-j\beta z}[1 - \hat{\Gamma}_L e^{j2\beta(z-L)}] \tag{7.49}$$

Equations (7.48) and (7.49) express voltage and current at any location z, away from the source, in terms of the unknown constant $\hat{V}_z^+$, and the load reflection coefficient.

In summary, the voltage and current at any location z, away from the source, can be obtained from

$$\hat{V}(z) = \hat{V}_z^+ e^{-j\beta z} + \hat{V}_z^- e^{j\beta z} \tag{7.50}$$

$$\hat{I}(z) = \frac{\hat{V}_z^+}{Z_C} e^{-j\beta z} - \frac{\hat{V}_z^-}{Z_C} e^{j\beta z} \tag{7.51}$$

or

$$\hat{V}(z) = \hat{V}_z^+ e^{-j\beta z}[1 + \hat{\Gamma}(z)] \tag{7.52}$$

$$\hat{I}(z) = \frac{\hat{V}_z^+}{Z_C} e^{-j\beta z}[1 - \hat{\Gamma}(z)] \tag{7.53}$$

or

$$\hat{V}(z) = \hat{V}_z^+ e^{-j\beta z}[1 + \hat{\Gamma}_L e^{j2\beta(z-L)}] \tag{7.54}$$

$$\hat{I}(z) = \frac{\hat{V}_z^+}{Z_C} e^{-j\beta z}[1 - \hat{\Gamma}_L e^{j2\beta(z-L)}] \tag{7.55}$$

The last set of equations is perhaps the most convenient since the load reflection coefficient, $\hat{\Gamma}_L$, can be obtained directly from Eq. (7.44) and the only unknown in this set is the constant $\hat{V}_z^+$.

The three sets of the equations, (7.50) through (7.55), can be used to determine the voltage and current at the input to the line, and at the load.

Letting $z = 0$, in Eqs. (7.50) through (7.55), we obtain the voltage and current at the input to the line as

$$\hat{V}(0) = \hat{V}_z^+ + \hat{V}_z^- \tag{7.56}$$

$$\hat{I}(0) = \frac{\hat{V}_z^+}{Z_C} - \frac{\hat{V}_z^-}{Z_C} \tag{7.57}$$

or

$$\hat{V}(0) = \hat{V}_z^+[1 + \hat{\Gamma}(0)] \tag{7.58}$$

$$\hat{I}(0) = \frac{\hat{V}_z^+}{Z_C}[1 - \hat{\Gamma}(0)] \tag{7.59}$$

or

$$\hat{V}(0) = \hat{V}_z^+ e^{-j\beta z}[1 + \hat{\Gamma}_L e^{-j2\beta L}] \tag{7.60}$$

$$\hat{I}(0) = \frac{\hat{V}_z^+}{Z_C} e^{-j\beta z}[1 - \hat{\Gamma}_L e^{-j2\beta L}] \tag{7.61}$$

Letting $z = L$, in Eqs. (7.50) through (7.55), we obtain the voltage and current at the load as

$$\hat{V}(L) = \hat{V}_z^+ e^{-j\beta L} + \hat{V}_z^- e^{j\beta L} \tag{7.62}$$

$$\hat{I}(L) = \frac{\hat{V}_z^+}{Z_C} e^{-j\beta L} - \frac{\hat{V}_z^-}{Z_C} e^{j\beta L} \tag{7.63}$$

or

$$\hat{V}(L) = \hat{V}_z^+ e^{-j\beta L}[1 + \hat{\Gamma}_L] \tag{7.64}$$

$$\hat{I}(L) = \frac{\hat{V}_z^+}{Z_C} e^{-j\beta L}[1 - \hat{\Gamma}_L] \tag{7.65}$$

Next, let us turn our attention to the undetermined constants $\hat{V}_z^+$ and $\hat{V}_z^-$. These constants can be determined from the knowledge of the voltage and current at the input to the line.

Equations (7.56) and (7.57) can be rewritten as

$$\hat{V}(0) = \hat{V}_z^+ + \hat{V}_z^- \tag{7.66}$$

$$Z_C \hat{I}(0) = \hat{V}_z^+ - \hat{V}_z^- \tag{7.67}$$

Adding Eqs. (7.66) and (7.67) gives

$$\hat{V}(0) + Z_C \hat{I}(0) = 2\hat{V}_z^+ \tag{7.68}$$

and thus

$$\hat{V}_z^+ = \frac{\hat{V}(0) + Z_C \hat{I}(0)}{2} \tag{7.69}$$

Subtracting Eq. (7.67) from Eq. (7.66) gives

$$\hat{V}(0) - Z_C \hat{I}(0) = 2\hat{V}_z^- \tag{7.70}$$

and thus

$$\hat{V}_z^- = \frac{\hat{V}(0) - Z_C \hat{I}(0)}{2} \tag{7.71}$$

The two undetermined constants, $\hat{V}_z^+$ and $\hat{V}_z^-$, can alternatively be obtained from the knowledge of the voltage and current at the load.

Equations (7.62) and (7.63) can be rewritten as

$$\hat{V}(L) = \hat{V}_z^+ e^{-j\beta L} + \hat{V}_z^- e^{j\beta L} \tag{7.72}$$

$$Z_C \hat{I}(L) = \hat{V}_z^+ e^{-j\beta L} - \hat{V}_z^- e^{j\beta L} \tag{7.73}$$

Adding Eqs. (7.72) and (7.73) gives

$$\hat{V}(L) + Z_C \hat{I}(L) = 2\hat{V}_z^+ e^{-j\beta L} \tag{7.74}$$

and thus

$$\hat{V}_z^+ = \frac{\hat{V}(L) + Z_C \hat{I}(L)}{2} e^{j\beta L} \tag{7.75}$$

Subtracting (7.73) from Eq. (7.72) gives

$$\hat{V}(L) - Z_C \hat{I}(L) = 2\hat{V}_z^- e^{j\beta L} \tag{7.76}$$

and thus

$$\hat{V}_z^- = \frac{\hat{V}(L) - Z_C \hat{I}(L)}{2} e^{-j\beta L} \tag{7.77}$$

Observation: To obtain the voltage or current at any location z, away from the source, we need the knowledge of the undetermined constants, $\hat{V}_z^+$ and $\hat{V}_z^-$ (or at least $\hat{V}_z^+$). To obtain the undetermined constant, $\hat{V}_z^+$ and $\hat{V}_z^-$, we need the knowledge of the voltage and current at the input to the line or at the load. We resolve this stalemate by introducing the concept of the input impedance to the line.

At any location z, away from the source, the input impedance to the line, $\hat{Z}_{in}(z)$, shown in Figure 7.7, is defined as the ratio of the total voltage to the total current at that point.

$$\hat{Z}_{in}(z) = \frac{\hat{V}(z)}{\hat{I}(z)} \tag{7.78}$$

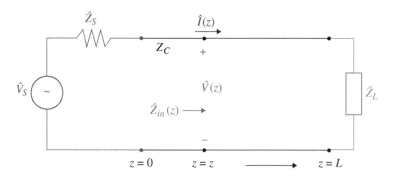

Figure 7.7   Input impedance to the line at any location z away from the source.

Since the total voltage and current at any location z away from the source can be obtained from three different sets of the equations, (7.50) through (7.55), it follows that the input impedance to the line, at any location z away from the source can be obtained from

$$\hat{Z}_{in}(z) = Z_C \frac{\hat{V}_z^+ e^{-j\beta z} + \hat{V}_z^- e^{j\beta z}}{\hat{V}_z^+ e^{-j\beta z} - \hat{V}_z^- e^{j\beta z}} \qquad (7.79)$$

or

$$\hat{Z}_{in}(z) = Z_C \frac{1 + \hat{\Gamma}(z)}{1 - \hat{\Gamma}(z)} \qquad (7.80)$$

or

$$\hat{Z}_{in}(z) = Z_C \frac{1 + \hat{\Gamma}_L e^{j2\beta(z-L)}}{1 - \hat{\Gamma}_L e^{j2\beta(z-L)}} \qquad (7.81)$$

Letting $z = 0$ in Eqs. (7.79)–(7.81), we obtain the *input impedance to the line at the input to the line* as

$$\hat{Z}_{in}(0) = Z_C \frac{\hat{V}_z^+ + \hat{V}_z^-}{\hat{V}_z^+ - \hat{V}_z^-} \qquad (7.82)$$

or

$$\hat{Z}_{in}(0) = Z_C \frac{1 + \hat{\Gamma}(0)}{1 - \hat{\Gamma}(0)} \qquad (7.83)$$

or

$$\hat{Z}_{in}(0) = Z_C \frac{1 + \hat{\Gamma}_L e^{-j2\beta L}}{1 - \hat{\Gamma}_L e^{-j2\beta L}} \qquad (7.84)$$

Since the constants, $\hat{V}_z^+$ and $\hat{V}_z^-$, are still unknown, in the calculations of the input impedance to the line at the input to the line, we are left with the remaining two equations, (7.83) and (7.84).

Since

$$\hat{\Gamma}(0) = \hat{\Gamma}_L e^{-j2\beta L} \qquad (7.85)$$

at this point, we effectively have just one equation, (7.84), to determine the input impedance to the line at the input to the line. Toward this end, we first determine the load reflection coefficient from

$$\Gamma_L = \frac{\hat{Z}_L - Z_C}{\hat{Z}_L + Z_C} \qquad (7.86)$$

and then use Eq. (7.84) to calculate the input impedance to the line *at the input to the line*.

There is one more useful set of formulas for obtaining the input impedance to the line at the input to the line. Using Eq. (7.86) in Eq. (7.84) we get

$$\hat{Z}_{in}(0) = Z_C \frac{1 + \frac{\hat{Z}_L - Z_C}{\hat{Z}_L + Z_C} e^{-j2\beta L}}{1 - \frac{\hat{Z}_L - Z_C}{\hat{Z}_L + Z_C} e^{-j2\beta L}} \qquad (7.87)$$

or

$$\hat{Z}_{in}(0) = Z_C \frac{\hat{Z}_L + \hat{Z}_L e^{-j2\beta L} + Z_C - Z_C e^{-j2\beta L}}{\hat{Z}_L - \hat{Z}_L e^{-j2\beta L} + Z_C + Z_C e^{-j2\beta L}} \tag{7.88}$$

or

$$\hat{Z}_{in}(0) = Z_C \frac{\hat{Z}_L(1 + e^{-j2\beta L}) + Z_C(1 - e^{-j2\beta L})}{\hat{Z}_L(1 - e^{-j2\beta L}) + Z_C(1 + e^{-j2\beta L})} \tag{7.89}$$

Now,

$$1 + e^{-j2\beta L} = e^{-j\beta L}(e^{j\beta L} + e^{-j\beta L}) \tag{7.90}$$

$$1 - e^{-j2\beta L} = e^{-j\beta L}(e^{j\beta L} - e^{-j\beta L}) \tag{7.91}$$

Utilizing Eqs. (7.90) and (7.91)in Eq. (7.89) we get

$$\hat{Z}_{in}(0) = Z_C \frac{\hat{Z}_L(e^{j\beta L} + e^{-j\beta L}) + Z_C(e^{j\beta L} - e^{-j\beta L})}{\hat{Z}_L(e^{j\beta L} - e^{-j\beta L}) + Z_C(e^{j\beta L} + e^{-j\beta L})} \tag{7.92}$$

or, using Euler's formulas

$$\hat{Z}_{in}(0) = Z_C \frac{\hat{Z}_L 2\cos(\beta L) + Z_C j2\sin(\beta L)}{\hat{Z}_L j2\sin(\beta L) + Z_C 2\cos(\beta L)} \tag{7.93}$$

leading to

$$\hat{Z}_{in}(0) = Z_C \frac{\hat{Z}_L \cos(\beta L) + jZ_C \sin(\beta L)}{Z_C \cos(\beta L) + j\hat{Z}_L \sin(\beta L)} \tag{7.94}$$

or equivalently,

$$\hat{Z}_{in}(0) = Z_C \frac{\hat{Z}_L + jZ_C \tan(\beta L)}{Z_C + j\hat{Z}_L \tan(\beta L)} \tag{7.95}$$

Thus, the input impedance to the line *at the input to the line* can be calculated either from Eq. (7.84) or (7.95).

At the input to the line, we have a situation depicted in Figure 7.8.

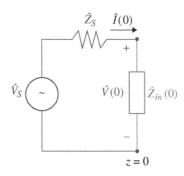

Figure 7.8   Equivalent circuit at the location $z = 0$.

It is apparent the voltage and current *at the input to the line* can be now obtained from

$$\hat{V}(0) = \frac{\hat{Z}_{in}(0)}{\hat{Z}_S + \hat{Z}_{in}(0)} \hat{V}_S \tag{7.96}$$

$$\hat{I}(0) = \frac{\hat{V}(0)}{\hat{Z}_{in}(0)} \tag{7.97}$$

Now, from the knowledge of $\hat{V}(0)$ and $\hat{I}(0)$ we can determine the constants $\hat{V}_z^+$ and $\hat{V}_z^-$ from

$$\hat{V}_z^+ = \frac{\hat{V}(0) + Z_C \hat{I}(0)}{2} \tag{7.98}$$

$$\hat{V}_z^- = \frac{\hat{V}(0) - Z_C \hat{I}(0)}{2} \tag{7.99}$$

At this point we can obtain the voltage, current, or impedance at any location z away from the source using previously derived equations.

In Section 7.1.1.4, we will analyze the circuit where we move from the load is located at $d = 0$ toward the source located at $d = L$ (Model 2).

7.1.1.4 Voltage, Current, and Input Impedance – Circuit Model 2

In this section we analyze Circuit Model 2, shown in Figure 7.9 [Adamczyk, 2023c].

Figure 7.9 shows a lossless transmission line with the characteristic impedance Z_C, driven by the source located at $d = L$ and terminated by the load located at $d = 0$.

The voltage, current, and input impedance now are a function of d, when moving from the load toward the source. Note that the input impedance to the line at any location d, $\hat{Z}_{in}(d)$, is always calculated looking toward the load, regardless whether we use Model 1 or Model 2.

The two distance variables are related by

$$d = L - z \quad \Rightarrow \quad z = L - d \tag{7.100}$$

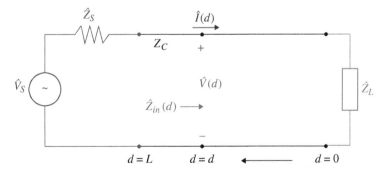

Figure 7.9 Circuit Model 2.

Recall: Using Model 1 the voltage and current at any location z away from the source were derived as

$$\hat{V}(z) = \hat{V}_z^+ e^{-j\beta z} + \hat{V}_z^- e^{j\beta z} \tag{7.101}$$

$$\hat{I}(z) = \frac{\hat{V}_z^+}{Z_C} e^{-j\beta z} - \frac{\hat{V}_z^-}{Z_C} e^{j\beta z} \tag{7.102}$$

Using the relations (7.100), in Eqs. (7.101) and (7.102), the voltage and current at any location d away from the load are

$$\hat{V}(d) = \hat{V}_z^+ e^{-j\beta(L-d)} + \hat{V}_z^- e^{j\beta(L-d)} \tag{7.103}$$

$$\hat{I}(d) = \frac{\hat{V}_z^+}{Z_C} e^{-j\beta(L-d)} - \frac{\hat{V}_z^-}{Z_C} e^{j\beta(L-d)} \tag{7.104}$$

or

$$\hat{V}(d) = \hat{V}_z^+ e^{-j\beta L} e^{j\beta d} + \hat{V}_z^- e^{j\beta L} e^{-j\beta d} \tag{7.105}$$

$$\hat{I}(d) = \frac{\hat{V}_z^+}{Z_C} e^{-j\beta L} e^{j\beta d} - \frac{\hat{V}_z^-}{Z_C} e^{j\beta L} e^{-j\beta d} \tag{7.106}$$

leading to

$$\hat{V}(d) = \hat{V}_d^+ e^{j\beta d} + \hat{V}_d^- e^{-j\beta d} \tag{7.107}$$

$$\hat{I}(d) = \frac{\hat{V}_d^+}{Z_C} e^{j\beta d} - \frac{\hat{V}_d^-}{Z_C} e^{-j\beta d} \tag{7.108}$$

where

$$\hat{V}_d^+ = \hat{V}_z^+ e^{-j\beta L} \tag{7.109}$$

$$\hat{V}_d^- = \hat{V}_z^- e^{j\beta L} \tag{7.110}$$

The solutions in Eqs. (7.107) and (7.108) consist of the forward- and backward-traveling waves. The forward-traveling voltage wave is described by

$$\hat{V}_f(d) = \hat{V}_d^+ e^{j\beta d} \tag{7.111}$$

while the backward-traveling voltage wave is given by

$$\hat{V}_b(d) = \hat{V}_d^- e^{-j\beta d} \tag{7.112}$$

Using these two waves, we define the voltage reflection coefficient at any location d, as the ratio of the backward-propagating wave to the forward-propagating wave

$$\hat{\Gamma}(d) = \frac{\hat{V}_b(d)}{\hat{V}_f(d)} = \frac{\hat{V}_d^- e^{-j\beta d}}{\hat{V}_d^+ e^{j\beta d}} \tag{7.113}$$

or

$$\hat{\Gamma}(d) = \frac{\hat{V}_d^-}{\hat{V}_d^+} e^{-j2\beta d} \tag{7.114}$$

From Eq. (7.114) we obtain

$$\hat{V}_d^- = \hat{\Gamma}(d)\hat{V}_d^+ e^{j2\beta d} \tag{7.115}$$

Utilizing Eq. (7.115) in Eq. (7.107) gives

$$\hat{V}(d) = \hat{V}_d^+ e^{j\beta d} + \hat{\Gamma}(d)\hat{V}_d^+ e^{j2\beta d} e^{-j\beta d} \tag{7.116}$$

or

$$\hat{V}(d) = \hat{V}_d^+ e^{j\beta d}[1 + \hat{\Gamma}(d)] \tag{7.117}$$

Utilizing Eq. (7.115) in Eq. (7.108) gives

$$\hat{I}(d) = \frac{\hat{V}_d^+}{Z_C} e^{j\beta d} - \frac{\hat{\Gamma}(d)\hat{V}_d^+ e^{j2\beta d}}{Z_C} e^{-j\beta d} \tag{7.118}$$

or

$$\hat{I}(d) = \frac{\hat{V}_d^+}{Z_C} e^{j\beta d}[1 - \hat{\Gamma}(d)] \tag{7.119}$$

Equations (7.117) and (7.119) express voltage and current at any location d, away from the load, in terms of the unknown constant $\hat{V}_d^+$ and the voltage reflection coefficient $\hat{\Gamma}(d)$, at any location d away from the load.

Let us return to this reflection coefficient, given by Eq. (7.114). Letting $d = 0$, we obtain the voltage reflection coefficient at the load

$$\hat{\Gamma}(0) = \hat{\Gamma}_L = \frac{\hat{V}_d^-}{\hat{V}_d^+} \tag{7.120}$$

Note that the load reflection coefficient can always be obtained directly from the knowledge of the load and the characteristic impedance of the line as

$$\hat{\Gamma}_L = \frac{\hat{Z}_L - Z_C}{\hat{Z}_L + Z_C} \tag{7.121}$$

Utilizing Eq. (7.120) in Eq. (7.114) we get

$$\hat{\Gamma}(d) = \hat{\Gamma}_L e^{-j2\beta d} \tag{7.122}$$

which expresses the voltage reflection coefficient at any location d, away from the load, in terms of the load reflection coefficient.

Equation (7.122) can be used to determine the voltage reflection coefficient at the input to the line, i.e. at $d = L$ (we will need it shortly),

$$\hat{\Gamma}(L) = \hat{\Gamma}_L e^{-j2\beta L} \tag{7.123}$$

Utilizing Eq. (7.122) in Eqs. (7.117) and (7.119) gives

$$\hat{V}(d) = \hat{V}_d^+ e^{j\beta d}[1 + \hat{\Gamma}_L e^{-j2\beta d}] \tag{7.124}$$

$$\hat{I}(d) = \frac{\hat{V}_d^+}{Z_C} e^{j\beta d}[1 - \hat{\Gamma}_L e^{-j2\beta d}] \tag{7.125}$$

Equations (7.124) and (7.125) express voltage and current at any location d, away from the load, in terms of the unknown constant $\hat{V}_d^+$, and the load reflection coefficient.

In summary, the voltage and current at any location d, away from the load, can be obtained from

$$\hat{V}(d) = \hat{V}_d^+ e^{j\beta d} + \hat{V}_d^- e^{-j\beta d} \tag{7.126}$$

$$\hat{I}(d) = \frac{\hat{V}_d^+}{Z_C} e^{j\beta d} - \frac{\hat{V}_d^-}{Z_C} e^{-j\beta d} \tag{7.127}$$

or

$$\hat{V}(d) = \hat{V}_d^+ e^{j\beta d}[1 + \hat{\Gamma}(d)] \tag{7.128}$$

$$\hat{I}(d) = \frac{\hat{V}_d^+}{Z_C} e^{j\beta d}[1 - \hat{\Gamma}(d)] \tag{7.129}$$

or

$$\hat{V}(d) = \hat{V}_d^+ e^{j\beta d}[1 + \hat{\Gamma}_L e^{-j2\beta d}] \tag{7.130}$$

$$\hat{I}(d) = \frac{\hat{V}_d^+}{Z_C} e^{j\beta d}[1 - \hat{\Gamma}_L e^{-j2\beta d}] \tag{7.131}$$

The last set of equations is perhaps the most convenient since the load reflection coefficient, $\hat{\Gamma}_L$, can be obtained directly from Eq. (7.125) and the only unknown in this set is the constant $\hat{V}_d^+$.

The three sets of Eqs. (7.126) through (7.131) can be used to determine the voltage and current at the load and at the input to the line.

Letting $d = 0$, in Eqs. (7.126) and (7.127) we obtain the voltage and current *at the load* as

$$\hat{V}(0) = \hat{V}_d^+ + \hat{V}_d^- \tag{7.132}$$

$$\hat{I}(0) = \frac{\hat{V}_d^+}{Z_C} - \frac{\hat{V}_d^-}{Z_C} \tag{7.133}$$

Letting $d = 0$, in Eqs. (7.130) and (7.131) we obtain the voltage and current *at the load* as

$$\hat{V}(0) = \hat{V}_d^+[1 + \hat{\Gamma}_L] \tag{7.134}$$

$$\hat{I}(0) = \frac{\hat{V}_d^+}{Z_C}[1 - \hat{\Gamma}_L] \tag{7.135}$$

Letting $d = L$, in Eqs. (7.126) through (7.131) we obtain the voltage and current *at the input to the line* as

$$\hat{V}(L) = \hat{V}_d^+ e^{j\beta L} + \hat{V}_d^- e^{-j\beta L} \tag{7.136}$$

$$\hat{I}(L) = \frac{\hat{V}_d^+}{Z_C} e^{j\beta L} - \frac{\hat{V}_d^-}{Z_C} e^{-j\beta L} \tag{7.137}$$

or

$$\hat{V}(L) = \hat{V}_d^+ e^{j\beta L}[1 + \hat{\Gamma}(L)] \tag{7.138}$$

$$\hat{I}(L) = \frac{\hat{V}_d^+}{Z_C} e^{j\beta L}[1 - \hat{\Gamma}(L)] \tag{7.139}$$

or

$$\hat{V}(L) = \hat{V}_d^+ e^{j\beta L}[1 + \hat{\Gamma}_L e^{-j2\beta L}] \tag{7.140}$$

$$\hat{I}(L) = \frac{\hat{V}_d^+}{Z_C} e^{j\beta L}[1 - \hat{\Gamma}_L e^{-j2\beta L}] \tag{7.141}$$

Next, let us turn our attention to the undetermined constants $\hat{V}_d^+$ and $\hat{V}_d^-$. These constants can be determined from the knowledge of the voltage and current *at the load*.

Equations (7.132) and (7.133) can be rewritten as

$$\hat{V}(0) = \hat{V}_d^+ + \hat{V}_d^- \tag{7.142}$$

$$Z_C \hat{I}(0) = \hat{V}_d^+ - \hat{V}_d^- \tag{7.143}$$

Adding Eqs. (7.142) and (7.143) gives

$$\hat{V}(0) + Z_C \hat{I}(0) = 2\hat{V}_d^+ \tag{7.144}$$

and thus

$$\hat{V}_d^+ = \frac{\hat{V}(0) + Z_C \hat{I}(0)}{2} \tag{7.145}$$

Subtracting Eq. (7.143) from Eq. (7.142) gives

$$\hat{V}(0) - Z_C \hat{I}(0) = 2\hat{V}_d^- \tag{7.146}$$

and thus

$$\hat{V}_d^- = \frac{\hat{V}(0) - Z_C \hat{I}(0)}{2} \tag{7.147}$$

These two undetermined constants, $\hat{V}_d^+$ and $\hat{V}_d^-$, can alternatively be obtained from the knowledge of the voltage and current at *the input to the line*.

Equations (7.136) and (7.137) can be rewritten as

$$\hat{V}(L) = \hat{V}_d^+ e^{j\beta L} + \hat{V}_d^- e^{-j\beta L} \tag{7.148}$$

$$Z_C \hat{I}(L) = \hat{V}_d^+ e^{j\beta L} - \hat{V}_d^- e^{-j\beta L} \tag{7.149}$$

Adding Eqs. (7.148) and (7.149) gives

$$\hat{V}(L) + Z_C \hat{I}(L) = 2\hat{V}_d^+ e^{j\beta L} \tag{7.150}$$

and thus

$$\hat{V}_d^+ = \frac{\hat{V}(L) + Z_C \hat{I}(L)}{2} e^{-j\beta L} \tag{7.151}$$

Subtracting Eq. (7.149) from Eq. (7.148) gives

$$\hat{V}(L) - Z_C \hat{I}(L) = 2\hat{V}_d^- e^{-j\beta L} \tag{7.152}$$

and thus

$$\hat{V}_d^- = \frac{\hat{V}(L) - Z_C \hat{I}(L)}{2} e^{j\beta L} \tag{7.153}$$

Observation: To obtain the voltage or current at any location d, away from the load, we need the knowledge of the undetermined constants, $\hat{V}_d^+$ and $\hat{V}_d^-$ (or at least $\hat{V}_d^+$). To obtain the undetermined constant, $\hat{V}_d^+$ and $\hat{V}_d^-$, we need the knowledge of the voltage and current at the input to the line, or at the load. We resolve this stalemate in a similar way we did in Model 1, by introducing the concept of the input impedance to the line.

At any location d, away from the load, the input impedance to the line, $\hat{Z}_{in}(d)$, shown in Figure 7.10, is defined as the ratio of the total voltage to the total current at that point.

$$\hat{Z}_{in}(d) = \frac{\hat{V}(d)}{\hat{I}(d)} \tag{7.154}$$

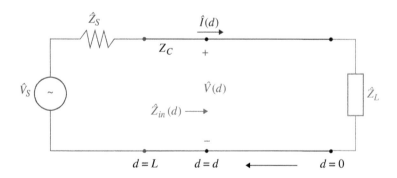

Figure 7.10   Input impedance to the line at any location d away from the load.

Since the total voltage and current at any location d away from the load can be obtained from the three different sets of the equations, (7.126) through (7.131), it follows that the input impedance to the line, at any location d away from the load can be obtained from

$$\hat{Z}_{in}(d) = Z_C \frac{\hat{V}_d^+ e^{j\beta d} + \hat{V}_d^- e^{-j\beta d}}{\hat{V}_d^+ e^{j\beta d} - \hat{V}_d^- e^{-j\beta d}} \tag{7.155}$$

or

$$\hat{Z}_{in}(d) = Z_C \frac{1 + \hat{\Gamma}(d)}{1 - \hat{\Gamma}(d)} \tag{7.156}$$

or

$$\hat{Z}_{in}(d) = Z_C \frac{1 + \hat{\Gamma}_L e^{-j2\beta d}}{1 - \hat{\Gamma}_L e^{-j2\beta d}} \tag{7.157}$$

Letting $d = L$ in Eqs. (7.155)–(7.157), we obtain the *input impedance to the line at the input to the line* as

$$\hat{Z}_{in}(L) = Z_C \frac{\hat{V}_d^+ e^{j\beta L} + \hat{V}_d^- e^{-j\beta L}}{\hat{V}_d^+ e^{j\beta L} - \hat{V}_d^- e^{-j\beta L}} \tag{7.158}$$

or

$$\hat{Z}_{in}(L) = Z_C \frac{1 + \hat{\Gamma}(L)}{1 - \hat{\Gamma}(L)} \tag{7.159}$$

or

$$\hat{Z}_{in}(L) = Z_C \frac{1 + \hat{\Gamma}_L e^{-j2\beta L}}{1 - \hat{\Gamma}_L e^{-j2\beta L}} \tag{7.160}$$

Since the constants, $\hat{V}_d^+$ and $\hat{V}_d^-$, are still unknown, in the calculations of the input impedance to the line at the input to the line, we are left with the remaining two equations, (7.159) and (7.160).
Since

$$\hat{\Gamma}(L) = \hat{\Gamma}_L e^{-j2\beta L} \tag{7.161}$$

at this point, we effectively have just one equation, (7.160), to determine the input impedance to the line at the input to the line. Toward this end, we first determine the load reflection coefficient from

$$\Gamma_L = \frac{\hat{Z}_L - Z_C}{\hat{Z}_L + Z_C} \tag{7.162}$$

and then use Eq. (7.160) to calculate the input impedance to the line *at the input to the line*.

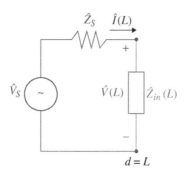

Figure 7.11   Equivalent circuit at the location $d = L$.

There is one more useful set of formulas for obtaining the input impedance to the line at the input to the line. Using Eq. (7.162) in Eq. (7.160) we get

$$\hat{Z}_{in}(L) = Z_C \frac{1 + \frac{\hat{Z}_L - Z_C}{\hat{Z}_L + Z_C} e^{-j2\beta L}}{1 - \frac{\hat{Z}_L - Z_C}{\hat{Z}_L + Z_C} e^{-j2\beta L}} \tag{7.163}$$

The RHS of Eq. (7.162) is equivalent to, [Adamczyk, Feb 2023b],

$$\hat{Z}_{in}(0) = Z_C \frac{\hat{Z}_L \cos(\beta L) + jZ_C \sin(\beta L)}{Z_C \cos(\beta L) + j\hat{Z}_L \sin(\beta L)} \tag{7.164}$$

or equivalently,

$$\hat{Z}_{in}(L) = Z_C \frac{\hat{Z}_L + jZ_C \tan(\beta L)}{Z_C + j\hat{Z}_L \tan(\beta L)} \tag{7.165}$$

At the input to the line, we have a situation depicted in Figure 7.11.

It is apparent the voltage and current *at the input to the line* can be now obtained from

$$\hat{V}(L) = \frac{\hat{Z}_{in}(L)}{\hat{Z}_S + \hat{Z}_{in}(L)} \hat{V}_S \tag{7.166}$$

$$\hat{I}(L) = \frac{\hat{V}(L)}{\hat{Z}_{in}(L)} \tag{7.167}$$

Now, from the knowledge of $\hat{V}(L)$ and $\hat{I}(L)$ we can determine the constants $\hat{V}_d^+$ and $\hat{V}_d^-$ from

$$\hat{V}_d^+ = \frac{\hat{V}(L) + Z_C \hat{I}(L)}{2} e^{-j\beta L} \tag{7.168}$$

$$\hat{V}_d^- = \frac{\hat{V}(L) - Z_C \hat{I}(L)}{2} e^{j\beta L} \tag{7.169}$$

At this point we can obtain the voltage, current, or impedance at any location d away from the load using previously derived equations.

7.1.2 Frequency-Domain Solution – Example

Consider the transmission line circuit shown in Figure 7.12.
Determine:

(a) Input impedance to the line at the input to the line.
(b) Input impedance to the line one wavelength away from the source.
(c) Input impedance to the line one wavelength away from the load.
(d) Voltage and current at the input to the line.
(e) Voltage and current at the load.
(f) Voltage and current one wavelength away from the source.
(g) Voltage and current one wavelength away from the load.

Solution:

Part a – Input Impedance to the Line at the Input to the Line:

First, we will use Eq. (7.84), repeated here:

$$\hat{Z}_{in}(0) = Z_C \frac{1 + \hat{\Gamma}_L e^{-j2\beta L}}{1 - \hat{\Gamma}_L e^{-j2\beta L}} \tag{7.170}$$

The load reflection coefficient is obtained from

$$\Gamma_L = \frac{\hat{Z}_L - Z_C}{\hat{Z}_L + Z_C} \tag{7.171}$$

Thus,

$$\Gamma_L = \frac{200 + j500 - 50}{200 + j500 + 50} = \frac{150 + j500}{250 + j500} \tag{7.172}$$

or

$$\Gamma_L = 0.92 + j0.16 = 0.933809\angle 9.86581° \tag{7.173}$$

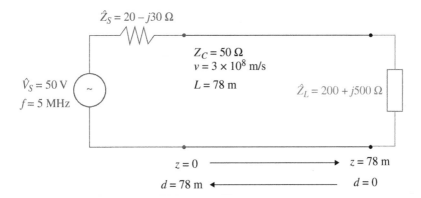

Figure 7.12 Transmission line circuit.

The phase constant is obtained from

$$\beta = \frac{\omega}{v} = \frac{2\pi \times 5 \times 10^6}{3 \times 10^8} \tag{7.174}$$

or

$$\beta = 0.1047 \, \frac{\text{rad}}{\text{m}} \tag{7.175}$$

Then,

$$-2\beta L = -(2)(0.1047)(L) = -(2)(0.1047)(78) = -16.332 \, \text{rad} \tag{7.176}$$

or

$$-2\beta L = -935.8234° = -215.8234° \tag{7.177}$$

Subsequently,

$$\hat{\Gamma}_L e^{-j2\beta L} = 0.933809 e^{j9.86581°} e^{-j215.8243°} \tag{7.178}$$

or

$$\hat{\Gamma}_L e^{-j2\beta L} = 0.933809 e^{-j205.95759°} = -0.839605 + j0.408734 \tag{7.179}$$

Then,

$$1 + \hat{\Gamma}_L e^{-j2\beta L} = 0.160395 + j0.408734 \tag{7.180}$$

and

$$1 - \hat{\Gamma}_L e^{-j2\beta L} = 1.839605 - j0.408734 \tag{7.181}$$

Finally,

$$\hat{Z}_{in}(0) = 50 \frac{0.160395 + j0.408734}{1.839605 - j0.408734} \tag{7.182}$$

or

$$\hat{Z}_{in}(0) = 1.8022 + j11.5097 = 11.6499\angle 81.1008° \tag{7.183}$$

To verify this solution, let's use Eq. (7.95), repeated here:

$$\hat{Z}_{in}(0) = Z_C \frac{\hat{Z}_L + jZ_C \tan(\beta L)}{Z_C + j\hat{Z}_L \tan(\beta L)} \tag{7.184}$$

The argument of the tangent function is

$$\beta L = (0.1047)(78) = 8.1666 \, \text{rad} = 467.9117° = 107.9117° \tag{7.185}$$

thus

$$\tan \beta L = -3.093893 \tag{7.186}$$

Therefore,

$$\hat{Z}_{in}(0) = 50 \frac{200 + j500 + j(50)(-3.093893)}{50 + j(200 + j500)(-3.093893)} \tag{7.187}$$

or
$$\hat{Z}_{in}(0) = 1.8022 + j11.50975 = 11.65\angle 81.1009° \; \Omega \tag{7.188}$$

which agrees with the previous result.

Part b – Input Impedance to the Line One Wavelength Away from the Source:

First, let's use the Eq. (7.81), repeated here, for Circuit Model 1:

$$\hat{Z}_{in}(z) = Z_C \frac{1 + \hat{\Gamma}_L e^{j2\beta(z-L)}}{1 - \hat{\Gamma}_L e^{j2\beta(z-L)}} \tag{7.189}$$

The wavelength is obtained from the relation

$$\beta = \frac{2\pi}{\lambda} \tag{7.190}$$

thus,

$$\lambda = \frac{2\pi}{\beta} = \frac{2\pi}{0.1047} = 60 \, \text{m} \tag{7.191}$$

and

$$2\beta(z - L) = (2)(0.1047)(60 - 78) = -3.7692 \, \text{rad} = -215.959252° \tag{7.192}$$

Subsequently,

$$\hat{\Gamma}_L e^{j2\beta(z-L)} = 0.933809 e^{j9.86581°} e^{-j215.959252°} \tag{7.193}$$

or

$$\hat{\Gamma}_L e^{j2\beta(z-L)} = 0.933809 e^{-j206.093442°} = -0.838625 + j0.410719 \tag{7.194}$$

Then,

$$1 + \hat{\Gamma}_L e^{j2\beta(z-L)} = 0.161375 + j0.410719 \tag{7.195}$$

and

$$1 - \hat{\Gamma}_L e^{j2\beta(z-L)} = 1.838625 - j0.410719 \tag{7.196}$$

Finally,

$$\hat{Z}_{in}(z = 60 \, \text{m}) = 50 \frac{0.161375 + j0.410719}{1.838625 - j0.410719} \tag{7.197}$$

or

$$\hat{Z}_{in}(z = 60 \, \text{m}) = 1.80346 + j11.57205 = 11.7117\angle 81.1419° \; \Omega \tag{7.198}$$

Alternatively, we could use Eq. (7.157), developed for the Circuit Model 2:

$$\hat{Z}_{in}(d) = Z_C \frac{1 + \hat{\Gamma}_L e^{-j2\beta d}}{1 - \hat{\Gamma}_L e^{-j2\beta d}} \tag{7.199}$$

The distance d from the load is $d = 18 \, \text{m}$, giving exactly the same result as the previous approach.

Another way of verifying these results is by using Eq. (7.79), developed for Circuit Model 1:

$$\hat{Z}_{in}(z) = Z_C \frac{\hat{V}_z^+ e^{-j\beta z} + \hat{V}_z^- e^{j\beta z}}{\hat{V}_z^+ e^{-j\beta z} - \hat{V}_z^- e^{j\beta z}} \tag{7.200}$$

or equation (7.155), developed for Circuit Model 2:

$$\hat{Z}_{in}(d) = Z_C \frac{\hat{V}_d^+ e^{j\beta d} + \hat{V}_d^- e^{-j\beta d}}{\hat{V}_d^+ e^{j\beta d} - \hat{V}_d^- e^{-j\beta d}} \tag{7.201}$$

In order to do that, we need to compute the undetermined constants $\hat{V}_z^+$ and $\hat{V}_z^-$ for Circuit Model 1, and $\hat{V}_d^+$ and $\hat{V}_d^-$ for Circuit Model 2. This two sets of constants are computed in the latter part of this example.

Part c – Input Impedance to the Line One Wavelength Away from the Load:

One wavelength away from the load we have $z = 18\,\text{m}$. Thus,

$$2\beta(z - L) = (2)(0.1047)(18 - 78) = -12.564\,\text{rad} = 0.135853° \tag{7.202}$$

Subsequently,

$$\hat{\Gamma}_L e^{j2\beta(z-L)} = 0.933809 e^{j9.86581°} e^{j0.135853°} \tag{7.203}$$

or

$$\hat{\Gamma}_L e^{j2\beta(z-L)} = 0.933809 e^{j10.001663°} = 0.919609 + j0.162179 \tag{7.204}$$

Then,

$$1 + \hat{\Gamma}_L e^{j2\beta(z-L)} = 1.919609 + j0.162179 \tag{7.205}$$

and

$$1 - \hat{\Gamma}_L e^{j2\beta(z-L)} = 0.080391 - j0.162179 \tag{7.206}$$

Finally,

$$\hat{Z}_{in}(z = 18\,\text{m}) = 50 \frac{1.919609 + j0.162179}{0.080391 - j0.162179} \tag{7.207}$$

or

$$\hat{Z}_{in}(z = 18\,\text{m}) = 195.358265 + j494.98025 = 532.137\angle 68.4619°\,\Omega \tag{7.208}$$

Alternatively, we could use Eq. (7.157), developed for the Circuit Model 2:

$$\hat{Z}_{in}(d) = Z_C \frac{1 + \hat{\Gamma}_L e^{-j2\beta d}}{1 - \hat{\Gamma}_L e^{-j2\beta d}} \tag{7.209}$$

The distance d from the load is $d = 60\,\text{m}$, giving exactly the same result.
Again, another way of verifying these results is by using Eqs. (7.79) and (7.155).

Part d – Voltage and Current at the Input to the Line:

The easiest way of calculating voltage and current at the input to the line is from Eqs. (7.96) and (7.97), repeated here:

$$\hat{V}(0) = \frac{\hat{Z}_{in}(0)}{\hat{Z}_S + \hat{Z}_{in}(0)} \hat{V}_S \tag{7.210}$$

$$\hat{I}(0) = \frac{\hat{V}(0)}{\hat{Z}_{in}(0)} \tag{7.211}$$

Thus,

$$\hat{V}(z = 0) = \frac{1.8022 + j11.5097}{20 - j30 + 1.8022 + j11.5097} \tag{7.212}$$

or

$$\hat{V}(z = 0) = -10.615956 + j17.390804 = 20.3749\angle121.401° \text{ V} \tag{7.213}$$

Then,

$$\hat{I}(z = 0) = \frac{20.3749\angle121.401°}{11.6499\angle81.1008°} \tag{7.214}$$

or

$$\hat{I}(z = 0) = 1.333852 + j1.131197 = 1.748933\angle40.3002° \text{ A} \tag{7.215}$$

Part e – Voltage and Current at the Load:

Voltage and current at the load can be obtained from Eqs. (7.64) and (7.65), repeated here:

$$\hat{V}(L) = \hat{V}_z^+ e^{-j\beta L}[1 + \hat{\Gamma}_L] \tag{7.216}$$

$$\hat{I}(L) = \frac{\hat{V}_z^+}{Z_C} e^{-j\beta L}[1 - \hat{\Gamma}_L] \tag{7.217}$$

Both of these equations involve the constant $\hat{V}_z^+$. The easiest approach is to use Eq. (7.98), repeated here:

$$\hat{V}_z^+ = \frac{\hat{V}(0) + Z_C\hat{I}(0)}{2} \tag{7.218}$$

Thus,

$$\hat{V}_z^+ = \frac{-10.615956 + j17.390804 + 50(1.333852 + j1.131197)}{2} \tag{7.219}$$

or

$$\hat{V}_z^+ = 28.038322 + j36.975327 = 46.4039\angle52.827° \tag{7.220}$$

Let's verify this result before using it. Toward this end, we could use Eq. (7.58), repeated here:

$$\hat{V}(0) = \hat{V}_z^+[1 + \hat{\Gamma}(0)] \tag{7.221}$$

From this equation we obtain $\hat{V}_z^+$ as

$$\hat{V}_z^+ = \frac{\hat{V}(0)}{1 + \hat{\Gamma}(0)} \qquad (7.222)$$

$\hat{\Gamma}(0)$ is obtained from Eq. (7.47), repeated here:

$$\hat{\Gamma}(0) = \hat{\Gamma}_L e^{-j2\beta L} \qquad (7.223)$$

This result was calculated earlier in Eq. (7.179) as

$$\hat{\Gamma}_L e^{-j2\beta L} = 0.933809 e^{-j205.95759°} \qquad (7.224)$$

$1 + \hat{\Gamma}(0)$ was calculated earlier in Eq. (7.180) as

$$1 + \hat{\Gamma}_L e^{-j2\beta L} = 0.160395 + j0.408734 = 0.439079\angle 68.574° \qquad (7.225)$$

Thus,

$$\hat{V}_z^+ = \frac{20.3749\angle 121.401°}{0.439079\angle 68.574°} = 46.403722\angle 52.827° \qquad (7.226)$$

which agrees with the result obtained in Eq. (7.220).

βL was calculated earlier in Eq. (7.185) as

$$\beta L = (0.1047)(78) = 8.1666\,\text{rad} = 467.9117° = 107.9117° \qquad (7.227)$$

Before evaluating Eqs. (7.216) and (7.217) let's evaluate two more expressions.

$$1 + \hat{\Gamma}_L = 1.92 + j0.16 = 1.92666\angle 4.76364° \qquad (7.228)$$

$$1 - \hat{\Gamma}_L = 0.08 - j0.16 = 0.178885\angle -63.4349° \qquad (7.229)$$

Thus, the voltage at the load is

$$\hat{V}(L = 78\,\text{m}) = (46.403722 e^{j52.827°})(e^{-j107.9117°})(1.92666^{j4.76364°}) \qquad (7.230)$$

or

$$\hat{V}(L = 78\,\text{m}) = 57.08323 - j68.80853 = 89.404195\angle -50.32106°\,\text{V} \qquad (7.231)$$

The current at the load is

$$\hat{I}(L = 78\,\text{m}) = \frac{(46.403722 e^{j52.827°})}{50}(e^{-j107.9117°})(0.178885 e^{-j63.4349°}) \qquad (7.232)$$

or

$$\hat{I}(L = 78\,\text{m}) = -0.079267 + j0.145873 = 0.1660186\angle -118.5196°\,\text{A} \qquad (7.233)$$

Part f – Voltage and Current One Wavelength from the Source:

First, let's use Circuit Model 1. Voltage and current can be obtained from Eqs. (7.54) and (7.55), repeated here:

$$\hat{V}(z) = \hat{V}_z^+ e^{-j\beta z}[1 + \hat{\Gamma}_L e^{j2\beta(z-L)}] \tag{7.234}$$

$$\hat{I}(z) = \frac{\hat{V}_z^+}{Z_C} e^{-j\beta z}[1 - \hat{\Gamma}_L e^{j2\beta(z-L)}] \tag{7.235}$$

One wavelength away from the source we have $z = 60\,\text{m}$. Thus,

$$-\beta z = -(0.1047)(60) = -6.282\,\text{rad} = -359.93209° = 0.06791° \tag{7.236}$$

$$2\beta(z - L) = (2)(0.1047)(60 - 78) = -3.7692\,\text{rad} = -215.959252° \tag{7.237}$$

$$\hat{\Gamma}_L e^{j2\beta(z-L)} = 0.933809 e^{-j206.093442°} = -0.838625 + j0.410719 \tag{7.238}$$

$$1 + \hat{\Gamma}_L e^{j2\beta(z-L)} = 0.161375 + j0.410719 = 0.441284\angle 68.5497° \tag{7.239}$$

$$1 - \hat{\Gamma}_L e^{j2\beta(z-L)} = 1.838625 - j0.410719 = 1.88394\angle -12.5922° \tag{7.240}$$

Thus, the voltage one wavelength away from the source is

$$\hat{V}(z = 60\,\text{m}) = (46.403722 e^{j52.827°})(e^{j0.06791°})(0.441284^{j68.5497°}) \tag{7.241}$$

or

$$\hat{V}(z = 60\,\text{m}) = 20.47722\angle 121.4876°\,\text{V} \tag{7.242}$$

The current one wavelength away from the source is

$$\hat{I}(z = 60\,\text{m}) = \frac{(46.403722 e^{j52.827°})}{50}(e^{j0.06791°})(1.88394 e^{-j12.5922°}) \tag{7.243}$$

or

$$\hat{I}(z = 60\,\text{m}) = 1.7484366\angle 40.3027°\,\text{A} \tag{7.244}$$

Let's verify these results using Circuit Model 2. Voltage and current can be obtained from Eqs. (7.130) and (7.131), repeated here:

$$\hat{V}(d) = \hat{V}_d^+ e^{j\beta d}[1 + \hat{\Gamma}_L e^{-j2\beta d}] \tag{7.245}$$

$$\hat{I}(d) = \frac{\hat{V}_d^+}{Z_C} e^{j\beta d}[1 - \hat{\Gamma}_L e^{-j2\beta d}] \tag{7.246}$$

This set of equations can be rewritten in an equivalent form as

$$\hat{V}(d) = \hat{V}_d^+[e^{j\beta d} + \hat{\Gamma}_L e^{-j\beta d}] \tag{7.247}$$

$$\hat{I}(d) = \frac{\hat{V}_d^+}{Z_C}[e^{j\beta d} - \hat{\Gamma}_L e^{-j\beta d}] \tag{7.248}$$

One wavelength away from the source we have $d = 18\,\text{m}$. Thus,

$$\beta d = (0.1047)(18) = 1.8846\,\text{rad} = 107.979626° \tag{7.249}$$

The constant $\hat{V}_d^+$ is related to the constant $\hat{V}_z^+$ by Eq. (7.109), repeated here

$$\hat{V}_d^+ = \hat{V}_z^+ e^{-j\beta L} \tag{7.250}$$

where
$$\beta L = (0.1047)(78) = 8.166\,\text{rad} = 467.9117° = 107.9117° \tag{7.251}$$

Thus,
$$\hat{V}_d^+ = (46.403722 e^{j52.827°})(e^{-j107.9117°}) = 46.403722 e^{-j55.0847°} \tag{7.252}$$

Additionally,

$$\hat{\Gamma}_L e^{-j\beta d} = (0.933809 e^{j9.86581°})(e^{-j107.979626°}) = 0.933809 e^{-j98.113816°} \tag{7.253}$$

$$e^{j\beta d} + \hat{\Gamma}_L e^{-j\beta d} = 0.441283 e^{j176.529°} \tag{7.254}$$

$$e^{j\beta d} - \hat{\Gamma}_L e^{-j\beta d} = 1.88394 e^{j95.3874°} \tag{7.255}$$

Thus, the voltage one wavelength away from the source is

$$\hat{V}(d = 18\,\text{m}) = (46.403722 e^{-j55.0847°})(0.441283^{j176.529°}) \tag{7.256}$$

or
$$\hat{V}(d = 18\,\text{m}) = 20.47717\angle121.4443°\,\text{V} \tag{7.257}$$

The current one wavelength away from the source is

$$\hat{I}(d = 18\,\text{m}) = \frac{(46.403722 e^{-j55.084713°})}{50}(1.88394 e^{j95.3874°}) \tag{7.258}$$

or
$$\hat{I}(d = 18\,\text{m}) = 1.748436\angle40.303°\,\text{A} \tag{7.259}$$

These results are consistent with the ones obtained using Circuit Model 1, and given by Eqs. (7.242) and (7.244).

Part g – Voltage and Current One Wavelength from the Load:

First, let's use Circuit Model 1. Voltage and current can be obtained from Eqs. (7.54) and (7.55), repeated here:

$$\hat{V}(z) = \hat{V}_z^+ e^{-j\beta z}[1 + \hat{\Gamma}_L e^{j2\beta(z-L)}] \tag{7.260}$$

$$\hat{I}(z) = \frac{\hat{V}_z^+}{Z_C} e^{-j\beta z}[1 - \hat{\Gamma}_L e^{j2\beta(z-L)}] \tag{7.261}$$

One wavelength away from the source we have $z = 18\,\text{m}$. Thus,

$$-\beta z = -(0.1047)(18) = -1.8846\,\text{rad} = -107.979626° \tag{7.262}$$

$$2\beta(z - L) = (2)(0.1047)(18 - 78) = -12.564\,\text{rad} = 0.1358° \tag{7.263}$$

$$\hat{\Gamma}_L e^{j2\beta(z-L)} = 0.933809 e^{j10.00161°} \tag{7.264}$$

$$1 + \hat{\Gamma}_L e^{j2\beta(z-L)} = 1.92646 e^{j4.8292°} \tag{7.265}$$

$$1 - \hat{\Gamma}_L e^{j2\beta(z-L)} = 0.181007 e^{-j63.6357°} \tag{7.266}$$

Thus, the voltage one wavelength away from the load is

$$\hat{V}(z = 18\,\text{m}) = (46.403722 e^{j52.827°})(e^{-j107.979626°})(1.92646^{j4.8292°}) \tag{7.267}$$

or

$$\hat{V}(z = 18\,\text{m}) = 89.394914\angle - 50.3234°\,\text{V} \tag{7.268}$$

The current one wavelength away from the load is

$$\hat{I}(z = 18\,\text{m}) = \frac{(46.403722 e^{j52.827°})}{50}(e^{-j107.979626°})(0.181007 e^{-j63.6357°}) \tag{7.269}$$

or

$$\hat{I}(z = 18\,\text{m}) = 0.167988\angle - 118.7453°\,\text{A} \tag{7.270}$$

Let's verify these results using Circuit Model 2. Voltage and current can be obtained from Eqs. (7.247) and (7.248), repeated here:

$$\hat{V}(d) = \hat{V}_d^+[e^{j\beta d} + \hat{\Gamma}_L e^{-j\beta d}] \tag{7.271}$$

$$\hat{I}(d) = \frac{\hat{V}_d^+}{Z_C}[e^{j\beta d} - \hat{\Gamma}_L e^{-j\beta d}] \tag{7.272}$$

One wavelength away from the source we have $d = 60\,\text{m}$. Thus,

$$\beta d = (0.1047)(60) = 6.282\,\text{rad} = -0.06791° \tag{7.273}$$

$$\hat{\Gamma}_L e^{-j\beta d} = (0.933809 e^{j9.86581°})(e^{j0.06791°}) = 0.933809 e^{j9.9337°} \tag{7.274}$$

$$e^{j\beta d} + \hat{\Gamma}_L e^{-j\beta d} = 1.92655 e^{j4.79642°} \tag{7.275}$$

$$e^{j\beta d} - \hat{\Gamma}_L e^{-j\beta d} = 0.179946 e^{-j63.536°} \tag{7.276}$$

Thus, the voltage one wavelength away from the load is

$$\hat{V}(d = 60\,\mathrm{m}) = (46.403722 e^{-j55.0847°})(1.92655 e^{j4.79642°}) \tag{7.277}$$

or

$$\hat{V}(d = 60\,\mathrm{m}) = 89.3991 \angle -50.2539° \,\mathrm{V} \tag{7.278}$$

The current one wavelength away from the source is

$$\hat{I}(d = 60\,\mathrm{m}) = \frac{(46.403722 e^{-j55.084713°})}{50}(0.179946 e^{-j63.536°}) \tag{7.279}$$

or

$$\hat{I}(d = 60\,\mathrm{m}) = 0.167003 \angle -118.5863° \,\mathrm{A} \tag{7.280}$$

These results are consistent with the ones obtained using Circuit Model 1, and given by Eqs. (7.268) and (7.270).

7.2 Smith Chart and Input Impedance to the Transmission Line

This section explains the creation of the Smith Chart, like the one shown in Figure 7.13 [Wikipidia, 2023].

Smith Chart is a graphical tool for designing transmission lines and RF matching circuits. It is based on a plot of the load reflection coefficient, as explained next.

7.2.1 Smith Chart Fundamentals

Recall a typical lossless transmission line model, shown in Figure 7.14.

In this model, the source is located at $z = 0$, and the load is located at $z = L$. In some applications it is convenient to use an alternative model, shown in Figure 7.15.

In this model, the load is located at $d = 0$, and the source is located at $d = L$. Note that, in either model, the input impedance to the line at any location is always calculated looking toward the load.

The load reflection coefficient, in either model, can be obtained directly from the knowledge of the load and the characteristic impedance of the line as

$$\hat{\Gamma}_L = \frac{\hat{Z}_L - Z_C}{\hat{Z}_L + Z_C} \tag{7.281}$$

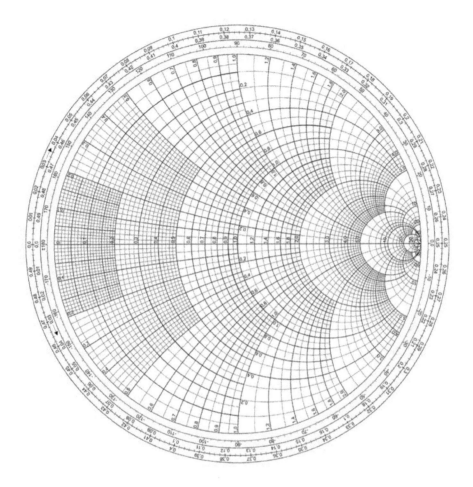

Figure 7.13   Sample Smith Chart.

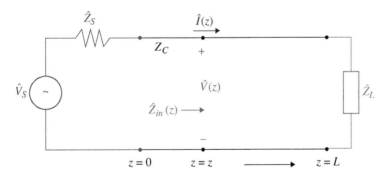

Figure 7.14   Transmission line Circuit Model 1.

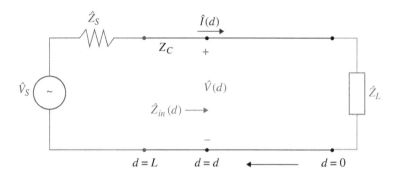

Figure 7.15 Transmission line Circuit Model 2.

Recall three special cases of the load reflection coefficient:

1. *Short-circuited line, $\hat{Z}_L = 0$*

$$\hat{\Gamma}_L = \frac{\hat{Z}_L - Z_C}{\hat{Z}_L + Z_C} = \frac{0 - Z_c}{0 + Z_c} = -1 \tag{7.282}$$

2. *Open-circuited line, $\hat{Z}_L = \infty$*

$$\hat{\Gamma}_L = \frac{\hat{Z}_L - Z_C}{\hat{Z}_L + Z_C} = \left(\frac{1 - \frac{Z_c}{\hat{Z}_L}}{1 + \frac{Z_c}{\hat{Z}_L}} \right)_{\hat{Z}_L = \infty} = 1 \tag{7.283}$$

3. *Matched line, $\hat{Z}_L = Z_C$*

$$\hat{\Gamma}_L = \frac{\hat{Z}_L - Z_C}{\hat{Z}_L + Z_C} = \frac{Z_C - Z_c}{Z_C + Z_c} = 0 \tag{7.284}$$

We will refer to these special cases shortly. Being a complex quantity, the load reflection coefficient can be expressed either in polar or rectangular form as

$$\hat{\Gamma}_L = \Gamma e^{j\theta} = \Gamma_r + j\Gamma_i \tag{7.285}$$

If we create a complex plane with a horizontal axis Γ_r, and a vertical axis Γ_i, then the load reflection coefficient will correspond to a unique point on that plane, as shown in Figure 7.16.

Magnitude of the load reflection coefficient is plotted as a directed line segment from the center of the plane. Angle is measured counterclockwise from the right-hand side (RHS) of the horizontal Γ_r, axis.

For passive loads, the magnitude of the load reflection coefficient is always

$$0 \leq \Gamma_L \leq 1 \tag{7.286}$$

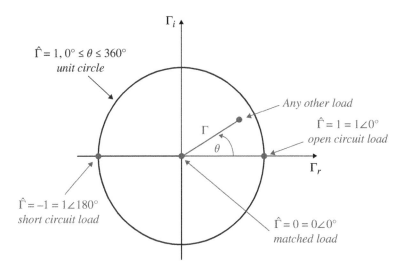

Figure 7.16   Load reflection coefficient and the complex Γ plane.

The magnitude of 0 (center of the complex plane) corresponds to a matched load, the magnitude of 1 with the angle of 0° represents an open circuit, while the magnitude of 1 with the angle of 180° represents a short circuit. Shown in Figure 7.16 is also a unit circle, which sets the boundary for all the points representing a passive load reflection coefficient.

Let's return to the load reflection coefficient given by Eq. (7.281) and divide the numerator and denominator by the characteristic impedance of the line, Z_C.

$$\hat{\Gamma}_L = \frac{\hat{Z}_L - Z_C}{\hat{Z}_L + Z_C} = \frac{\dfrac{\hat{Z}_L}{Z_C} - \dfrac{Z_C}{Z_C}}{\dfrac{\hat{Z}_L}{Z_C} + \dfrac{Z_C}{Z_C}} = \frac{\dfrac{\hat{Z}_L}{Z_C} - 1}{\dfrac{\hat{Z}_L}{Z_C} + 1} \tag{7.287}$$

or

$$\hat{\Gamma}_L = \frac{\hat{z}_L - 1}{\hat{z}_L + 1} \tag{7.288}$$

where the small letter $\hat{z}_L$ denotes the normalized load impedance, i.e. $\hat{z}_L = \hat{Z}_L/Z_C$. Smith Chart is based on the plot of the load reflection coefficient utilizing this normalized load impedance, as explained next.

Equation (7.288) expresses the load reflection coefficient in terms of the normalized load impedance. This equation can be solved for the normalized load impedance in terms of the load reflection coefficient as follows.

Multiplying both sides by the denominator of the RHS, we obtain

$$\hat{\Gamma}_L(\hat{z}_L + 1) = \hat{z}_L - 1 \tag{7.289}$$

or

$$\hat{z}_L \hat{\Gamma}_L + \hat{\Gamma}_L = \hat{z}_L - 1 \tag{7.290}$$

and thus

$$\hat{\Gamma}_L + 1 = \hat{z}_L - \hat{z}_L\hat{\Gamma}_L \qquad (7.291)$$

leading to

$$\hat{\Gamma}_L + 1 = \hat{z}_L(1 - \hat{\Gamma}_L) \qquad (7.292)$$

and finally

$$\hat{z}_L = \frac{1 + \hat{\Gamma}_L}{1 - \hat{\Gamma}_L} \qquad (7.293)$$

Both the normalized load impedance and the load reflection coefficient are complex quantities and as such they can be expressed in terms of their real and imaginary parts. That is,

$$\hat{\Gamma}_L = \Gamma_r + j\Gamma_i \qquad (7.294)$$

$$\hat{z}_L = r_L + jx_L \qquad (7.295)$$

where r_L is the normalized load resistance and x_L is the normalized load reactance.

Utilizing Eqs. (7.294) and (7.295) in Eq. (7.293) we obtain

$$r_L + jx_L = \frac{(1 + \Gamma_r) + j\Gamma_i}{(1 - \Gamma_r) - j\Gamma_i} \qquad (7.296)$$

For now, let's focus on the RHS of Eq. (7.296). Multiplying numerator and denominator by the complex conjugate of the denominator we get

$$RHS = \frac{(1 + \Gamma_r) + j\Gamma_i}{(1 - \Gamma_r) - j\Gamma_i} \times \frac{(1 - \Gamma_r) + j\Gamma_i}{(1 - \Gamma_r) + j\Gamma_i} \qquad (7.297)$$

or

$$RHS = \frac{[(1 + \Gamma_r) + j\Gamma_i][(1 - \Gamma_r) + j\Gamma_i]}{(1 - \Gamma_r)^2 + \Gamma_i^2} \qquad (7.298)$$

Multiplying out the terms in the numerator and grouping the real and imaginary parts gives

$$RHS = \frac{1 - \Gamma_r^2 - \Gamma_i^2 + j(\Gamma_i + \Gamma_r\Gamma_i + \Gamma_i - \Gamma_i\Gamma_r)}{(1 - \Gamma_r)^2 + \Gamma_i^2} \qquad (7.299)$$

or

$$RHS = \frac{1 - \Gamma_r^2 - \Gamma_i^2}{(1 - \Gamma_r)^2 + \Gamma_i^2} + j\frac{2\Gamma_i}{(1 - \Gamma_r)^2 + \Gamma_i^2} \qquad (7.300)$$

This result can be substituted for the RHS of Eq. (7.296) resulting in

$$r_L + jx_L = \frac{1 - \Gamma_r^2 - \Gamma_i^2}{(1 - \Gamma_r)^2 + \Gamma_i^2} + j\frac{2\Gamma_i}{(1 - \Gamma_r)^2 + \Gamma_i^2} \qquad (7.301)$$

Equating the real and imaginary parts we get

$$r_L = \frac{1 - \Gamma_r^2 - \Gamma_i^2}{(1 - \Gamma_r)^2 + \Gamma_i^2} \tag{7.302}$$

$$x_L = \frac{2\Gamma_i}{(1 - \Gamma_r)^2 + \Gamma_i^2} \tag{7.303}$$

For now, let's focus on Eq. (7.302). Multiplying both sides by the denominator of the RHS, we obtain

$$r_L[(1 - \Gamma_r)^2 + \Gamma_i^2] = 1 - \Gamma_r^2 - \Gamma_i^2 \tag{7.304}$$

or

$$r_L(1 - 2\Gamma_r + \Gamma_r^2 + \Gamma_i^2) = 1 - \Gamma_r^2 - \Gamma_i^2 \tag{7.305}$$

Multiplying out and rearranging we get

$$r_L\Gamma_r^2 + \Gamma_r^2 - 2r_L\Gamma_r + r_L\Gamma_i^2 + \Gamma_i^2 = 1 - r_L \tag{7.306}$$

or

$$\Gamma_r^2(r_L + 1) - 2r_L\Gamma_r + \Gamma_i^2(r_L + 1) = 1 - r_L \tag{7.307}$$

Dividing both sides by $(1 + r_L)$ we have

$$\Gamma_r^2 - \frac{2r_L\Gamma_r}{1 + r_L} + \Gamma_i^2 = \frac{1 - r_L}{1 + r_L} \tag{7.308}$$

Adding $\left(\frac{r_L}{1+r_L}\right)^2$ to both sides results in

$$\Gamma_r^2 - \frac{2r_L\Gamma_r}{1 + r_L} + \left(\frac{r_L}{1 + r_L}\right)^2 + \Gamma_i^2 = \frac{1 - r_L}{1 + r_L} + \left(\frac{r_L}{1 + r_L}\right)^2 \tag{7.309}$$

leading to

$$\left(\Gamma_r - \frac{r_L}{1 + r_L}\right)^2 + \Gamma_i^2 = \frac{(1 - r_L)(1 + r_L) + r_L^2}{(1 + r_L)^2} \tag{7.310}$$

or

$$\left(\Gamma_r - \frac{r_L}{1 + r_L}\right)^2 + \Gamma_i^2 = \frac{1 + r_L - r_L - r_L^2 + r_L^2}{(1 + r_L)^2} \tag{7.311}$$

and finally

$$\left(\Gamma_r - \frac{r_L}{1 + r_L}\right)^2 + \Gamma_i^2 = \left(\frac{1}{1 + r_L}\right)^2 \tag{7.312}$$

Now, let's look at Eq. (7.303). Multiplying both sides by the denominator of the RHS, we obtain

$$x_L[(1 - \Gamma_r)^2 + \Gamma_i^2] = 2\Gamma_i \tag{7.313}$$

or

$$x_L(1 - 2\Gamma_r + \Gamma_r^2 + \Gamma_i^2) = 2\Gamma_i \tag{7.314}$$

Multiplying out and rearranging we get

$$x_L\Gamma_r^2 - x_L 2\Gamma_r + x_L\Gamma_i^2 - 2\Gamma_i = -x_L \tag{7.315}$$

Dividing both sides by x_L we have

$$\Gamma_r^2 - 2\Gamma_r + \Gamma_i^2 - \frac{2\Gamma_i}{x_L} = -1 \tag{7.316}$$

Adding $\left(1 + \frac{1}{x_L^2}\right)^2$ to both sides results in

$$\Gamma_r^2 - 2\Gamma_r + 1 + \Gamma_i^2 - \frac{2\Gamma_i}{x_L} + \frac{1}{x_L^2} = -1 + 1 + \frac{1}{x_L^2} \tag{7.317}$$

leading to

$$(\Gamma_r - 1)^2 = \left(\Gamma_i - \frac{1}{x_L}\right)^2 = \left(\frac{1}{x_L}\right)^2 \tag{7.318}$$

Equations (7.312) and (7.318) have the form of

$$(x - h)^2 + (y - k)^2 = b^2 \tag{7.319}$$

which is an equation describing a circle in the xy plane. This circle has a radius b and is centered at $(x, y) = (h, k)$.

Thus, Eq. (7.312), repeated here

$$\left(\Gamma_r - \frac{r_L}{1 + r_L}\right)^2 + \Gamma_i^2 = \left(\frac{1}{1 + r_L}\right)^2 \tag{7.320}$$

describes a circle in the $\Gamma_r\Gamma_i$ plane. This circle has a radius

$$radius = \frac{1}{1 + r_L} \tag{7.321}$$

and is centered at

$$(\Gamma_r, \Gamma_i) = \left(\frac{r_L}{1 + r_L}, 0\right) \tag{7.322}$$

We refer to this circle as the *resistance circle*.

Similarly, Eq. (7.318), repeated here

$$(\Gamma_r - 1)^2 = \left(\Gamma_i - \frac{1}{x_L}\right)^2 = \left(\frac{1}{x_L}\right)^2 \tag{7.323}$$

describes a circle in the $\Gamma_r\Gamma_i$ plane. This circle has a radius

$$radius = \frac{1}{x_L} \tag{7.324}$$

and is centered at

$$(\Gamma_r, \Gamma_i) = \left(1, \frac{1}{x_L}\right) \tag{7.325}$$

We refer to this circle as the *reactance circle*.

7.2.1.1 Resistance Circles

Let us calculate the radii and centers of the resistance circles for typical values of the normalized resistance r_L [Ulaby, 1999]; this is shown in Figure 7.17.

Figure 7.18 shows the plots of these circles on the complex Γ plane.

Observations: All circles pass through the point $(\Gamma_r, \Gamma_i) = (1,0)$. The largest circle is for $r_L = 0$ (which is the unit circle corresponding to $\Gamma = 1$). All circles lie within the bounds of $\Gamma = 1$ unit circle.

Let us identify some of these circles on the actual Smith Chart; this is shown in Figure 7.19.

The values of the normalized resistances corresponding to these circles are shown on the horizontal axis (and other places) of the Smith Chart as shown in Figure 7.20.

7.2.1.2 Reactance Circles

Let us calculate the radii and centers of the reactance circles for typical values of the normalized reactance x_L [Ulaby, 1999]. Note that unlike the normalized resistance (which is always non-negative), the normalized reactance can be positive (inductive load) or negative (capacitive load). Thus, Eq. (7.323) represents two families of circles, as shown in Figures 7.21 and 7.22.

Of interest to us is the part of a given reactance circle that falls within the bounds of $\Gamma = 1$ unit circle.

Normalized resistance (r_L)	Radius $\left(\dfrac{1}{1+r_L}\right)$	Center $\left(\dfrac{r_L}{1+r_L}, 0\right)$
0	1	(0,0)
1/2	2/3	(1/3, 0)
1	1/2	(1/2, 0)
2	1/3	(2/3, 0)
5	1/6	(5/6, 0)
∞	0	(1, 0)

Figure 7.17   Radii and centers of resistance circles.

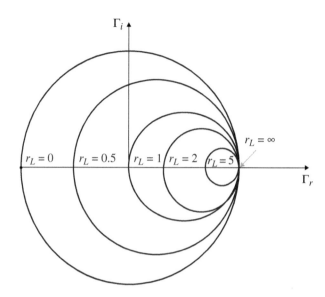

Figure 7.18   Typical *r* circles.

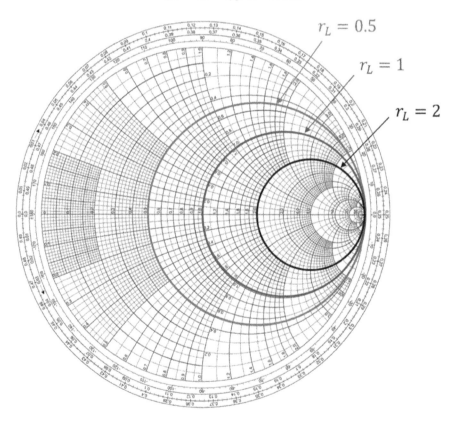

Figure 7.19   Selected *r*-circles on the Smith Chart.

Figure 7.20   Normalized resistance values on the Smith Chart.

Normalized resistance (x_L)	Radius $\left(\frac{1}{x_L}\right)$	Center $\left(1, \frac{1}{x_L}\right)$
0	∞	(1, ∞)
±1/2	2	(1, ±2)
±1	1	(1, ±1)
±2	1/2	(1, ±1/2)
±5	1/5	(1, ±1/5)
±∞	0	(1, 0)

Figure 7.21   Radii and Centers of x-Circles.

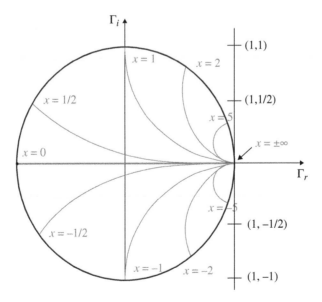

Figure 7.22   Typical *x*-circles.

Observations: The centers of all the reactance circles lie on the vertical $\Gamma_r = 1$ line. All reactance circles also pass through the $(\Gamma_r, \Gamma_i) = (1,0)$ point (just like the r_L circles).

Let us identify some of these partial circles on the actual Smith Chart; this is shown in Figure 7.23.

The values of the normalized reactances corresponding to these partial circles are shown on the perimeter circle (and other places) of the Smith Chart, as shown in Figure 7.24.

When we superimpose the resistance and reactance circles onto each other, we obtain Smith Chart shown in Figure 7.13. The intersection of any *r*-circle with any *x*-circle corresponds to a normalized load impedance, as shown in Figure 7.25.

7.2.2 Input Impedance to the Transmission Line

Let's return to the transmission line Circuit Model 2, shown in Figure 7.26.

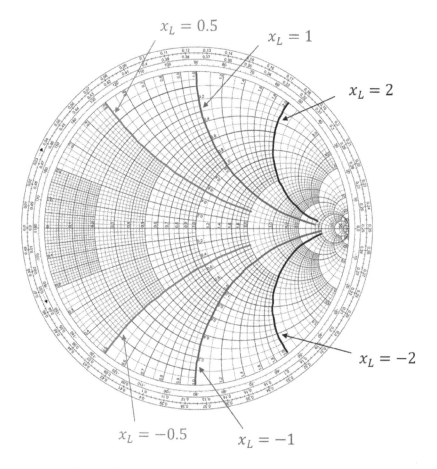

Figure 7.23 Parts of the selected *x*-circles on the Smith Chart.

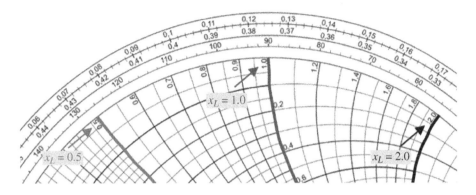

Figure 7.24 Normalized reactance values on the Smith Chart.

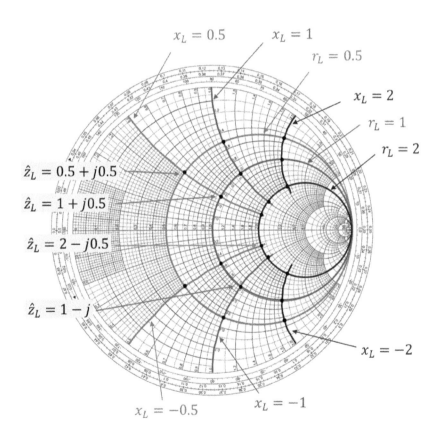

Figure 7.25 Normalized load impedance on the Smith Chart.

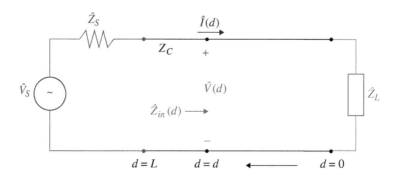

Figure 7.26   Transmission line Circuit Model 2.

The input impedance to the line at any location d away from the load can be obtained from Eq. (7.156), repeated here

$$\hat{Z}_{in}(d) = Z_C \frac{1 + \hat{\Gamma}(d)}{1 - \hat{\Gamma}(d)} \tag{7.326}$$

where $\hat{\Gamma}(d)$ is the voltage reflection coefficient at any location d, away from the load, and can be expressed in terms of the load reflection coefficient by (7.122) repeated here

$$\hat{\Gamma}(d) = \hat{\Gamma}_L e^{-j2\beta d} \tag{7.327}$$

The load reflection coefficient can be expressed in terms of its magnitude and angle as

$$\hat{\Gamma}_L = \Gamma_L e^{-j\theta} \tag{7.328}$$

Utilizing Eq. (7.328) in Eq. (7.327) we obtain

$$\hat{\Gamma}(d) = \hat{\Gamma}_L e^{j\theta} e^{-j2\beta d} = \hat{\Gamma}_L e^{j(\theta - 2\beta d)} \tag{7.329}$$

We refer to $\hat{\Gamma}(d)$ as the phase-shifted load reflection coefficient. Note that the phase-shifted load reflection coefficient has the same magnitude as the load reflection coefficient, but the phase of $\hat{\Gamma}(d)$ is shifted by $2\beta d$ relative to the phase of $\hat{\Gamma}_L$.

On the Smith Chart, obtaining $\hat{\Gamma}(d)$ from $\hat{\Gamma}_L$ means keeping the magnitude, Γ_L, constant and decreasing the phase by $2\beta d$. Phase decrease corresponds to the clockwise rotation on the Smith Chart. This is shown in Figure 7.27.

Let's return to Eq. (7.326). Dividing both sides by the characteristic impedance of the line, Z_C, we obtain the *normalized* input impedance to the line at any location d away from the load,

$$\hat{z}_{in}(d) = \frac{\hat{Z}_{in}(d)}{Z_C} \tag{7.330}$$

or

$$\hat{z}_{in}(d) = \frac{1 + \hat{\Gamma}(d)}{1 - \hat{\Gamma}(d)} \tag{7.331}$$

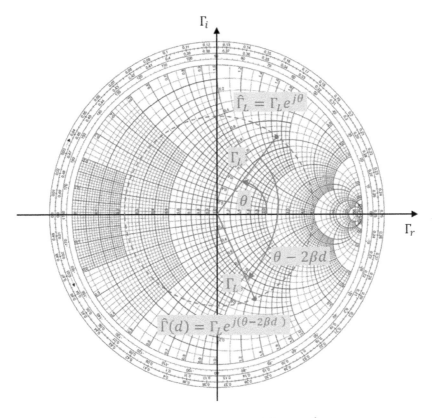

Figure 7.27   Transformation of $\hat{\Gamma}_L$ into $\hat{\Gamma}(d)$.

Let's compare this expression with the one for normalized load impedance, given by Eq. (7.293), repeated here

$$\hat{z}_L = \frac{1 + \hat{\Gamma}_L}{1 - \hat{\Gamma}_L} \tag{7.332}$$

The mathematical form of these two equations is the same! If $\hat{\Gamma}_L$ is replaced by $\hat{\Gamma}(d)$ then $\hat{z}_L$ gets replaced by $\hat{z}_{in}(d)$.

On the Smith Chart obtaining $\hat{z}_{in}(d)$ from $\hat{z}_L$ means keeping the magnitude of $\hat{z}_L$ constant and decreasing the phase by $2\beta d$. This is shown in Figure 7.28.

When using Smith Chart, the distance d is expressed in terms of wavelengths. A complete rotation around the Smith Chart corresponds to a phase change of $360°$ or 2π radians.

$$2\beta d = 2\pi \tag{7.333}$$

The phase constant β is related to the wavelength λ by

$$\beta = \frac{2\pi}{\lambda} \tag{7.334}$$

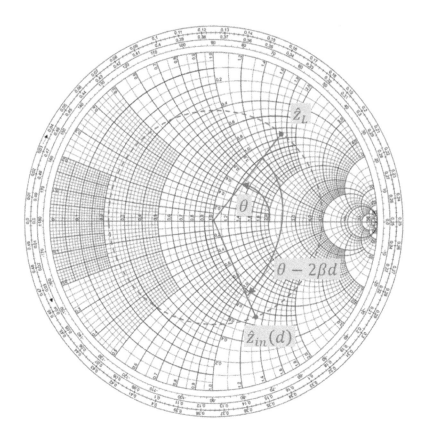

Figure 7.28   Transformation of $\hat{z}_L$ into $\hat{z}_{in}(d)$.

Thus,

$$\frac{2\pi}{\lambda}d = \pi \tag{7.335}$$

from which

$$d = \frac{\lambda}{2} \tag{7.336}$$

This means that one complete rotation around the Smith Chart corresponds to the distance change of $\frac{\lambda}{2}$. The rotation around Smith Chart, either in degrees or wavelengths, is denoted on the outer three scales around its perimeter, as shown in Figure 7.29.

We have shown how to obtain $\hat{z}_{in}(d)$, the normalized input impedance to the line at a distance d away from the load. What about $\hat{z}_{in}(z)$, the normalized input impedance to the line at a distance z away from the source?

$\hat{z}_{in}(z)$ can be easily obtained from $\hat{z}_{in}(d)$ since the two distance variables are related by,

$$d = L - z \tag{7.337}$$

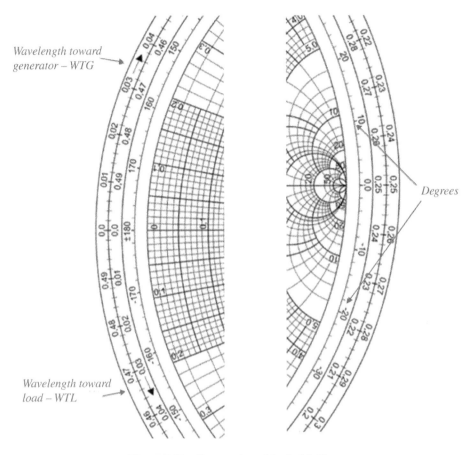

Figure 7.29   Outer scales of the Smith Chart.

Thus, if we are given z, the distance from the source, we can calculate d, the corresponding distance from the load, and use the procedure just described to obtain the input impedance to the line at that location.

In summary, to determine the input impedance to the line at any distance from the load or the source, using Smith Chart, the following steps need to be taken:

1. Calculate the load reflection coefficient

$$\Gamma_L = \frac{\hat{Z}_L - Z_C}{\hat{Z}_L + Z_C} \tag{7.338}$$

2. Obtain the normalized load impedance

$$\hat{z}_L = \frac{1 + \hat{\Gamma}_L}{1 - \hat{\Gamma}_L} = r_L + jx_L \tag{7.339}$$

3. Locate this impedance on the Smith Chart at the intersection of the resistance and reactance circles.
4. Calculate the distance from the load, d, or the source, z, in terms of wavelengths.
5. To obtain the (normalized) input impedance to the line at a distance d from the load, $\hat{z}_{in}(d)$, move clockwise on the constant radius circle from z_L.
6. Multiply the normalized input impedance by the characteristic impedance of the line, Z_C, to obtain the actual input impedance

$$\hat{Z}_{in}(d) = Z_C \hat{z}_{in}(d) \tag{7.340}$$

7.3 Standing Waves and VSWR

Consider the transmission line circuit shown in Figure 7.30 [Adamczyk, 2017b].

A sinusoidal voltage source $\hat{V}_S$ with its source impedance $\hat{Z}_S$ drives a lossless transmission line with characteristic impedance Z_C, terminated in a resistive load R_L.

The magnitudes of voltage and current along the line at any distance d away from the load are [Adamczyk, 2017a],

$$\left|\hat{V}(d)\right| = \left|\hat{V}^+\right|\left|\left[1 + \hat{\Gamma}_L e^{-j2\beta d}\right]\right| \tag{7.341}$$

$$\left|\hat{I}(d)\right| = \left|\frac{\hat{V}^+}{Z_C}\right|\left|\left[1 - \hat{\Gamma}_L e^{-j2\beta d}\right]\right| \tag{7.342}$$

where $|\hat{V}(d)|$ denotes the amplitude of the sinusoidal voltage wave, β is the phase constant of the wave, and $\hat{\Gamma}_L$ is the load reflection coefficient given by

$$\hat{\Gamma}_L = \frac{\hat{Z}_L - Z_C}{\hat{Z}_L + Z_C} \tag{7.343}$$

In the circuit shown in Figure 7.30, we have $\hat{Z}_L = R_L$

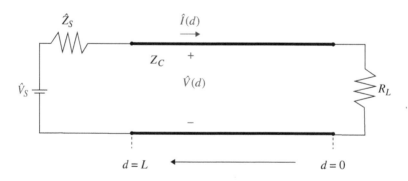

Figure 7.30 Transmission line driven by a sinusoidal source.

There are four important cases of special interest that we will investigate:

1. The load is a short circuit, $\hat{Z}_L = R_L = 0$
2. The load is an open circuit, $\hat{Z}_L = R_L = \infty$
3. The load is matched, $\hat{Z}_L = R_L = Z_C$
4. Arbitrary resistive load R

Case 1: Short-circuited load $\hat{Z}_L = R_L = 0$

The load reflection coefficient in this case is

$$\hat{\Gamma}_L = -1 \tag{7.344}$$

In this case, Eqs. (7.341) and (7.342) reduce to [Adamczyk, 2017a],

$$\left|\hat{V}(d)\right| = 2\left|\hat{V}^+\right||\sin(\beta d)| \tag{7.345}$$

$$\left|\hat{V}(d)\right| = 2\left|\frac{\hat{V}^+}{Z_C}\right||\cos(\beta d)| \tag{7.346}$$

The phase constant β can be expressed in terms of the wavelength λ as

$$\beta = \frac{2\pi}{\lambda} \tag{7.347}$$

Substituting Eq. (7.347) into Eqs. (7.345) and (7.346) we get

$$\left|\hat{V}(d)\right| = 2\left|\hat{V}^+\right|\left|\sin(2\pi\frac{d}{\lambda})\right| \tag{7.348}$$

$$\left|\hat{V}(d)\right| = 2\left|\frac{\hat{V}^+}{Z_C}\right|\left|\cos(2\pi\frac{d}{\lambda})\right| \tag{7.349}$$

The magnitudes of the voltage and current waves for a short-circuited load are shown in Figure 7.31.

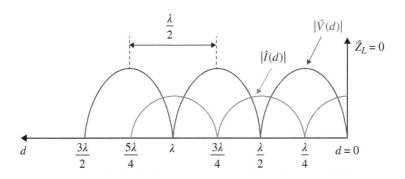

Figure 7.31 Magnitudes of the voltage and current for a short-circuited load.

We observe the following:

1. The voltage is zero at the load and at distances from the load which are multiples of a half wavelength.
2. The current is maximum at the load and is zero at distances from the load that are odd multiples of a quarter wavelength.
3. The corresponding points are separated by one-half wavelength.

Case 2: Open-circuited load $\hat{Z}_L = \infty$

The load reflection coefficient in this case is

$$\hat{\Gamma}_L = 1 \tag{7.350}$$

In this case, Eqs. (7.341) and (7.342) reduce to [Adamczyk, 2017a],

$$\left|\hat{V}(d)\right| = 2\left|\hat{V}^+\right|\left|\cos(2\pi\frac{d}{\lambda})\right| \tag{7.351}$$

$$\left|\hat{V}(d)\right| = 2\left|\frac{\hat{V}^+}{Z_C}\right|\left|\sin(2\pi\frac{d}{\lambda})\right| \tag{7.352}$$

The magnitudes of the voltage and current waves for an open-circuited load are shown in Figure 7.32.
We observe the following:

1. The current is zero at the load and at distances from the load which are multiples of a half wavelength.
2. The voltage is maximum at the load and is zero at distances from the load that are odd multiples of a quarter wavelength.
3. The corresponding points are separated by one-half wavelength.

In both cases, the voltage and current waves do not travel as the time advances, but stay where they are, only oscillating in time between the stationary zeros. In other words, they do not represent a traveling wave in either direction. The resulting wave is termed a *standing wave*.

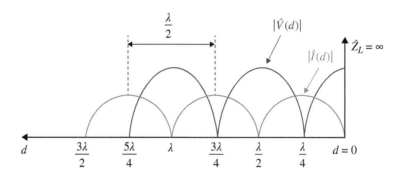

Figure 7.32   Magnitudes of the voltage and current for an open-circuited load.

Case 3: Matched load $\hat{Z}_L = Z_C$

The load reflection coefficient in this case is

$$\hat{\Gamma}_L = 0 \tag{7.353}$$

In this case, Eqs. (7.341) and (7.342) reduce to [Adamczyk, 2017a],

$$\left|\hat{V}(d)\right| = \left|\hat{V}^+\right| \tag{7.354}$$

$$\left|\hat{V}(d)\right| = \left|\frac{\hat{V}^+}{Z_C}\right| \tag{7.355}$$

The magnitudes of the voltage and current waves for a matched load are shown in Figure 7.33.

We observe that the voltage and current magnitudes are constant along the line.

Case 4: Arbitrary resistive load R

The magnitudes of the voltage and current waves for an arbitrary load given by Eqs. (7.341) and (7.342) are shown in Figure 7.34.

We observe that the locations of the voltage and current maxima and minima are determined by the actual load impedance, but again, adjacent, corresponding points on each waveform are separated by one-half-wavelength.

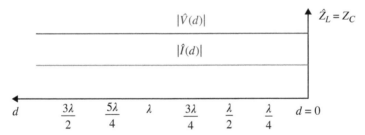

Figure 7.33 Magnitudes of the voltage and current for a matched load.

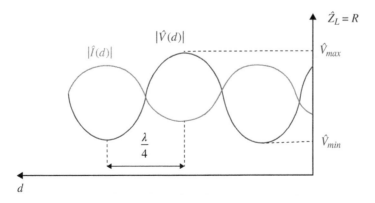

Figure 7.34 Magnitudes of the voltage and current for an arbitrary load.

In all cases, except for the matched load, the magnitudes of the voltage and current vary along the line. This variation is quantitatively described by the *voltage standing wave ratio (VSWR)* defined as

$$VSWR = \left| \frac{\hat{V}_{max}}{\hat{V}_{min}} \right| \tag{7.356}$$

VSWR can also be expressed in terms of the magnitude of the load reflection coefficient as

$$VSWR = \frac{1 + \left| \hat{\Gamma}_L \right|}{1 - \left| \hat{\Gamma}_L \right|} \tag{7.357}$$

when the load is short-circuited or open-circuited, we have

$$VSWR = \infty, \quad \hat{Z}_L = 0 \tag{7.358}$$

$$VSWR = \infty, \quad \hat{Z}_L = \infty \tag{7.359}$$

When the load is matched, we have

$$VSWR = 1, \quad \hat{Z}_L = Z_C \tag{7.360}$$

In general,

$$1 \leq VSWR \leq \infty \tag{7.361}$$

We will return to the VSWR in Section 8.11.1, when measuring the impedance of a log periodic antenna.

7.4 Laboratory Exercises

7.4.1 Input Impedance to Transmission Line – Smith Chart

Objective: This laboratory exercise uses Smith Chart to determine the input impedance to the transmission line at different distance from the source and the load.

7.4.2 Laboratory Procedure – Smith Chart

Consider the transmission line circuit shown in Figure 7.12, and repeated here in Figure 7.35.
Use Smith Chart to determine:

(a) Input impedance to the line at the input to the line.
(b) Input impedance to the line one wavelength away from the source.
(c) Input impedance to the line one wavelength away from the load.

Compare the solutions to the ones obtained analytically in Section 7.1.2.

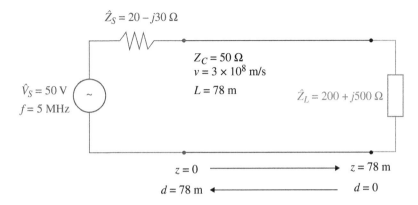

$\hat{Z}_S = 20 - j30\ \Omega$

$Z_C = 50\ \Omega$
$v = 3 \times 10^8$ m/s
$L = 78$ m

$\hat{V}_S = 50$ V
$f = 5$ MHz

$\hat{Z}_L = 200 + j500\ \Omega$

$z = 0$ $z = 78$ m
$d = 78$ m $d = 0$

Figure 7.35   Transmission line circuit.

References

Bogdan Adamczyk. *Foundations of Electromagnetic Compatibility*. John Wiley & Sons, Ltd, Chichester, UK, 2017a. ISBN 9781119120810. doi: https://doi.org/10.1002/9781119120810.

Bogdan Adamczyk. Standing Waves on Transmission Lines and VSWR Measurements. *In Compliance Magazine*, November 2017b. URL https://incompliancemag.com/article/standing-waves-on-transmission-lines-and-vswr-measurements/.

Bogdan Adamczyk. EM Waves, Voltage, and Current Waves. *In Compliance Magazine*, April 2021. URL https://incompliancemag.com/article/em-waves-voltage-and-current-waves/.

Bogdan Adamczyk. Concept of a Phasor in Sinusoidal Steady State Analysis. *In Compliance Magazine*, December 2022. URL https://incompliancemag.com/article/concept-of-a-phasor-in-sinusoidal-steady-state-analysis/.

Bogdan Adamczyk. Sinusoidal Steady State Analysis of Transmission Lines Part I: Transmission Line Model, Equations and Their Solutions, and the Concept of the Input Impedance to the Line. *In Compliance Magazine*, January 2023a. URL https://incompliancemag.com/article/sinusoidal-steady-state-analysis-of-transmission-lines-part-1/.

Bogdan Adamczyk. Sinusoidal Steady State Analysis of Transmission Lines Part II: Voltage, Current, and Input Impedance Calculations –Circuit Model 1. *In Compliance Magazine*, February 2023b. URL https://incompliancemag.com/article/sinusoidal-steady-state-analysis-of-transmission-lines-part2/.

Bogdan Adamczyk. Sinusoidal Steady State Analysis of Transmission Lines Part III: Voltage, Current, and Input Impedance Calculations –Circuit Model 2. *In Compliance Magazine*, March 2023c. URL https://incompliancemag.com/article/sinusoidal-steady-state-analysis-of-transmission-lines-part3/.

Fawwaz T. Ulaby. *Fundamentals of Applied Electromagnetics*. Prentice-Hall, Upper Saddle River, New Jersey, 1999. ISBN 0-13-011554-1.

Wikipidia. Sample Smith Chart, 2023.

Chapter 8: Antennas and Radiation

8.1 Bridge Between the Transmission Line Theory and Antennas

In this section we will use the theory of the standing waves on transmission lines discussed in Chapter 7 to build a bridge between the transmission line theory and the fundamental antenna structure of a dipole antenna. Consider a standing wave pattern in lossless two-wire transmission line terminated in an open, shown in Figure 8.1.

When the incident wave arrives at the open-circuited load, it undergoes a complete reflection. The incident and reflected waves combine to create a pure standing wave pattern as shown in Figure 8.1. The current reflection coefficient at an open-circuited load is -1, and the current in each wire undergoes a 180 degree phase reversal between adjoining half-cycles (this is shown by the reversal of the arrow directions).

The current in a half-cycle of one wire is of the same magnitude but 180 degrees out-of-phase from that in the corresponding half-cycle of the other wire. If the spacing between two wires is very small, $s \ll \lambda$, the fields radiated by the current of each wire are canceled by those of the other. Effectively, there is no radiation from this transmission line.

Now, let's flare the terminal section of the transmission line, as shown in Figure 8.2 [Balanis, 2005].

It is reasonable to assume that the current distribution is essentially unaltered in form in each of the wires of the transmission line. Since the two wires are no longer parallel and close to each other, the radiated fields do not cancel each other and there is a net radiation from the flared section. Continuing the flaring process we arrive at the structure shown in Figure 8.3.

The fields radiated by the two vertical parts of the transmission line will reinforce each other, as long as the total length of the flared section is $0 < l < \lambda$. The maximum radiation (in the direction shown in Figure 8.3, i.e. broadside to this antenna) occurs for the total length of both vertical parts equal to $l = \frac{\lambda}{2}$.

This is shown in Figure 8.4.

The radiating structure shown in Figure 8.4 is referred to as a *half-wave dipole*. Half-wave dipole radiated fields are obtained from the fields of the Hertzian dipole antenna discussed next.

Principles of Electromagnetic Compatibility: Laboratory Exercises and Lectures, First Edition. Bogdan Adamczyk.
© 2024 John Wiley & Sons Ltd. Published 2024 by John Wiley & Sons Ltd.
Companion website: www.wiley.com/go/principlesofelectromagneticcompatibility

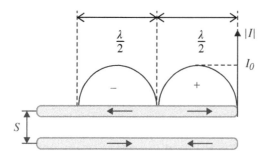

Figure 8.1 Standing wave pattern in a transmission line terminated in an open.

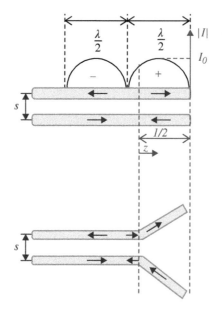

Figure 8.2 Transmission line with terminal section flared.

8.2 Electric (Hertzian) Dipole Antenna

Hertzian dipole antenna is a fictitious (infinitely small) antenna that serves as a building block in the study of more complex antennas and their properties. We will briefly introduce a Hertzian dipole, and use it to study several antenna topics relevant to EMC: the half-wave dipole and quarter-wave monopole antennas, antenna arrays, log-periodic antenna, as well as the radiation from the common-mode and differential-mode currents (discussed in Chapter 9).

Hertzian dipole antenna is shown in Figure 8.5.

Hertzian dipole consists of a short linear wire carrying a sinusoidal current $\hat{I}_0$. The length of the wire l, $l \ll \lambda$, where λ is the wavelength corresponding to the frequency of the sinusoidal current. The wire is positioned symmetrically at the origin of the Cartesian coordinate system and oriented along the z axis. A spherical coordinate system is usually

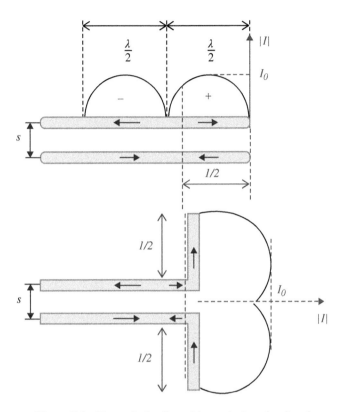

Figure 8.3 Transmission line with terminal section flared.

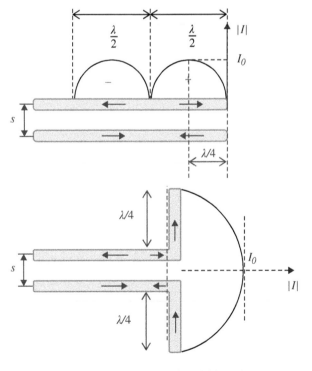

Figure 8.4 Maximum radiation broadside to the antenna.

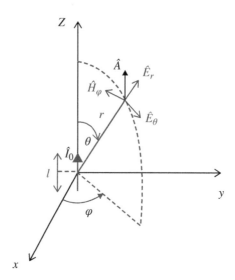

Figure 8.5   Hertzian dipole antenna.

superimposed on the Cartesian system, as it is better suited for the antenna radiation pattern study.

The Hertzian dipole fields at a distance r from the origin can be obtained from the vector magnetic potential **A** shown in Figure 8.5.

The complete fields of the Hertzian dipole were derived in [Adamczyk, 2017] and are given by [Paul, 2006]

$$\hat{E}_r = 2\frac{I_0 l}{4\pi}\eta_0\beta_0^2\cos\theta\left(\frac{1}{\beta_0^2 r^2} - j\frac{1}{\beta_0^3 r^3}\right)e^{-j\beta_0 r} \tag{8.1}$$

$$\hat{E}_\theta = \frac{I_0 l}{4\pi}\eta_0\beta_0^2\sin\theta\left(\frac{1}{\beta_0 r} + \frac{1}{\beta_0^2 r^2} - j\frac{1}{\beta_0^3 r^3}\right)e^{-j\beta_0 r} \tag{8.2}$$

$$\hat{E}_\phi = 0 \tag{8.3}$$

$$\hat{H}_r = 0 \tag{8.4}$$

$$\hat{H}_\theta = 0 \tag{8.5}$$

$$\hat{H}_\phi = \frac{I_0 l}{4\pi}\beta_0^2\sin\theta\left(j\frac{1}{\beta_0 r} + \frac{1}{\beta_0^2 r^2}\right)e^{-j\beta_0 r} \tag{8.6}$$

Hertzian dipole fields described by Eqs. (8.1) through (8.6) apply at any distance r from the antenna. They can be simplified at a "large enough" distance from the antenna. To determine what "large enough" corresponds to consider a positive number x.

$$\text{if } x < 1 \Rightarrow \frac{1}{x} < \frac{1}{x^2} < \frac{1}{x^3} \tag{8.7}$$

$$\text{if } x > 1 \Rightarrow \frac{1}{x} > \frac{1}{x^2} > \frac{1}{x^3} \tag{8.8}$$

Thus, for very small x, $x \ll 1$, the term $1/x^2$ will dominate the term $1/x$, and the term $1/x^3$ will dominate the term $1/x^2$.

On the other hand, for large x, $x \gg 1$, the terms $1/x^2$ and $1/x^3$ will be negligible compared to the term $1/x$.

Now let

$$x = \beta r \tag{8.9}$$

Thus, at a small distance from the antenna (referred to as the *near field*)

$$\beta r < 1 \Rightarrow \frac{1}{\beta r} < \frac{1}{(\beta r)^2} < \frac{1}{(\beta r)^3} \tag{8.10}$$

At a large distance from the antenna (referred to as the *far field*)

$$\beta r > 1 \Rightarrow \frac{1}{\beta r} > \frac{1}{(\beta r)^2} > \frac{1}{(\beta r)^3} \tag{8.11}$$

This leads to the *far-field* expressions for the Hertzian dipole

$$\hat{E}_\theta = j\eta_0\beta_0 \frac{I_0 l}{4\pi} \sin\theta \frac{e^{-j\beta_0 r}}{r} \tag{8.12}$$

$$\hat{H}_\phi = j\beta_0 \frac{I_0 l}{4\pi} \sin\theta \frac{e^{-j\beta_0 r}}{r} \tag{8.13}$$

with the remaining field components being zero (negligible).

8.2.1 Wave Impedance and Far-Field Criterion

With the Hertzian dipole wave described by Eqs. (8.1) through (8.6) we associate *wave impedance*, defined as

$$\hat{Z}_{w,e} = \frac{\hat{E}_\theta}{\hat{H}_\phi} \tag{8.14}$$

Using Eqs. (8.2) and (8.6) in Eq. (8.14) we get

$$\hat{Z}_{w,e} = \frac{\frac{I_0 l}{4\pi}\eta_0\beta_0^2 \sin\theta \left(\frac{1}{\beta_0 r} + \frac{1}{\beta_0^2 r^2} - j\frac{1}{\beta_0^3 r^3} \right) e^{-j\beta_0 r}}{\frac{I_0 l}{4\pi}\beta_0^2 \sin\theta \left(j\frac{1}{\beta_0 r} + \frac{1}{\beta_0^2 r^2} \right) e^{-j\beta_0 r}} \tag{8.15}$$

or

$$\hat{Z}_{w,e} = \frac{\eta_0 \left(\frac{1}{\beta_0 r} + \frac{1}{\beta_0^2 r^2} - j\frac{1}{\beta_0^3 r^3} \right)}{\left(j\frac{1}{\beta_0 r} + \frac{1}{\beta_0^2 r^2} \right)} \tag{8.16}$$

Multiplying the numerator and denominator by $j(\beta_0 r)^3$ we obtain

$$\hat{Z}_{w,e} = \frac{\eta_0 \left(\frac{1}{\beta_0 r} + \frac{1}{\beta_0^2 r^2} - j\frac{1}{\beta_0^3 r^3} \right)}{\left(j\frac{1}{\beta_0 r} + \frac{1}{\beta_0^2 r^2} \right)} \frac{j(\beta_0 r)^3}{j(\beta_0 r)^3} \tag{8.17}$$

or

$$\hat{Z}_{w,e} = \eta_0 \frac{-(\beta_0 r)^2 + j(\beta_0 r) + 1}{-(\beta_0 r)^2 + j(\beta_0 r)} \tag{8.18}$$

Letting

$$\beta_0 = \frac{2\pi}{\lambda_0} \tag{8.19}$$

we obtain

$$\hat{Z}_{w,e} = \frac{\hat{E}_\theta}{\hat{H}_\phi} = \eta_0 \frac{1 - \left(\frac{2\pi r}{\lambda_0}\right)^2 + j\left(\frac{2\pi r}{\lambda_0}\right)}{-\left(\frac{2\pi r}{\lambda_0}\right)^2 + j\left(\frac{2\pi r}{\lambda_0}\right)} \tag{8.20}$$

We can evaluate this expression at different distances (in terms of the wavelength) from the antenna, for instance $r = \frac{\lambda_0}{2\pi}$, $r = \lambda_0$, $r = 3\lambda_0$. When evaluated at $r = 3\lambda_0$, the wave impedance becomes

$$\hat{Z}_{w,e} = 375.93\angle - 0.01° \cong \eta_0 \tag{8.21}$$

The result in Eq. (8.21) leads to the far-field criterion for the Hertzian dipole (and other wire-type antennas):

$$r_{far\,field} = 3\lambda_0 \tag{8.22}$$

8.2.2 Wave Impedance in the Near Field

Let's return to the wave impedance given by Eq. (8.16), repeated here

$$\hat{Z}_{w,e} = \frac{\eta_0 \left(\frac{1}{\beta_0 r} + \frac{1}{\beta_0^2 r^2} - j\frac{1}{\beta_0^3 r^3}\right)}{\left(j\frac{1}{\beta_0 r} + \frac{1}{\beta_0^2 r^2}\right)} \tag{8.23}$$

At small distance from the antenna, $\beta r \ll 1$, the term $1/(\beta r)^2$ will dominate the term $1/(\beta r)$, and the term $1/(\beta r)^3$ will dominate the term $1/(\beta r)^2$.

Thus, the wave impedance in Eq. (8.23) can be approximated by

$$\hat{Z}_{w,e} \approx \eta_0 \frac{-j\frac{1}{\beta_0^3 r^3}}{\frac{1}{\beta_0^2 r^2}} \tag{8.24}$$

or

$$\hat{Z}_{w,e} \approx \eta_0 \left(-\frac{j}{\beta_0 r}\right) \tag{8.25}$$

The magnitude of this wave impedance is

$$|\hat{Z}_{w,e}| \approx \frac{\eta_0}{\beta_0 r} \tag{8.26}$$

In the very near field

$$\beta_0 r \ll 1 \implies |\hat{Z}_{w,e}| \gg \eta_0 \tag{8.27}$$

For that reason, we refer to the electric dipole as a high-impedance source. Since

$$\beta_0 = \frac{2\pi}{\lambda_0} \tag{8.28}$$

we have

$$|\hat{Z}_{w,e}| \approx \frac{\eta_0}{\frac{2\pi}{\lambda_0}r} = \frac{120\pi}{\frac{2\pi}{\lambda_0}r} \tag{8.29}$$

or

$$|\hat{Z}_{w,e}| \approx 60\frac{\lambda_0}{r} \tag{8.30}$$

We will use this expression when discussing near-field shielding for electric sources in Chapter 11.

8.3 Magnetic Dipole Antenna

Magnetic dipole, shown in Figure 8.6, consists of a small thin circular wire loop of radius a, carrying a current I_0 positioned in the xy plane, with the center of the loop at $z = 0$.

The complete fields of the magnetic dipole are given by [Paul, 2006],

$$\hat{E}_r = 0 \tag{8.31}$$

$$\hat{E}_\theta = 0 \tag{8.32}$$

$$\hat{E}_\phi = -j\frac{\omega\mu_0 I_0 a^2 \beta_0^2}{4} \sin\theta \left(j\frac{1}{\beta_0 r} + \frac{1}{\beta_0^2 r^2} \right) e^{-j\beta_0 r} \tag{8.33}$$

$$\hat{H}_r = -j2\frac{\omega\mu_0 I_0 a^2 \beta_0^2}{4\eta_0} \cos\theta \left(\frac{1}{\beta_0^2 r^2} - j\frac{1}{\beta_0^3 r^3} \right) e^{-j\beta_0 r} \tag{8.34}$$

$$\hat{H}_\theta = j\frac{\omega\mu_0 I_0 a^2 \beta_0^2}{4\eta_0} \sin\theta \left(j\frac{1}{\beta_0 r} + \frac{1}{\beta_0^2 r^2} - j\frac{1}{\beta_0^3 r^3} \right) e^{-j\beta_0 r} \tag{8.35}$$

$$\hat{H}_\phi = 0 \tag{8.36}$$

The *far-field* expressions are given by

$$\hat{E}_\phi = \frac{\omega\mu_0 I_0 a^2 \beta_0}{4} \sin\theta \frac{e^{-j\beta_0 r}}{r} \tag{8.37}$$

$$\hat{H}_\theta = -\frac{\omega\mu_0 I_0 a^2 \beta_0}{4\eta_0} \sin\theta \frac{e^{-j\beta_0 r}}{r} \tag{8.38}$$

with the remaining field components being zero (negligible).

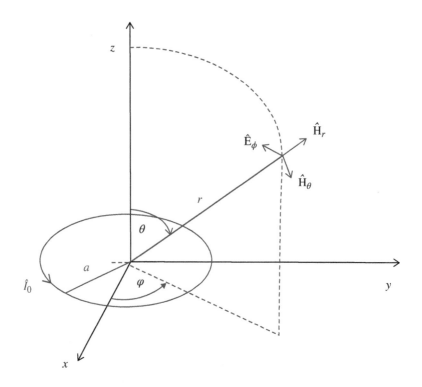

Figure 8.6   Magnetic dipole antenna.

8.3.1 Wave Impedance and Far-Field Criterion

The wave impedance for magnetic dipole is defined as

$$\hat{Z}_{w,m} = \frac{\hat{E}_\phi}{\hat{H}_\theta}$$
(8.39)

Using Eqs. (8.33) and (8.35) in Eq. (8.39) we get

$$\hat{Z}_{w,m} = \frac{-j\frac{\omega\mu_0 I_0 a^2 \beta_0^2}{4}\sin\theta\left(j\frac{1}{\beta_0 r} + \frac{1}{\beta_0^2 r^2}\right)e^{-j\beta_0 r}}{j\frac{\omega\mu_0 I_0 a^2 \beta_0^2}{4\eta_0}\sin\theta\left(j\frac{1}{\beta_0 r} + \frac{1}{\beta_0^2 r^2} - j\frac{1}{\beta_0^3 r^3}\right)e^{-j\beta_0 r}}$$
(8.40)

or

$$\hat{Z}_{w,m} = -\eta_0\frac{\left(j\frac{1}{\beta_0 r} + \frac{1}{\beta_0^2 r^2}\right)}{\left(j\frac{1}{\beta_0 r} + \frac{1}{\beta_0^2 r^2} - j\frac{1}{\beta_0^3 r^3}\right)}$$
(8.41)

Multiplying the numerator and denominator by $j(\beta_0 r)^3$ we obtain

$$\hat{Z}_{w,m} = -\eta_0 \frac{\left(j\frac{1}{\beta_0 r} + \frac{1}{\beta_0^2 r^2}\right)}{\left(j\frac{1}{\beta_0 r} + \frac{1}{\beta_0^2 r^2} - j\frac{1}{\beta_0^3 r^3}\right)} \frac{j(\beta_0 r)^3}{j(\beta_0 r)^3} \tag{8.42}$$

or

$$\hat{Z}_{w,m} = -\eta_0 \frac{(j\beta_0^2 r^2 + \beta_0 r)}{(j\beta_0^2 r^2 + \beta_0 r - j)} \tag{8.43}$$

leading to

$$\hat{Z}_{w,m} = -\eta_0 \frac{\beta_0 r + j\beta_0^2 r^2}{\beta_0 r + j(\beta_0^2 r^2 - 1)} \tag{8.44}$$

Letting

$$\beta_0 = \frac{2\pi}{\lambda_0} \tag{8.45}$$

we obtain

$$\hat{Z}_{w,m} = \frac{\hat{E}_\phi}{\hat{H}_\theta} = -\eta_0 \frac{\left(\frac{2\pi r}{\lambda_0}\right) + j\left(\frac{2\pi r}{\lambda_0}\right)^2}{\left(\frac{2\pi r}{\lambda_0}\right) + j\left[\left(\frac{2\pi r}{\lambda_0}\right)^2 - 1\right]} \tag{8.46}$$

Evaluating Eq. (8.46) at $r = 3\lambda_0$ we get

$$\hat{Z}_{w,m} = -\eta_0 \frac{6\pi + j(6\pi)^2}{6\pi + j[(6\pi)^2 - 1]} - \eta_0 \frac{6\pi + j36\pi^2}{6\pi + j36\pi^2 - 1} \cong -\eta_0 \tag{8.47}$$

Thus, the magnitude of the wave impedance

$$|\hat{Z}_{w,m}| \cong \eta_0 \tag{8.48}$$

We have arrived again at the far-field criterion as

$$r_{far\,field} = 3\lambda_0 \tag{8.49}$$

8.3.2 Wave Impedance in the Near Field

Let's return to the wave impedance given by Eq. (8.41), repeated here

$$\hat{Z}_{w,m} = -\eta_0 \frac{\left(j\frac{1}{\beta_0 r} + \frac{1}{\beta_0^2 r^2}\right)}{\left(j\frac{1}{\beta_0 r} + \frac{1}{\beta_0^2 r^2} - j\frac{1}{\beta_0^3 r^3}\right)} \tag{8.50}$$

At small distance from the antenna, $\beta r \ll 1$, the term $1/(\beta r)^2$ will dominate the term $1/(\beta r)$, and the term $1/(\beta r)^3$ will dominate the term $1/(\beta r)^2$.

Thus, the wave impedance in Eq. (8.50) can be approximated by

$$\hat{Z}_{w,m} \approx -\eta_0 \frac{\frac{1}{\beta_0^2 r^2}}{\frac{1}{-j\beta_0^3 r^3}} \tag{8.51}$$

or

$$\hat{Z}_{w,m} \approx j\eta_0 \beta_0 r \tag{8.52}$$

The magnitude of this wave impedance is

$$|\hat{Z}_{w,m}| \approx \eta_0 \beta_0 r \tag{8.53}$$

In the very near field

$$\beta_0 r \ll 1 \;\Rightarrow\; |\hat{Z}_{w,m}| \ll \eta_0 \tag{8.54}$$

For that reason, we refer to the magnetic dipole as a low-impedance source. Since

$$\beta_0 = \frac{2\pi}{\lambda_0} \tag{8.55}$$

we have

$$|\hat{Z}_{w,m}| \approx (120\pi)\left(\frac{2\pi}{\lambda_0}\right) r = 240\pi^2 \frac{r}{\lambda_0} = 2368.7 \frac{r}{\lambda_0} \tag{8.56}$$

or

$$|\hat{Z}_{w,m}| \approx 2369 \frac{r}{\lambda_0} \tag{8.57}$$

We will use this expression when discussing near-field shielding for magnetic sources in Chapter 11.

8.4 Half-Wave Dipole and Quarter-Wave Monopole Antennas

8.4.1 Half-Wave Dipole Antenna

Half-wave dipole antenna is shown in Figure 8.7.

Half-wave dipole consists of a thin wire fed or excited at the midpoint by a voltage source. The total length of a dipole equals half-wave length. Each leg of a dipole has a length equal to the quarter of a wavelength.

In many practical applications the half-wave dipole antenna is connected to an RF source via a coaxial cable, as shown in Figure 8.8.

Ideally, the current from the source flows as a *conduction* current, $\hat{I}_C$, along the transmission line, through the upper arm of the antenna structure, then it flows as a *displacement* current, $\hat{I}_D$, through a parasitic capacitance between the antenna arms,

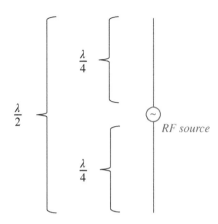

Figure 8.7   Half-wave dipole antenna driven by an RF source.

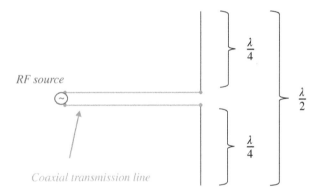

Figure 8.8   Half-wave dipole antenna connected to an RF source by a coaxial cable.

then it returns to the source as a conduction current through the lower arm and the coaxial cable. This is shown in Figure 8.9.

Figure 8.10 shows the details of the current flow in the center conductor and the (ideally) the shield of the coaxial cable (see Section 8.5 for detailed explanation).

With respect to the geometry shown in Figure 8.11, the half-wave dipole *far* electric field, $\hat{E}_\theta$, and *far* magnetic field, $\hat{H}_\phi$, are given by [Paul, 2006],

$$\hat{E}_\theta = j\frac{\eta_0 I e^{-j\beta_0 r}}{2\pi r}F(\theta) \tag{8.58}$$

$$\hat{H}_\phi = j\frac{I e^{-j\beta_0 r}}{2\pi r}F(\theta) \tag{8.59}$$

where I is the amplitude of the conduction current, η_0 is the intrinsic impedance of free space, and β_0 is the phase constant related to the wavelength in free space by

$$\beta_0 = \frac{2\pi}{\lambda_0} \tag{8.60}$$

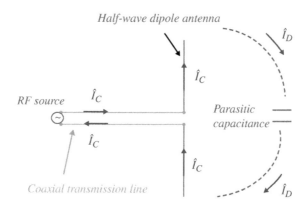

Figure 8.9   Ideal half-wave dipole antenna current flow.

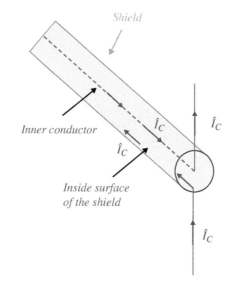

Figure 8.10   Coaxial cable ideal current flow.

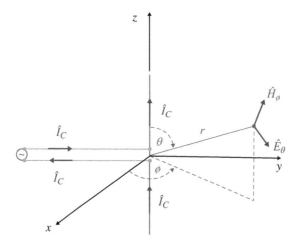

Figure 8.11   Half-wave dipole far fields, $\hat{E}_\theta$ and $\hat{H}_\phi$.

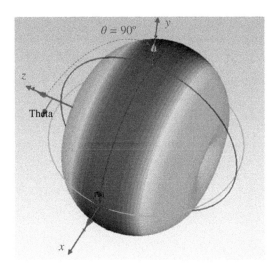

Figure 8.12   Half-wave dipole radiation pattern.

$F(\theta)$ is the *space pattern* given by

$$F(\theta) = \frac{\cos(\frac{\pi}{2}\cos\theta)}{\sin\theta} \tag{8.61}$$

The fields are at their maximum broadside to the antenna, i.e. for $\theta = 90°$ when $F(\theta) = 1$.
Figure 8.12 shows the radiation pattern of a half-wave dipole.

8.4.2 Quarter-Wave Monopole Antenna

Quarter-wave monopole can be obtained from a half-wave dipole by replacing one of the arms of the dipole by an infinite ground plane, as shown in Figure 8.13.

Infinite ground plane is, of course, not realistic; a practical quarter-wave antenna has a finite-dimension ground plane, like the one shown in Figure 8.14.

The radiation pattern of a quarter-wave monopole *above the ground plane* is the same as that for the half-wave dipole.

8.5 Balanced–Unbalanced Antenna Structures and Baluns

This section explains the operation of a balun that transforms an unbalanced antenna structure into a balanced one. Simple and intuitive models of balanced and unbalanced structures are presented and the formation of a sleeve (bazooka) balun is detailed.

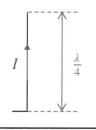

I $\frac{\lambda}{4}$

Infinite ground plane

Figure 8.13   Quarter-wave monopole antenna model.

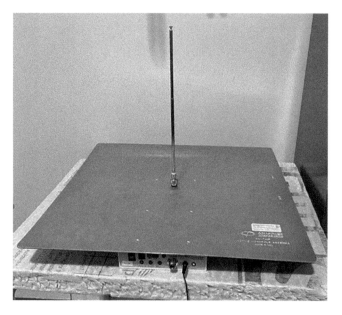

Figure 8.14   Practical quarter-wave monopole antenna.

The balanced–unbalanced antenna structures and the operation of the balun are explained using the half-dipole antenna and transmission line theory [Adamczyk, 2019].

8.5.1 Balanced and Unbalanced Half-Wave Dipole Antenna

Consider the ideal half-dipole antenna, shown in Figure 8.15. It is assumed that the medium surrounding the antenna is free space, free of obstacles, and extending to infinity in all directions.

Under these assumptions, this ideal antenna is an inherently balanced structure. The sinusoidal current $\hat{I}_F$ at any point on the upper arm is the same in magnitude as the current $\hat{I}_R$ at the corresponding position on the lower arm (same distance from the RF voltage source). This ideal case assumes that there are no other metallic or conducting objects in

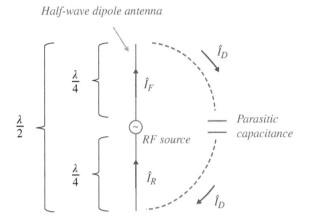

Figure 8.15 Balanced half-wave dipole antenna.

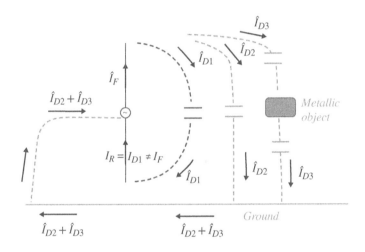

Figure 8.16 Unbalanced half-wave dipole antenna structure.

the vicinity of the antenna, so the "forward" conduction current $\hat{I}_F$ is the same in magnitude as the displacement current $\hat{I}_D$, which in turn is the same in magnitude as the "return" conduction current $\hat{I}_R$.

If there are any metallic or conducting objects (ground) in the vicinity of the antenna, this balance condition is violated, as illustrated in a simplified model shown in Figure 8.16.

The forward current, $\hat{I}_F$, upon leaving the upper arm returns to the source through several paths, only one of them being the lower antenna arm. The magnitude of the current in the upper antenna arm is no longer equal to the magnitude of the lower arm current, $\hat{I}_R = \hat{I}_{F1} \neq \hat{I}_F$. The antenna structure in no longer balanced.

Now, let's consider another ideal balanced antenna structure shown in Figure 8.17. The half-wave dipole antenna is connected to an RF source via a coaxial cable.

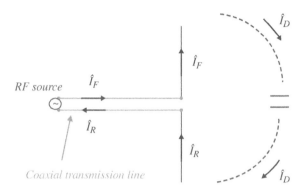

Figure 8.17   Half-wave dipole connected to a coax cable.

Again, let's assume that the medium surrounding the antenna is free space, free of obstacles, and extending to infinity in all directions. Also, let's assume that the forward current travels from the RF source on the inner coax conductor, then through the upper antenna arm, continues as a displacement current, and returns on the lower antenna arm and the coax shield conductor to the source.

The forward and the return currents are equal in magnitude and the antenna structure is balanced. If there are any metallic or conducting objects (ground) in the vicinity of the antenna, this balance condition is violated, in a similar manner as shown earlier in Figure 8.16.

In order for the structure shown in Figure 8.17 to be balanced, the entire return current must travel on the inner surface of the shield, as shown in Figure 8.18.

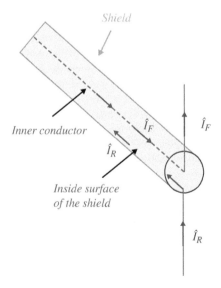

Figure 8.18   (Ideal) shield return current.

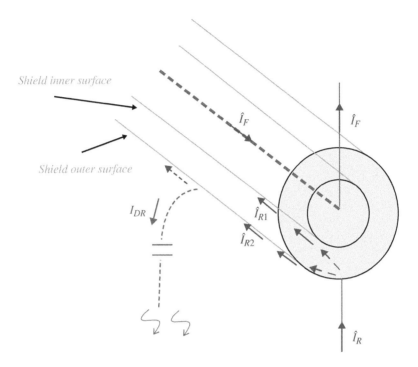

Figure 8.19 Unwanted radiation for the return current on the outside of the shield.

If any part of the return current is allowed to flow on the outside surface of the shield (current $\hat{I}_{R2}$ in Figure 8.19), then some of this current will return to the source through a parasitic capacitance as a displacement current (current $\hat{I}_{DR}$ in Figure 8.19). This results in an unbalanced structure and an unwanted radiation.

The unwanted current on the outside of the shield, $\hat{I}_{R2}$, and subsequently the displacement current $\hat{I}_{DR}$, are in effect the common-mode currents (see Chapter 9 for common-mode currents). This is shown in Figure 8.20.

8.5.2 Sleeve (Bazooka) Balun

The usual way of addressing this unbalance due to a coaxial cable is the use of a balun, which is an acronym for balanced to unbalanced. By using a balun we can transform an unbalanced structure into a balanced one.

The intent of the balun is to increase the impedance to ground seen by the outside shield current, and thus to minimize (ideally eliminate) the outer shield current. Thus, ideally, the return current in the lower antenna arm would equal the return current on the inside of the shield.

A common balun for the coax – dipole antenna structure is a sleeve or bazooka balun shown in Figure 8.21.

The sleeve length is $\lambda/4$ and it is grounded to the outer shield surface at its perimeter as shown in Figure 8.21 (point D). The upper arm of the dipole antenna is connected to

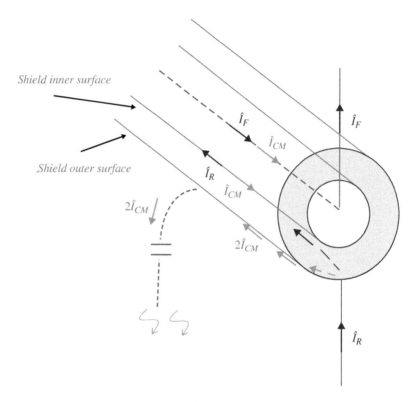

Figure 8.20 Common-mode shield current.

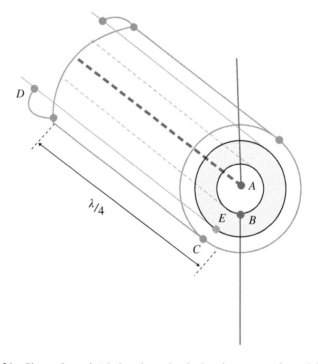

Figure 8.21 Sleeve (bazooka) balun short-circuited to the outer surface of the shield.

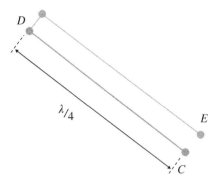

Figure 8.22 Quarter-wave transmission line terminated in a short.

the inner coax conductor (point A). The lower arm is connected to the inside of the shield conductor (point B).

With such connections a transmission line is created where the sleeve constitutes one conductor and the outside of the coax constitutes the other conductor of the transmission line. This transmission line is short-circuited at one end, as shown in Figure 8.22.

In order to explain the effectiveness of the sleeve balun we need to use transmission line theory from Chapter 7.

8.5.3 Input Impedance to the Transmission Line

Consider the transmission line of length L, shown in Figure 8.23.

The transmission line is driven by a sinusoidal source and is terminated by a complex impedance $\hat{Z}_L$. The characteristic impedance of the line is Z_C. The input impedance to the line at the input to the line is given by Eq. (7.165), repeated here

$$\hat{Z}_{IN}(L) = Z_C \frac{\hat{Z}_L + jZ_C \tan(\beta L)}{Z_C + j\hat{Z}_L \tan(\beta L)} \tag{8.62}$$

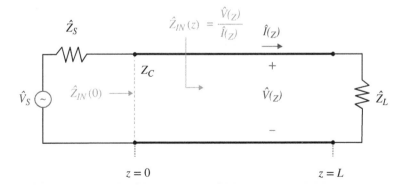

Figure 8.23 Transmission line driven by a sinusoidal source.

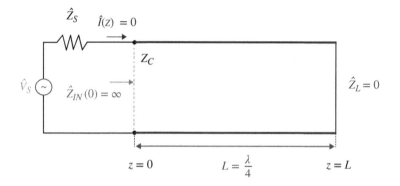

Figure 8.24   Transmission line of length $\lambda \mid 4$ terminated with a short.

where the phase constant β is related to the wavelength λ by

$$\beta = \frac{2\pi}{\lambda} \tag{8.63}$$

Let's assume that the line length is $\lambda/4$, and the line is terminated with a short as shown in Figure 8.24. (Note that this is the same configuration as the sleeve balun shown in Figure 8.22).

Now the input impedance to this quarter-wavelength line terminated in a short is

$$\hat{Z}_{IN}(0) = Z_C \frac{\hat{Z}_L + jZ_C \tan(\beta L)}{Z_C + j\hat{Z}_L \tan(\beta L)} = Z_C \frac{jZ_C \tan(\frac{2\pi}{\lambda})(\frac{\lambda}{4})}{Z_C} \tag{8.64}$$

or

$$\hat{Z}_{IN}(0) = jZ_C \tan(\frac{\pi}{2}) = \infty \tag{8.65}$$

Thus, the input impedance of the quarter-wavelength transmission line terminated by a short is infinite. This means that no current will flow into this transmission line, as shown in Figure 8.24.

8.5.4 Quarter-Wavelength Sleeve Balun

Since the quarter-wave long transmission line terminated with a short creates an infinite impedance at its input, all current flows on the inside surface of the shield. This is shown in Figure 8.25. Now this structure become balanced.

Sleeve (bazooka) baluns are the narrow-band baluns since work well only at the frequency where its length is one-quarter wavelength.

There are many other types of baluns; they all attempt to prevent (choke) the common-mode current from flowing on the outside of the shield. One such alternative is a ferrite toroid choke balun shown in Figure 8.26.

In Chapter 9 we will show that in an ideal common-mode choke, the self-inductance of each winding is equal to the mutual inductance between the windings, i.e. $L = M$.

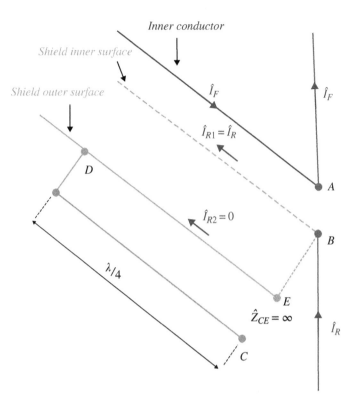

Figure 8.25   Balanced balun-coax-antenna structure.

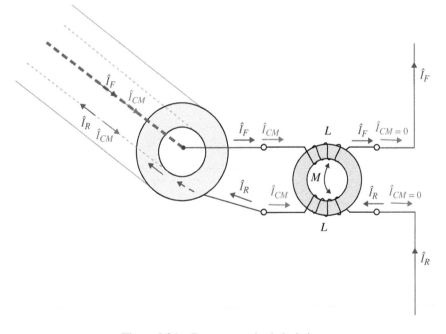

Figure 8.26   Common-mode choke balun.

The impedance seen by the differential-mode current in each winding $(\hat{I}_F, \hat{I}_R)$ is

$$\hat{Z}_{DM} = j\omega(L - M) = 0 \tag{8.66}$$

The impedance seen by the common-mode current in each winding is

$$\hat{Z}_{CM} = j\omega(L + M) = \infty \tag{8.67}$$

Thus, in an ideal case, the choke is transparent to the differential-mode currents and blocks the common-mode currents. Ferrite-core baluns provide a high common-mode impedance over a broader frequency range.

8.6 Sleeve Dipole Antenna Design and Build

This section describes a practical approach to the design and build of a sleeve dipole antenna mandated by a custom EMC standard [Adamczyk and Pearson, 2019]. The antenna was required to operate over the frequency range of 220 to 225 MHz specified in the standard. An additional requirement of having an input impedance close to 50 Ω was imposed to ensure good antenna efficiency when used with 50 Ω characteristic impedance test equipment. We begin our discussion by reviewing a symmetrically fed half-wave dipole antenna followed by the asymmetrically driven dipole, and finally focusing on a design and build of a sleeve dipole antenna.

8.6.1 Symmetrically Driven Half-Wave Dipole Antenna

Symmetrically driven half-wave dipole consists of a thin wire fed or excited at the midpoint by a voltage source. The total length of a dipole equals half-wave length. Each leg of a dipole has a length equal to the quarter of a wavelength, as shown in Figure 8.27a. Often the voltage source is connected to the antenna via transmission line, as shown in Figure 8.27b.

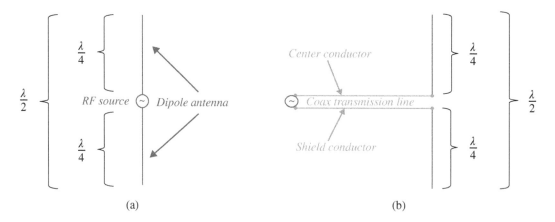

(a)   (b)

Figure 8.27 Symmetrical dipole antenna (a) driven directly (b) driven via coaxial transmission line.

To maximize the radiated power, the antenna should be matched to the transmission line. The input impedance of the (ideal) symmetrical half-wave dipole is

$$\hat{Z}_L = 73 + j42.5 \ \Omega \tag{8.68}$$

The impedance of practical dipoles depends on the thickness-to-length ratio of its elements, with typical values of the real part in the range $60 - 70 \ \Omega$.

To maximize the power delivered to the antenna its length is usually shortened to make the impedance purely real. This length shortening can be performed through iterative physical length reduction and impedance measurements until the desired characteristics are obtained. This time-consuming process can be reduced through software simulations if available.

The half-wave dipole antenna is one of the simplest, yet practical antennas but its impedance is very frequency sensitive, i.e. it has a narrow bandwidth. The input impedance of the half-wave dipole can be reduced by moving the fed point off-center and the bandwidth can be increased by the addition of a sleeve, as discussed next.

8.6.2 Asymmetrically Driven Dipole Antenna and a Sleeve Dipole

We can reduce the input impedance of a dipole by moving the fed point off-center; this is shown in Figure 8.28a. The creation of a sleeve, shown in Figures 8.28b,c, can increase the bandwidth up to more than an octave and fine-tune the input impedance.

Figure 8.29 shows the details of the antenna connections.

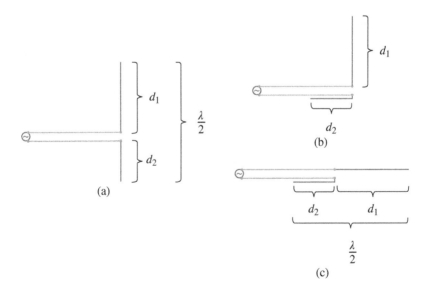

Figure 8.28   (a) Asymmetrical dipole antenna, (b) creation of a sleeve, and (c) model of a sleeve antenna.

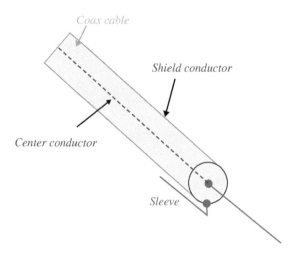

Figure 8.29   More detailed model of a sleeve antenna.

8.6.3 Sleeve Dipole Antenna Design

The desired operating frequency of the antenna was 220 − 225 MHz, and the desired input impedance was 50 Ω, so that it could be matched to a 50 Ω coaxial transmission line.

A simple way to begin the design process is to start with the symmetrical sleeve antenna where both antenna elements are equal to each other, each being $\lambda/4$. This length can be determined from

$$l = \frac{\lambda}{4} = \frac{v}{4f_0} \tag{8.69}$$

where v is the velocity of signal propagation in a coax and $f_0 = 222.5$ MHz.

Subsequently one could build such an antenna and measure its response. It is safe to say that the antenna's response would not satisfy the design requirements. The next step would be iteratively adjusting the length of the antenna elements to create an asymmetrical antenna and repeat the testing until the desired response is obtained.

Needless to say this is an ineffective and time-consuming process. With the help of a simulation software one can arrive at the proper antenna dimension in a more efficient way.

8.6.4 Sleeve Dipole Antenna Design Through Simulation

A 3D model of a sleeve dipole was drawn in CST Microwave Studio. Electrical material properties were assigned. Images of the antenna used in simulations are shown in Figures 8.30 and 8.31.

Assuming a symmetrical antenna the starting length of each arm was determined as

$$l = \frac{\lambda}{4} \approx \frac{3 \times 10^8}{4 \times 222.5 \times 10^6} = 337 \text{ mm} \tag{8.70}$$

It is not necessary to use the exact speed of the signal propagation, as the antenna length will be determined using an optimizer program, as discussed later. The simulation

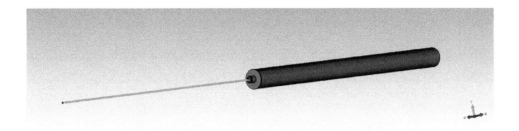

Figure 8.30 Sleeve dipole elements without enclosure.

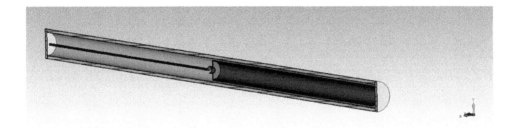

Figure 8.31 Cutaway of antenna 3D with PVC enclosure.

was performed from DC to 300 MHz. The resulting reflection coefficient magnitude (s_{11} parameter magnitude) is shown in Figures 8.32 and 8.33.

Figure 8.32 reveals that the antenna resonates at 199 MHz which is not within the antenna's desired operating frequency of 220 to 225 MHz.

In order to determine the optimal antenna length the simulators optimizer was run with a goal of minimum s_{11} magnitude in the frequency range 202 – 225 MHz. The antenna

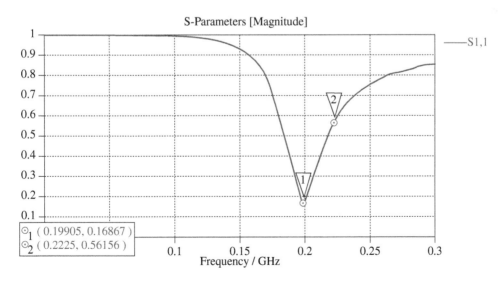

Figure 8.32 Linear magnitude of s_{11} parameter – 1/4 wavelength elements.

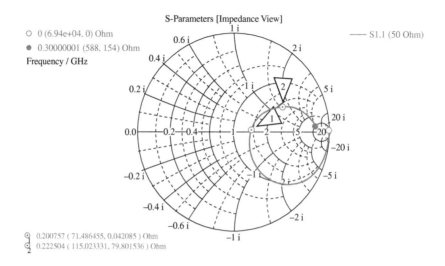

Figure 8.33 Smith chart of s_{11} parameter – 1/4 wavelength elements.

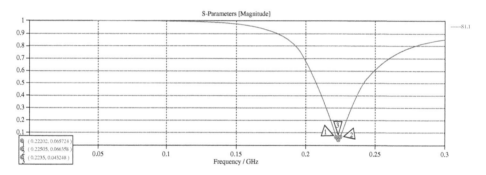

Figure 8.34 Linear magnitude of s_{11} parameter – optimized element lengths.

element lengths were set as free parameters, not necessarily equal, in the range of 150–340 mm.

The optimizer found the optimum element lengths to be 218 mm for the tube element and 330 mm for the wire element. A plot of the magnitude of the antennas s_{11} is shown in Figure 8.34. This plot reveals the minimum magnitude at 223.5 MHz, which is in the desired range.

The impedance is real valued (around 54Ω) at 223.5 MHz as shown by the Smith chart in Figure 8.35. It was also observed that this simulation with asymmetrical elements has a wider bandwidth than the simulation with symmetrical elements.

8.6.5 Construction and Tuning of a Sleeve Dipole

The simulation results were used to create a physical antenna shown in Figures 8.36 and 8.37.

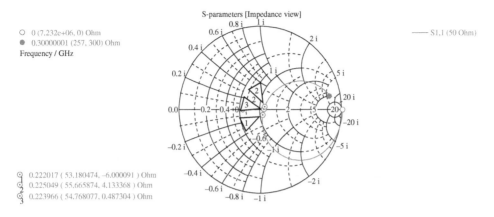

Figure 8.35   Smith chart of s_{11} parameter – 1/4 wavelength elements.

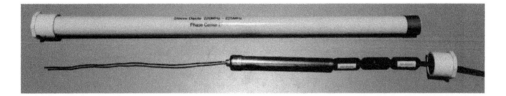

Figure 8.36   Sleeve dipole removed from enclosure.

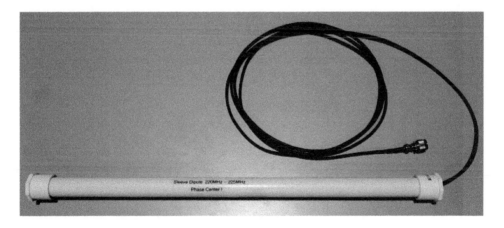

Figure 8.37   Sleeve dipole inside enclosure.

The antenna was constructed using a piece of copper pipe, copper end cap, a BNC female to solder-cup bulkhead, and a length of 18 AWG solid core wire with PVC insulation. The enclosure was made from PVC pipe and end caps. The feedline is a $50\,\Omega$ coaxial cable with $50\,\Omega$ BNC male connectors. The antenna was constructed using the

Figure 8.38   VNA antenna measurement setup.

element lengths found by the optimizer. The wire element was 330 mm and the copper tubing element was 218 mm.

To limit the common-mode current on the antenna feedline, three ferrite chokes were placed on the antennas feedline as shown in Figure 8.36. The sleeve dipoles reflection coefficient (s_{11} parameter magnitude) was measured using a vector network analyzer (Figure 8.38).

While the input impedance of the antenna requirement was satisfied (low reflection coefficient at resonance), the resonant frequency of the antenna was found to be at 210.6 MHz (instead of the target 222.5 MHz), as revealed by the plot in Figure 8.39.

To determine the target length of the antenna elements the following relationship was used:

$$\frac{f_{measured}}{f_{target}} = \frac{l_{target}}{l_{measured}} \tag{8.71}$$

The target lengths were determined as

$$l_{target} = l_{measured}\frac{f_{measured}}{f_{target}} = 218 \text{ mm } \frac{210.6 \times 10^6}{2222.5 \times 10^6} = 206 \text{ mm } (sleeve) \tag{8.72}$$

$$l_{target} = l_{measured}\frac{f_{measured}}{f_{target}} = 330 \text{ mm } \frac{210.6 \times 10^6}{2222.5 \times 10^6} = 312 \text{ mm } (arm) \tag{8.73}$$

Lengths of wire and copper tubing were cut from the antenna to achieve the target length. The antenna's s_{11} parameter magnitude was measured again with the VNA, as shown in Figure 8.40.

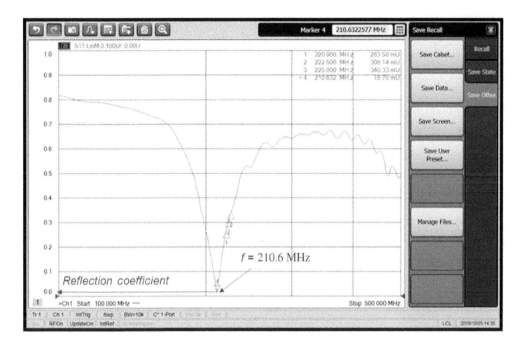

Figure 8.39 s_{11} magnitude of un-tuned antenna.

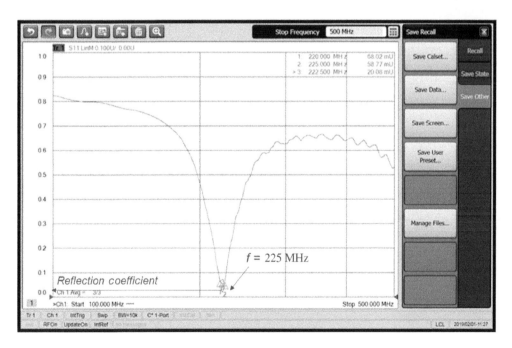

Figure 8.40 s_{11} magnitude of tuned antenna.

This figure reveals that both the redesigned antennas meet input impedance requirement of being close to $50\,\Omega$ and its resonant frequency is in the vicinity of 225 MHz.

8.7 Antennas Arrays

Hertzian dipole, half-wave dipole, and quarter-wave monopole antennas radiated equally in any plane perpendicular to the antenna axis, i.e. they radiate equally in any ϕ direction. In EMC radiated emissions and immunity measurements it is often desired to focus the radiated signal in a particular direction.

This can be accomplished by forming an antenna that consists of an assembly (an array) of the radiating elements. Log-periodic antenna, discussed next, consists of an array of the half-wave dipoles antennas.

8.8 Log-Periodic Antenna

Log-periodic antenna is one of the most common antennas used in EMC measurements. It is a broadband antenna and it is usually used for the measurement of the radiated emissions from 200 MHz to 1 GHz [Adamczyk, 2020a].

The log-periodic antenna consists of a sequence of parallel linear dipoles forming a coplanar dipole array, as shown in Figure 8.41 [Adamczyk, 2020a].

The dipole lengths increase along the antenna while keeping the angle α constant. The lengths l and the spacing R of adjacent elements increase logarithmically as defined by the inverse ratio

$$\frac{1}{\tau} = \frac{l_{n+1}}{l_n} = \frac{R_{n+1}}{R_n} \tag{8.74}$$

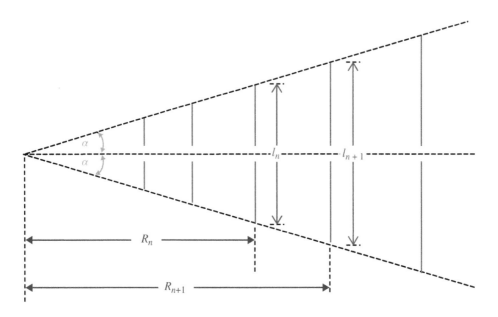

Figure 8.41 Log-periodic dipole array.

Figure 8.42   Log-periodic antenna is fed at the small end of the structure.

Figure 8.43   Log-periodic antenna for radiated emissions measurements.

All the dipole elements of the log-periodic antennas are connected and fed at the small end of the structure. Figure 8.42 shows such a connection for a PCB log-periodic antenna.

Figure 8.43 shows a log-periodic antenna used for EMC radiated emissions measurements in the frequency range 300 MHz–1 GHz.

The details of the coaxial line connections for this antenna are shown in Figure 8.44.

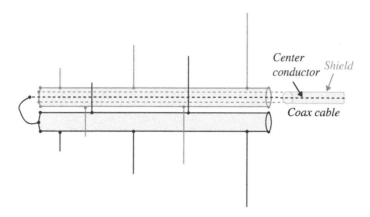

Figure 8.44   Connection details of a log-periodic antenna fed by a coaxial cable.

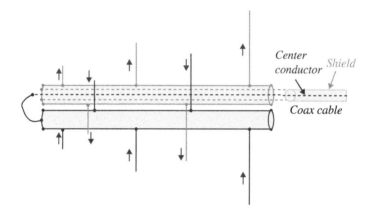

Figure 8.45   Current flow in the dipole array elements.

The dipole connections shown in Figure 8.44 create the current flow in the radiating elements as shown in Figure 8.45.

Note that the currents in the adjacent elements are $180°$ out of phase. At any given frequency only a fraction of the antenna is used (where the dipoles are $\lambda/2$ long). Since the phase between the adjacent closely spaced elements shorter than $\lambda/2$ is opposite, the fields radiated by them tend to cancel out. The elements longer than $\lambda/2$ present a large impedance and carry only a small current. Often, a stub is placed at the back of the antenna (to suppress reflections) as shown in Figure 8.46.

The phase reversal between the short elements produces a phase progression so that the energy is beamed in the direction of the shorter elements. Radiation pattern of the log-periodic antenna is shown in Figure 8.47.

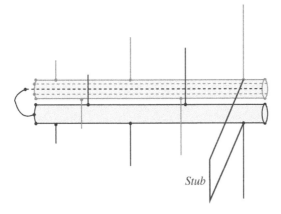

Figure 8.46 Stub at the end of the log-periodic antenna.

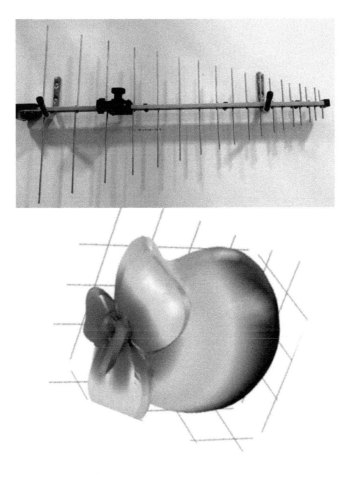

Figure 8.47 Radiation pattern of the log-periodic antenna.

8.9 Biconical Antenna

Biconical antenna, shown in Figure 8.48, is another type of a dipole antenna commonly used for the measurement of the radiated emissions from 30 MHz to 200 or 300 GHz [Adamczyk, 2020a].

Biconical (bicon) antenna belongs to the dipole family of antennas and is classified as an intermediate bandwidth antenna.

Radiation pattern of the bicon antenna is shown in Figure 8.49.

Note that this pattern resembles the one of a dipole antenna in Figure 8.12. The difference is that the half-wave dipole is a narrow-band antenna and the bicon is a broadband antenna.

Figure 8.48   Biconical antenna.

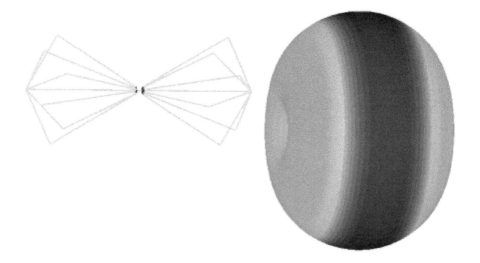

Figure 8.49   Radiation pattern of the bicon antenna.

Figure 8.50 Bicon antenna with a balun.

Normally, an EMC bicon antenna has a balun [Adamczyk, 2019], shown in Figure 8.50, attached at the input to the antenna to make it a balanced structure.

8.10 Antenna Impedance and VSWR

Consider the model of an antenna system in the receiving mode shown in Figure 8.51 [Adamczyk, 2020b].

Spectrum analyzer is matched to the coaxial cable (thus, there are no reflections at the receiver). If the antenna's radiation resistance were 50 Ω over the measurement frequency range then the voltage induced at the base of the antenna would appear at the spectrum analyzer (assuming no cable loss).

If the antenna's resistance differed from 50 Ω then some of the power received by the antenna would be reflected back or reradiated and the reading at the spectrum analyzer

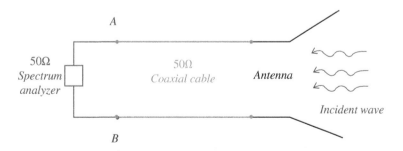

Figure 8.51 Antenna in the receiving mode.

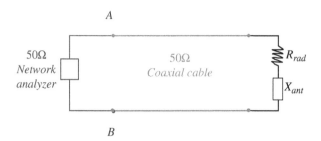

Figure 8.52   Antenna model.

would be lower. It is therefore very useful to know the impedance of the antenna over its measurement range.

One very good indicator of the antenna impedance is obtained by measuring VSWR of the antenna. The antenna can be represented by its circuit model consisting of the radiation resistance R_{rad} and its reactance X_{ant}, as shown in Figure 8.52.

If, in a given frequency range, the antenna's impedance is purely resistive and equals $50\,\Omega$ then the VSWR reading will be 1. The more the impedance of the antenna differs from $50\,\Omega$, the higher the VSWR reading.

Figure 8.53 shows the VNA setup and VNA calibration for the log-periodic antenna impedance and VSWR measurements [Adamczyk, 2020b].

A calibration of the network analyzer was performed inside the semianechoic chamber at the end of the cable that connects to the antenna using short, open, and load calibration standards. The calibration was performed with the cable positioned as close as possible to the final antenna-measurement configuration.

The impedance and VSWR measurements for the log-periodic antenna were performed with the antenna in both the horizontal and vertical orientations (polarization).

Figure 8.53 shows the impedance and VSWR of the log-periodic antenna in a horizontal polarization, while Figure 8.54 shows the results for the vertical polarization. The measurements were taken in the frequency range of 190 MHz–7 GHz.

Figure 8.53   Impedance and VSWR for log-periodic antenna in horizontal polarization.

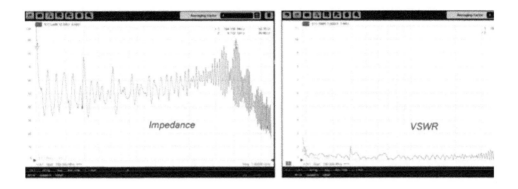

Figure 8.54   Impedance and VSWR for log-periodic antenna in vertical polarization.

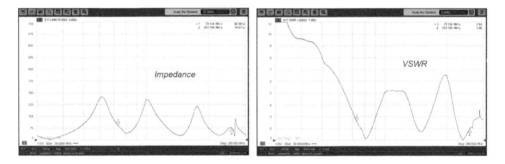

Figure 8.55   Impedance and VSWR for bicon antenna in vertical polarization.

Note that the impedance is close to 50 Ω while the VSWR is close to the value of one, which is a very desirable antenna behavior for EMC measurements. This is not the case for the bicon antenna, as demonstrated next.

The impedance and VSWR measurements for the bicon antenna were performed with the antenna in both the horizontal and vertical orientations.

Figure 8.55 shows the impedance VSWR for the bicon antenna in a vertical polarization. The measurements were taken in the frequency range of 30–300 MHz.

Note that, unlike the log-periodic antenna, the impedance and VSWR of the bicon antenna vary considerably over its intended frequency range.

8.11 Laboratory Exercises

This section contains two different laboratory exercises. Section 8.11.1 is devoted to the log-periodic and bicon antennas VSWR and impedance measurements. In Section 8.11.2 we describe how to build a small loop antenna.

8.11.1 Log-Periodic and Bicon Antenna Impedance and VSWR Measurements

Objective: In this laboratory exercise we use a network analyzer to measure the impedance and VSWR of a log-periodic and biconical antennas, over their intended frequency ranges.

8.11.1.1 Laboratory Equipment and Supplies

Network Analyzer: KEYSIGHT ENA Network Analyzer E5080A (or equivalent), shown in Figure 8.56.

   Log-periodic Antenna – Compower Model AL-100 (or equivalent), shown in Figure 8.57
   Biconical Antenna – Compower – Model AB-900 (or equivalent), shown in Figure 8.58
   Coaxial cable assembly – N Male to N Male Cable RG-393 Coax 48 Inch (or equivalent), shown in Figure 8.59. Fairview Microwave Part number: FMC0101393LF-48.

8.11.1.2 Laboratory Procedure

Note: The measurements in Steps 1 and 2 do not need to be taken inside a semi anechoic chamber.

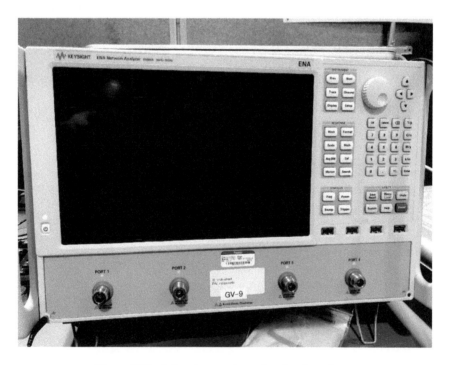

Figure 8.56   Laboratory equipment: network analyzer.

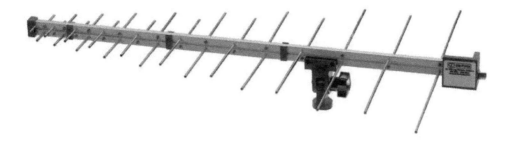

Figure 8.57 Compower log-periodic antenna.

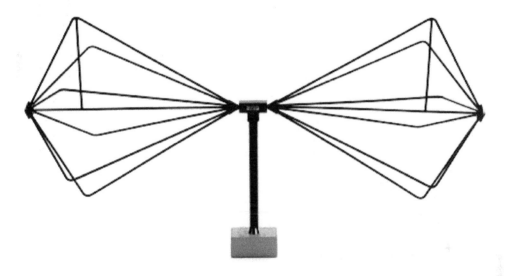

Figure 8.58 Compower Bicon antenna.

Step 1: Connect the log-periodic antenna to the network analyzer. Measure its impedance and VSWR over the frequency range 300–1000 MHz. Comment on the results.

Step 2: Connect the biconical antenna to the network analyzer. Measure its impedance and VSWR over the frequency range 25–200 MHz. Comment on the results.

8.11.2 Loop Antenna Construction

Objective: This laboratory exercise is devoted to the construction of a small loop antenna [Koeller, 2023], that will subsequently be used in Chapters 10 and 11 laboratory exercises.

Figure 8.59   Coaxial cable assembly.

Figure 8.60   Magnetic wire.

8.11.2.1 Equipment List

Magnetic Wire, 3-in Long, shown in Figure 8.60.
Supplier: Remington Industries, 14H200P.125,
Digi-Key Part Number: 2328-14H200P.125-ND.

Adapter Coaxial Connector, shown in Figure 8.61.
Supplier: Amphenol RF, 31-219-RFX,
Digi-Key Part Number: ARFX1069-ND.

Figure 8.61   Adapter coaxial connector.

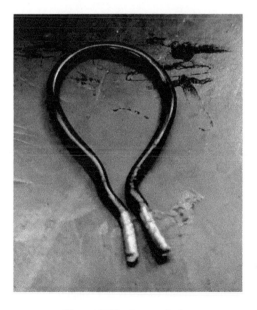

Figure 8.62   Loop of wire.

Figure 8.63   Loop of wire soldered to the connector center pin.

Figure 8.64   Loop of wire soldered to the connector wall.

8.11.2.2 Laboratory Procedure

Step 1: Cut about 3 inches of the wire, scrape about a quarter of an inch on both ends, and form a loop, as shown in Figure 8.62.

Step 2: Solder one end of the loop to the center pin of the connector, as shown in Figure 8.63.

Step 3: Solder the other end of the loop to the wall of the connector, as shown in Figure 8.64.

References

Bogdan Adamczyk. *Foundations of Electromagnetic Compatibility*. John Wiley & Sons, Ltd, Chichester, UK, 2017. ISBN 9781119120810. doi: https://doi.org/10.1002/9781119120810.

Bogdan Adamczyk. Balanced-Unbalanced Antenna Structures and Baluns. *In Compliance Magazine*, October 2019. URL https://incompliancemag.com/article/balanced-unbalanced-antenna-structures-and-baluns/.

Bogdan Adamczyk. Dipole-Type Antennas in EMC Testing −Part I: Antenna Models and Construction. *In Compliance Magazine*, June 2020a. URL https://incompliancemag.com/article/dipole-type-antennas-in-emc-testing-part1/.

Bogdan Adamczyk. Dipole-Type Antennas in EMC Testing −Part II: Antenna Parameters and Measurements. *In Compliance Magazine*, July 2020b. URL https://incompliancemag.com/article/dipole-type-antennas-in-emc-testing-part-ii/.

Bogdan Adamczyk and Alex Pearson. Sleeve Dipole Antenna Design and Build. *In Compliance Magazine*, November 2019. URL https://incompliancemag.com/article/sleeve-dipole-antenna-design-and-build/.

Constantine A. Balanis. *Antenna Theory - Analysis and Design*. Wiley-Interscience, Hoboken, New Jersey, 2005. ISBN 978-0-471-66782-7.

Nick Koeller. Loop Antenna Designer, 2023.

Clayton R. Paul. *Introduction to Electromagnetic Compatibility*. John Wiley & Sons, Inc., Hoboken, New Jersey, 2nd edition, 2006. ISBN 978-0-471-75500-5.

Chapter 9: Differential- and Common-Mode Currents and Radiation

9.1 Differential- and Common-Mode Currents

Consider a typical circuit model shown in Figure 9.1.

If the fields generated by the forward current cancel the fields of the return current and no other circuits, or sources, or coupling paths are present, then the forward current equals the return current. In virtually any practical circuit a different scenario takes place, as shown in Figure 9.2.

$\hat{I}_D$ is referred to as the *differential-mode (DM) current* while $\hat{I}_C$ is referred to as the *common-mode (CM) current*. The DM currents are usually the functional currents, they are equal in magnitude and of opposite directions. The CM (unwanted) currents are equal in magnitude and of the same direction.

In the analysis of the DM and CM currents we often use the circuit model shown in Figure 9.3, where, in addition to the DM and CM currents, we show the *total* currents $\hat{I}_1$ and $\hat{I}_2$ flowing in the same direction. The reason for this is that it is easier to apply the classical circuit theory to the total currents than it is to the individual currents. Once the equations are developed for the total currents, we simply substitute the differential- and/or common-mode currents for the total currents in the derived expressions. This approach will be demonstrated in Section 9.2 when discussing a common-mode choke.

The total currents $\hat{I}_1$ and $\hat{I}_2$ are related to the DM and CM currents by

$$\hat{I}_1 = \hat{I}_C + \hat{I}_D \tag{9.1}$$

$$\hat{I}_2 = \hat{I}_C - \hat{I}_D \tag{9.2}$$

Adding and subtracting Eqs. (9.1) and (9.2) gives

$$\hat{I}_1 + \hat{I}_2 = 2\hat{I}_C \tag{9.3}$$

$$\hat{I}_1 - \hat{I}_2 = 2\hat{I}_D \tag{9.4}$$

Thus, in terms of the total currents, the DM and CM currents can be expressed as

$$\hat{I}_D = \frac{1}{2}\left(\hat{I}_1 - \hat{I}_2\right) \tag{9.5}$$

$$\hat{I}_C = \frac{1}{2}\left(\hat{I}_1 + \hat{I}_2\right) \tag{9.6}$$

Principles of Electromagnetic Compatibility: Laboratory Exercises and Lectures, First Edition. Bogdan Adamczyk.
© 2024 John Wiley & Sons Ltd. Published 2024 by John Wiley & Sons Ltd.
Companion website: www.wiley.com/go/principlesofelectromagneticcompatibility

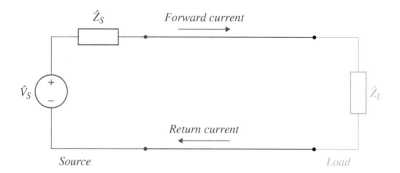

Figure 9.1   A typical circuit model.

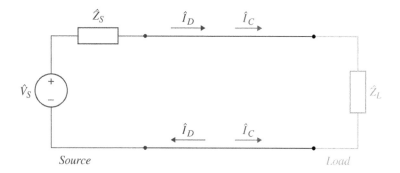

Figure 9.2   A realistic circuit model.

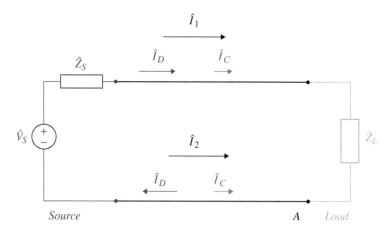

Figure 9.3   Circuit model showing the total currents.

Returning to Figure 9.3 we note that something is missing, since Kirchhoff's Current Law (KCL) seems to be violated. Current into the node A is not equal to the current out of the node A:

$$\hat{I}_D + 2\hat{I}_C \neq \hat{I}_D \tag{9.7}$$

What happened to the total common-mode current $2\hat{I}_C$ entering node A? Where is its return path? One possible return path is shown in Figure 9.4, where the common-mode currents return to the source as displacement currents through a parasitic capacitance.

9.1.1 Common-Mode Current Creation

There are many scenarios under which common-mode currents can be created; some of them are easier than others to understand or explain. In this section we present perhaps the most intuitive and easy scenario using a typical circuit, like the one shown in Figure 9.5 [Adamczyk, 2019].

In this idealized circuit the forward current consists only of the differential current $\hat{I}_D$ flowing from the source to the load, while the return current consists only of the differential current $\hat{I}_D$ equal in magnitude and flowing in the opposite direction from the load to the source. Let's keep in mind that these are not DC currents. These are high-frequency RF currents that propagate as current (and voltage) waves, and have differential-mode EM waves associated with them, as shown in Figure 9.6.

Now consider the scenario where the circuit shown in Figure 9.6 is in the far-field of another source that generates uniform plane EM waves, as shown in Figure 9.7.

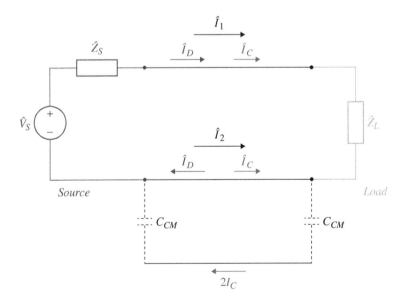

Figure 9.4 Common-mode currents return path.

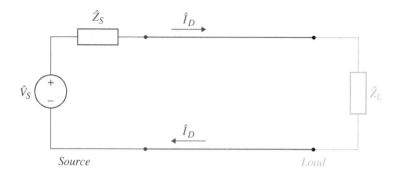

Figure 9.5 Circuit with differential currents only.

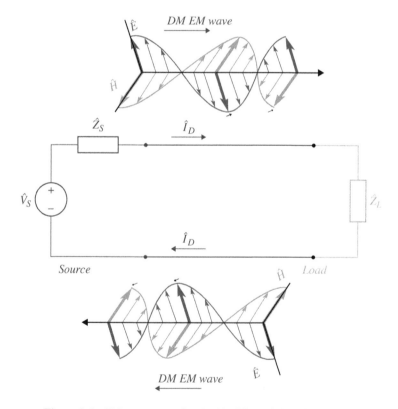

Figure 9.6 EM waves associated with differential-mode currents.

Note that the waves generated by the far-field source propagate in the same direction. When the far-field waves arrive at the original circuit we have a scenario shown in Figure 9.8.

Just like the differential-mode current has a DM wave associated with it, the CM wave has a corresponding CM current, as shown in Figure 9.8. When drawing electrical circuits,

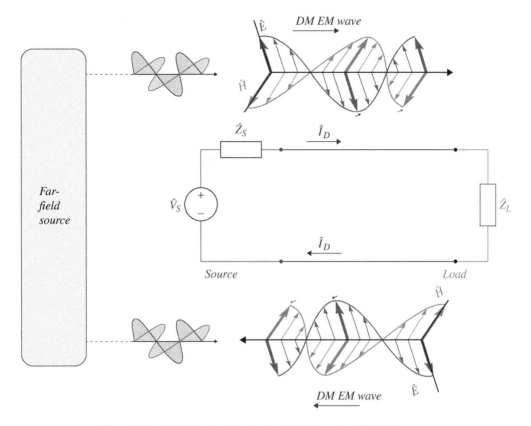

Figure 9.7   Original circuit in the far field of another EM-field source.

we normally do not show the associated EM waves, and thus a typical DM/CM current circuit is shown simply as the circuit in Figure 9.9.

9.2 Common-Mode Choke

A common-mode choke, shown in Figure 9.10 consists of a pair of wires carrying currents $\hat{I}_1$ and $\hat{I}_2$ wound around a ferromagnetic core.

As we shall see, the common-mode choke (ideally) blocks the common-mode currents and has no effect on the differential-mode currents. The currents shown in Figure 9.10 and the total current flowing in each wire are shown in Figure 9.11.

Let's investigate the effect of the choke on the CM- and DM-mode currents. The circuit model of the choke is shown in Figure 9.12, where L_1 and L_2 are the self-inductances of each winding and M is the mutual inductance [Paul, 2006 Adamczyk, 2017].

$$\hat{Z}_1 = \frac{\hat{V}_1}{\hat{I}_1} = \frac{j\omega L \hat{I}_1 + j\omega M \hat{I}_2}{\hat{I}_1} \tag{9.8}$$

$$\hat{Z}_2 = \frac{\hat{V}_2}{\hat{I}_2} = \frac{j\omega L \hat{I}_2 + j\omega M \hat{I}_1}{\hat{I}_2} \tag{9.9}$$

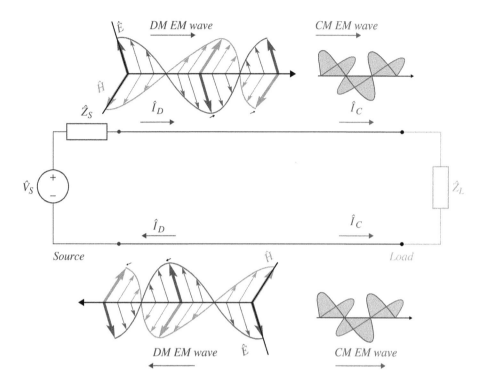

Figure 9.8   CM waves and associated CM currents.

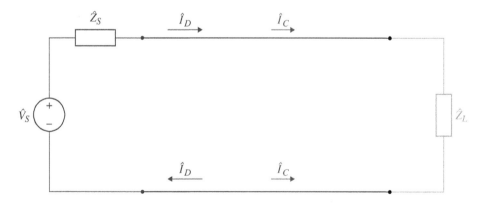

Figure 9.9   Designation of DM and CM currents.

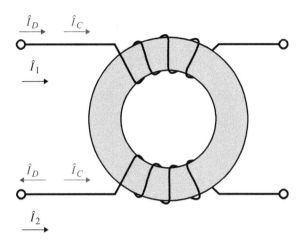

Figure 9.10 Common-mode choke.

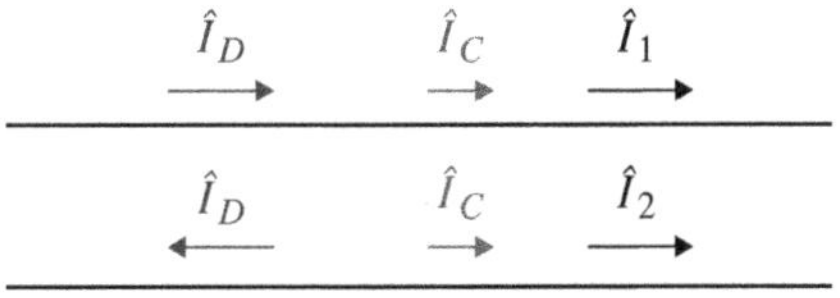

$$\hat{I}_D \qquad \hat{I}_C \qquad \hat{I}_1$$

$$\hat{I}_D \qquad \hat{I}_C \qquad \hat{I}_2$$

Figure 9.11 DM-, CM-, and total currents flowing in each wire.

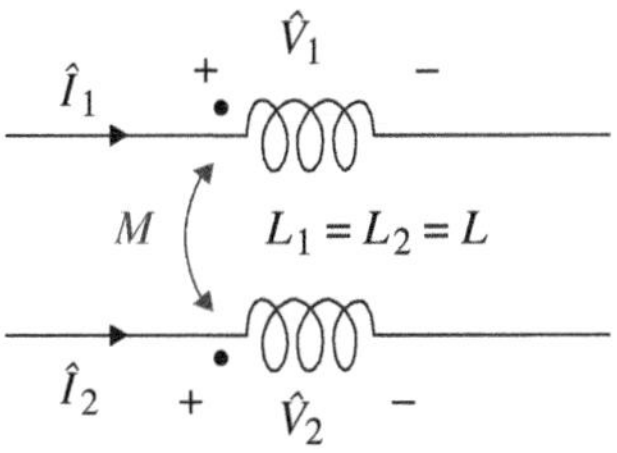

$$\hat{V}_1$$

$$\hat{I}_1 \qquad +$$

$$M \qquad L_1 = L_2 = L$$

$$\hat{I}_2 \qquad + \qquad \hat{V}_2$$

Figure 9.12 Circuit model of the CM choke.

To determine the effect of the choke on the DM currents let

$$\hat{I}_1 = \hat{I}_D \tag{9.10}$$

$$\hat{I}_2 = -\hat{I}_D \tag{9.11}$$

Using Eqs. (9.10) and (9.11) in Eq. (9.8) we get

$$\hat{Z}_{DM} = \frac{\hat{V}_1}{\hat{I}_1} = \frac{j\omega L\hat{I}_1 + j\omega M\hat{I}_2}{\hat{I}_1} = \frac{j\omega L\hat{I}_D - j\omega M\hat{I}_D}{\hat{I}_D} = j\omega(L - M) \tag{9.12}$$

Similarly, using Eqs. (9.10) and (9.11) in Eq. (9.9) we get

$$\hat{Z}_{DM} = \frac{\hat{V}_2}{\hat{I}_2} = \frac{j\omega L\hat{I}_2 + j\omega M\hat{I}_1}{\hat{I}_2} = \frac{-j\omega L\hat{I}_D + j\omega M\hat{I}_D}{-\hat{I}_D} = j\omega(L - M) \tag{9.13}$$

Thus, the impedance seen by the DM current in each winding is

$$\hat{Z}_{DM} = j\omega(L - M) \tag{9.14}$$

In the ideal case, where $L = M$, we have

$$\hat{Z}_{DM} = 0 \tag{9.15}$$

Thus, the ideal CM choke is transparent to the DM currents, i.e. it does not affect them at all, over the entire frequency range. Equivalently, the differential-mode insertion loss of an ideal CM choke should be 0 dB over the entire frequency range. Non-ideal, realistic CM chokes exhibit the DM insertion loss similar to the one shown in Figure 9.13 [Murata, 2023].

Next, let's determine the effect of the choke on CM currents. Toward this end let

$$\hat{I}_1 = \hat{I}_D \tag{9.16}$$

$$\hat{I}_2 = -\hat{I}_D \tag{9.17}$$

Using Eqs. (9.16) and (9.17) in Eq. (9.8) we get

$$\hat{Z}_{CM} = \frac{\hat{V}_1}{\hat{I}_1} = \frac{j\omega L\hat{I}_1 + j\omega M\hat{I}_2}{\hat{I}_1} = \frac{j\omega L\hat{I}_C + j\omega M\hat{I}_C}{\hat{I}_C} = j\omega(L + M) \tag{9.18}$$

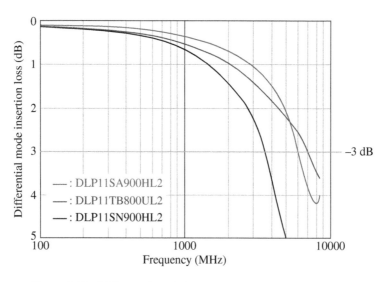

Figure 9.13   Differential-mode insertion loss of Murata CM chokes.

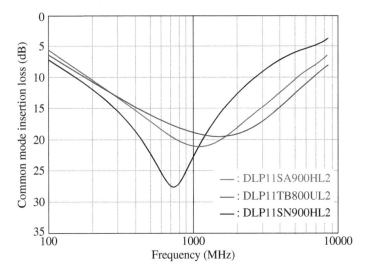

Figure 9.14   Common-mode insertion loss of Murata CM chokes.

Similarly, using Eqs. (9.16) and (9.17) in Eq. (9.9) we get

$$\hat{Z}_{CM} = \frac{\hat{V}_2}{\hat{I}_2} = \frac{j\omega L\hat{I}_2 + j\omega M\hat{I}_1}{\hat{I}_2} = \frac{j\omega L\hat{I}_C + j\omega M\hat{I}_C}{-\hat{I}_C} = j\omega(L+M) \tag{9.19}$$

Thus, the impedance seen by the DM current in each winding is

$$\hat{Z}_{CM} = j\omega(L+M) \tag{9.20}$$

Thus, the CM choke inserts an inductance $L + M$ in each winding, and consequently, it tends to block CM currents. Ideally, this total inductance should be very large over the entire frequency range. Equivalently, the common-mode insertion loss of an ideal CM choke should be very large over the entire frequency range. Non-ideal, realistic CM chokes exhibit the CM insertion loss similar to the one shown in Figure 9.14 [Murata, 2023].

9.3 Differential-Mode and Common-Mode Radiation

Differential- and common-mode radiation can be modeled as the radiation from two Hertzian dipoles driven by a noise voltage. Let's begin with the DM radiation. Consider the scenario shown in Figure 9.15 where two linear antennas (conductors 1 and 2) placed along the *x* axis, carry the differential-mode currents along the *z*-direction.

The maximum radiated field is broadside to the antenna (in the *xy*-plane, where $\theta = 90°$) and in the *z*-direction, as shown. Note that the radiated fields due to both conductors are of opposite directions, giving a small total radiated field as shown. This total radiated field at the observation point in the far field can be obtained by superimposing the fields due to each antenna.

Now, consider the scenario shown in Figure 9.16 where two linear antennas carry the common-mode currents.

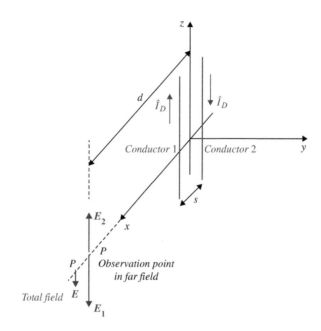

Figure 9.15   DM currents and the associated fields.

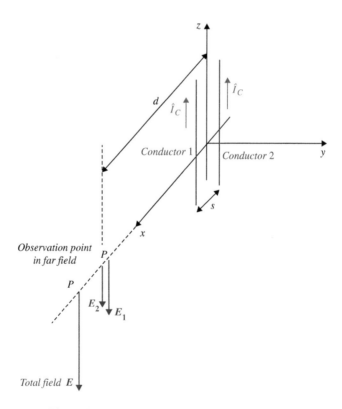

Figure 9.16   CM currents and the associated fields.

The radiated fields due to both conductors are of same directions, thus reinforcing each other to give the total radiated field as shown.

It should be noted that the CM currents could be several orders of magnitude smaller than the DM currents, yet the radiation from them could exceed the regulatory limits.

The total radiated field at the observation point the far field can be obtained by treating each of the conductors as a small dipole antenna and superimposing the fields [Paul, 2006].

In order to calculate the DM and CM radiation consider the scenario shown in Figure 9.17 [Adamczyk, 2017].

The two linear antennas shown are placed along the x axis and carry the currents into the z direction. We will determine the radiated field due to the both antennas broadside to them (i.e. in the xy plane, or for $\theta = 90°$).

The total radiated electric field at an observation point P in the far field will be the sum of the field of each conductor

$$\hat{E}_\theta = \hat{E}_{\theta 1} + \hat{E}_{\theta 2} \tag{9.21}$$

Treating each conductor as a Hertzian dipole and utilizing Eq. (8.12) we have

$$\hat{E}_{\theta 1} = j\eta_0\beta_0\frac{\hat{I}_1 l}{4\pi}\sin\theta\frac{e^{-j\beta_0 r_1}}{r_1} \tag{9.22}$$

$$\hat{E}_{\theta 1} = j\eta_0\beta_0\frac{\hat{I}_2 l}{4\pi}\sin\theta\frac{e^{-j\beta_0 r_2}}{r_2} \tag{9.23}$$

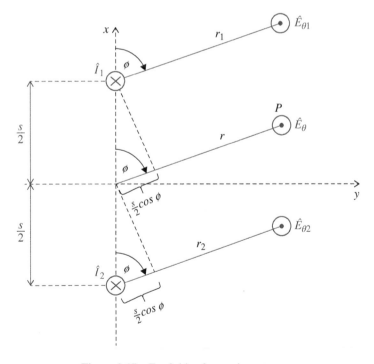

Figure 9.17   Far fields of two wire antennas.

Thus, the total field in Eq. (9.21) equals

$$\hat{E}_\theta = j\eta_0\beta_0\frac{\hat{I}_1 l}{4\pi}\sin\theta\frac{e^{-j\beta_0 r_1}}{r_1} + j\eta_0\beta_0\frac{\hat{I}_2 l}{4\pi}\sin\theta\frac{e^{-j\beta_0 r_2}}{r_2} \tag{9.24}$$

or

$$\hat{E}_\theta = j\eta_0\beta_0\frac{l}{4\pi}\sin\theta\left(\frac{\hat{I}_1 e^{-j\beta_0 r_1}}{r_1} + \frac{\hat{I}_2 e^{-j\beta_0 r_2}}{r_2}\right) \tag{9.25}$$

Using the parallel-ray approximation [Adamczyk, 2017], we have

$$r_1 = r - \frac{s}{2}\cos\phi \tag{9.26}$$

$$r_2 = r + \frac{s}{2}\cos\phi \tag{9.27}$$

Substituting Eqs. (9.26) and (9.27) into the exponential phase terms in Eq. (9.25) and substituting

$$r_1 = r \tag{9.28}$$

$$r_2 = r \tag{9.29}$$

into the distance term in the denominators in Eq. (9.25) we obtain

$$\hat{E}_\theta = j\eta_0\beta_0\frac{l}{4\pi}\sin\theta\left[\frac{\hat{I}_1 e^{-j\beta_0(r-\frac{s}{2}\cos\phi)}}{r} + \frac{\hat{I}_2 e^{-j\beta_0(r+\frac{s}{2}\cos\phi)}}{r}\right] \tag{9.30}$$

Broadside to the antennas, $\sin\theta = 1$, and therefore, the total radiated field is

$$\hat{E}_\theta = j\frac{l}{4\pi}\eta_0\beta_0\frac{e^{-j\beta_0 r}}{r}\left(\hat{I}_1 e^{j\beta_0\frac{s}{2}\cos\phi} + \hat{I}_2 e^{-j\beta_0\frac{s}{2}\cos\phi}\right) \tag{9.31}$$

The maximum radiation will occur in the plane of the wires and on the line perpendicular to the conductors, thus for $\phi = 0°$ or $\phi = 180°$ [Paul, 2006]. Using $\phi = 180°$ in Eq. (9.31) we obtain

$$\hat{E}_\theta = j\frac{l}{4\pi}\eta_0\beta_0\frac{e^{-j\beta_0 r}}{r}\left(\hat{I}_1 e^{-j\beta_0\frac{s}{2}} + \hat{I}_2 e^{j\beta_0\frac{s}{2}}\right) \tag{9.32}$$

Next, we will apply Eq. (9.32) to the differential- and common-mode currents shown in Figures 9.15 and 9.16, respectively.

9.3.1 Differential-Mode Radiation

Letting

$$\hat{I}_1 = \hat{I}_D \tag{9.33}$$

$$\hat{I}_2 = -\hat{I}_D \tag{9.34}$$

and replacing the distance r by d (taken from the midpoint between the conductors) in Eq. (9.32) we obtain

$$\hat{E}_\theta = j\frac{l}{4\pi}\eta_0\beta_0\frac{e^{-j\beta_0 d}}{d}\left(\hat{I}_D e^{-j\beta_0\frac{s}{2}} - \hat{I}_D e^{j\beta_0\frac{s}{2}}\right) \tag{9.35}$$

or

$$\hat{E}_\theta = j\frac{l}{4\pi}\eta_0\beta_0\frac{\hat{I}_D e^{-j\beta_0 d}}{d}\left(e^{-j\beta_0\frac{s}{2}} - e^{j\beta_0\frac{s}{2}}\right) \tag{9.36}$$

or

$$\hat{E}_\theta = -j\frac{l}{4\pi}\eta_0\beta_0\frac{\hat{I}_D e^{-j\beta_0 d}}{d}j2\sin\left(\beta_0\frac{s}{2}\right) \tag{9.37}$$

and thus

$$\hat{E}_\theta = \frac{l}{2\pi}\eta_0\beta_0\frac{\hat{I}_D e^{-j\beta_0 d}}{d}\sin\left(\beta_0\frac{s}{2}\right) \tag{9.38}$$

Utilizing

$$\beta_0 = \frac{2\pi}{\lambda_0} \tag{9.39}$$

in Eq. (9.38) we obtain

$$\hat{E}_\theta = \frac{l}{2\pi}\eta_0\frac{2\pi}{\lambda_0}\frac{\hat{I}_D e^{-j\beta_0 d}}{d}\sin\left(\frac{2\pi}{\lambda_0}\frac{s}{2}\right) \tag{9.40}$$

or

$$\hat{E}_\theta = \frac{l\eta_0}{\lambda_0}\frac{\hat{I}_D e^{-j\beta_0 d}}{d}\sin\left(\pi\frac{s}{\lambda_0}\right) \tag{9.41}$$

If we assume that the spacing between the dipole antennas is much smaller than the wavelength, i.e.

$$s \ll \lambda_0 \ or \ \frac{s}{\lambda_0} \ll 1 \tag{9.42}$$

we can use the approximation

$$\sin\left(\pi\frac{s}{\lambda_0}\right) \approx \pi\frac{s}{\lambda_0} \tag{9.43}$$

Using the approximation (9.43) in Eq. (9.41) we arrive at

$$\hat{E}_\theta = \frac{l\eta_0}{\lambda_0} \frac{\hat{I}_D e^{-j\beta_0 d}}{d} \pi \frac{s}{\lambda_0} \tag{9.44}$$

Now,

$$\lambda_0 = \frac{v_0}{f} = \frac{3 \times 10^8}{f} \tag{9.45}$$

$$\eta_0 = 120\pi \tag{9.46}$$

Using Eqs. (9.45) and (9.46) in Eq. (9.44) produces

$$\hat{E}_\theta = \frac{120\pi l f}{3 \times 10^8} \frac{\hat{I}_D e^{-j\beta_0 d}}{d} \pi \frac{s f}{3 \times 10^8} \tag{9.47}$$

or

$$\hat{E}_\theta = \frac{120\pi^2}{9 \times 10^{16}} f^2 l s \hat{I}_D \frac{e^{-j\beta_0 d}}{d} \tag{9.48}$$

or

$$\hat{E}_\theta = -131.59 \times 10^{-16} f^2 l s \hat{I}_D \frac{e^{-j\beta_0 d}}{d} \tag{9.49}$$

The magnitude of the total field is

$$E_\theta = 131.59 \times 10^{-16} f^2 I_D \frac{ls}{d} \tag{9.50}$$

This result is valid when the two Hertzian dipoles (two short wires) are in free space. Most of the EMC measurements of radiation are made over a ground plane which is a reflective surface that can increase the measured radiation as much as 6 dB, or a factor of two [Ott, 2009]. In this case, we double the result in Eq. (9.50) to arrive at

$$E_\theta = 263.18 \times 10^{-16} f^2 I_D \frac{ls}{d} \tag{9.51}$$

Consider the loop of differential-mode current shown in Figure 9.18.
If we replace the product ls in Eq. (9.51) by the area $A = l \times s$ we obtain

$$E_\theta = 263.18 \times 10^{-16} f^2 I_D \frac{A}{d} \tag{9.52}$$

This is the same result as the one presented in Ott [2009] for radiation from a small loop of current. It can be used to determine the maximum loop area that will not exceed a specified emission limit at a given distance, frequency, and current.

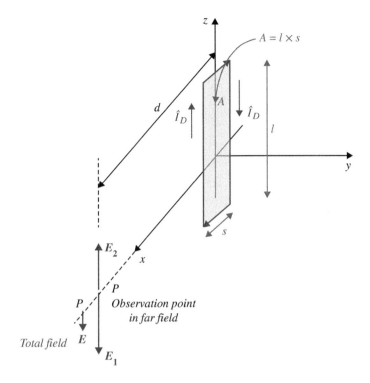

Figure 9.18 Loop of DM current.

9.3.2 Common-Mode Radiation

Letting

$$\hat{I}_1 = \hat{I}_C \tag{9.53}$$

$$\hat{I}_2 = \hat{I}_C \tag{9.54}$$

and replacing the distance r by d (taken from the midpoint between the conductors) in Eq. (9.32) we obtain

$$\hat{E}_\theta = j\frac{l}{4\pi}\eta_0\beta_0\frac{e^{-j\beta_0 d}}{d}\left(\hat{I}_C e^{-j\beta_0\frac{s}{2}} + \hat{I}_C e^{j\beta_0\frac{s}{2}}\right) \tag{9.55}$$

or

$$\hat{E}_\theta = j\frac{l}{4\pi}\eta_0\beta_0\frac{\hat{I}_C e^{-j\beta_0 d}}{d}\left(e^{-j\beta_0\frac{s}{2}} + e^{j\beta_0\frac{s}{2}}\right) \tag{9.56}$$

or

$$\hat{E}_\theta = -j\frac{l}{4\pi}\eta_0\beta_0\frac{\hat{I}_C e^{-j\beta_0 d}}{d}2\cos\left(\beta_0\frac{s}{2}\right) \tag{9.57}$$

or

$$\hat{E}_\theta = -j\frac{l}{2\pi}\eta_0\beta_0\frac{\hat{I}_C e^{-j\beta_0 d}}{d}\cos\left(\beta_0\frac{s}{2}\right) \tag{9.58}$$

Utilizing

$$\beta_0 = \frac{2\pi}{\lambda_0} \tag{9.59}$$

in Eq. (9.58) we obtain

$$\hat{E}_\theta = -j\frac{l}{2\pi}\eta_0\frac{2\pi}{\lambda_0}\frac{\hat{I}_C e^{-j\beta_0 d}}{d}\cos\left(\frac{2\pi}{\lambda_0}\frac{s}{2}\right) \tag{9.60}$$

or

$$\hat{E}_\theta = -j\frac{l\eta_0}{\lambda_0}\frac{\hat{I}_C e^{-j\beta_0 d}}{d}\cos\left(\pi\frac{s}{\lambda_0}\right) \tag{9.61}$$

If we assume that the spacing between the dipole antennas is much smaller than the wavelength, i.e.

$$s \ll \lambda_0 \text{ or } \frac{s}{\lambda_0} \ll 1 \tag{9.62}$$

we can use the approximation

$$\cos\left(\pi\frac{s}{\lambda_0}\right) \approx 1 \tag{9.63}$$

Using the approximation (9.63) in Eq. (9.61) we arrive at

$$\hat{E}_\theta = -j\frac{l\eta_0}{\lambda_0}\frac{\hat{I}_C e^{-j\beta_0 d}}{d} \tag{9.64}$$

Again,
Now,

$$\lambda_0 = \frac{v_0}{f} = \frac{3\times 10^8}{f} \tag{9.65}$$

$$\eta_0 = 120\pi \tag{9.66}$$

Using Eqs. (9.65) and (9.66) in Eq. (9.64) produces

$$\hat{E}_\theta = -j\frac{120\pi l f}{3\times 10^8}\frac{\hat{I}_C e^{-j\beta_0 d}}{d} \tag{9.67}$$

or

$$\hat{E}_\theta = -j125.66\times 10^{-8} f l\hat{I}_C\frac{e^{-j\beta_0 d}}{d} \tag{9.68}$$

The magnitude of the total field is

$$E_\theta = 125.66 \times 10^{-8} f I_C \frac{l}{d} \qquad (9.69)$$

Solving Eq. (9.69) for the CM current gives

$$I_C = \frac{E_\theta \times d \times 10^8}{125.66 \times f \times l} = 795,798.2 \frac{E_\theta \times d}{f \times l} \qquad (9.70)$$

If the CM current is expressed in μA, frequency in MHz, and the electric field strength is expressed in μV/m, then Eq. (9.70) becomes

$$I_C = 0.79 \frac{E_\theta \times d}{f \times l} \qquad (9.71)$$

This result can be used to determine the maximum allowable CM current at a given frequency, distance, and regulatory emission limit.

9.4 Laboratory Exercises

9.4.1 Differential-Mode and Common-Mode Current Measurement

Objective: The objective of this laboratory is to measure the differential-mode and common-mode currents.

Note: This laboratory exercise uses a SMPS designed in Chapter 12.

9.4.2 Laboratory Equipment and Supplies

DC Power Supply: BK Precision1760A (or equivalent), shown in Figure 9.19.

Banana Cables: Tektronix MDO3104 (or equivalent), shown in Figure 9.20.
Test Lead Banana to Banana 36" (2 sets)
Supplier: Ponoma Electronics
Digi-Key Part Number: 501-1423-ND

RF Current Probe: Fischer F-33-1 (or equivalent), shown in Figure 9.21.
SMPS: A step-down 24–3.3 V DC Switched-Mode Power Supply, designed and built in laboratory exercise in Chapter 12; shown in Figure 9.22.

9.4.3 Laboratory Procedure – Differential-Mode and Common-Mode Current Measurements

This section describes the differential-mode and common-mode currents measurements, made with a step-down, 12–5 V, SMPS [Adamczyk, 2020].

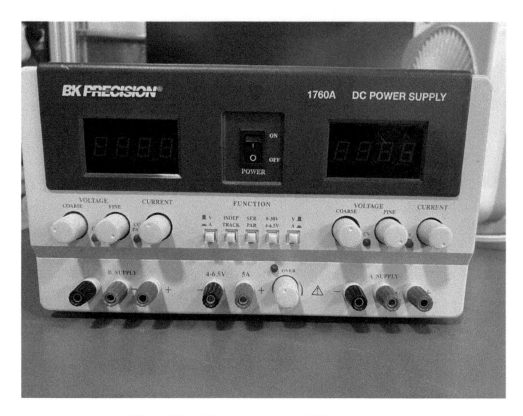

Figure 9.19   Laboratory equipment: DC power supply.

Figure 9.20   Laboratory equipment: banana cables.

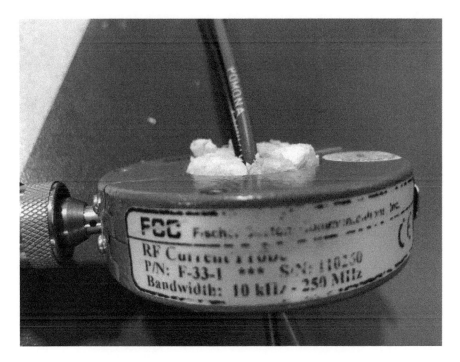

Figure 9.21 Laboratory equipment: RF current probe.

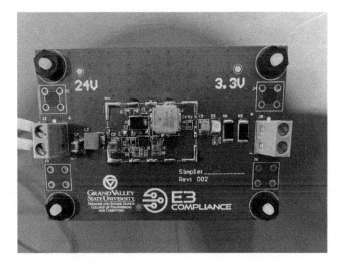

Figure 9.22 Laboratory equipment: SMPS.

Laboratory Exercise: After reading this section, replicate this laboratory exercise with your own 24–3.3 V SMPS designed in Chapter 12.

Figure 9.23 shows the test setup to measure the differential- and common-mode currents.

The SMPS used in this experiment is a step-down (buck), 12–5 V DC, switching at 420 kHz. The CM currents are measured with the current probe, where both the power and ground wires were placed inside the current probe, as shown in Figure 9.24.

With both wires inside the probe, the differential current fields (ideally) cancel each other, and the current probe measures only the common-mode currents. To be precise, it (ideally) measures twice the value of the CM current, i.e. $2I_C$. The measurement results are shown in Figure 9.25 and summarized in Figure 9.26. The measurement values are the relative values in the units of dBμV.

Next, let's measure the total currents. The total currents are measured with two different setups: current probe over the ground wire and the current probe over the power wire, as shown in Figure 9.27.

Note that in these setups, we measure the magnitudes of the total currents $\hat{I}_1$ and $\hat{I}_2$ given by Eqs. (9.1) and (9.2) and shown in Figure 9.2. As we shall see, in this experiment, the CM currents exist at the different frequencies than the DM currents, and thus, the total current measurements can be used to extract the DM currents. That's why we measured the CM currents first.

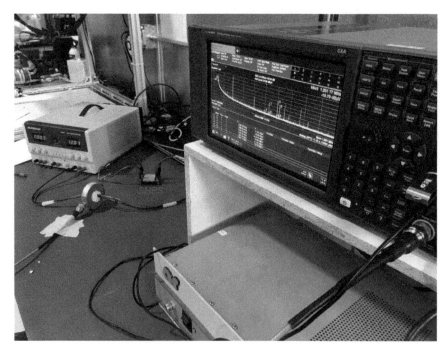

Figure 9.23   Measurement setup.

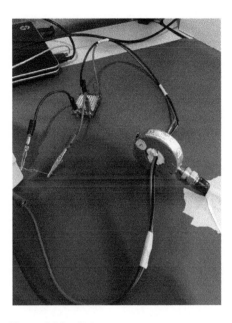

Figure 9.24   CM-current measurement setup.

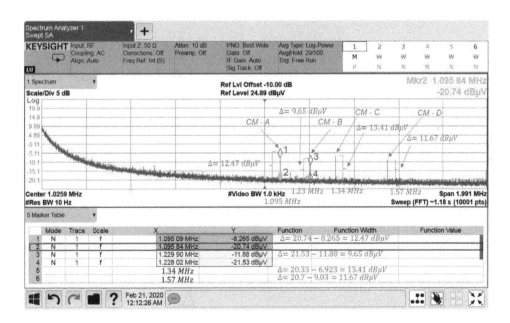

Figure 9.25   CM-current measurement results.

Common-mode current	Frequency (MHz)	Magnitude (dBµV)
CM-A	1.095	12.47
CM-B	1.23	9.65
CM-C	1.34	13.41
CM-D	1.57	11.67

Figure 9.26   Tabulated CM-current measurement results.

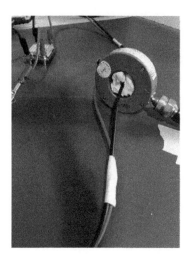 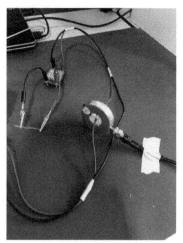

Figure 9.27   Total and DM-current measurements.

The measurement results with the probe over the ground line are shown in Figure 9.28, while the results for the power line are shown in Figure 9.29. Both results are summarized in Figure 9.30. The measurement values are the relative values in the units of dBµV.

Observations: Except for the frequency of 114.8 kHz, the DM currents appear at the harmonics of the switching frequency. The ground and the power wire total measurements capture both the DM and the CM currents. The CM currents are not as predictable as the DM currents. Note that the ground-wire CM-current is present at point *A* in Figure 9.28, but it is not present at that frequency on the power wire in Figure 9.29. Another CM current at a lower frequency, at point *K*, appears in Figure 9.29, and it was not present at that frequency in Figure 9.28.

As stated earlier, replicate this laboratory exercise with your own 24–3.3 V SMPS designed in Chapter 12.

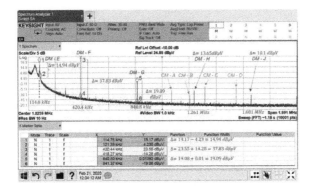

Figure 9.28 Total and DM-current measurement results – ground wire.

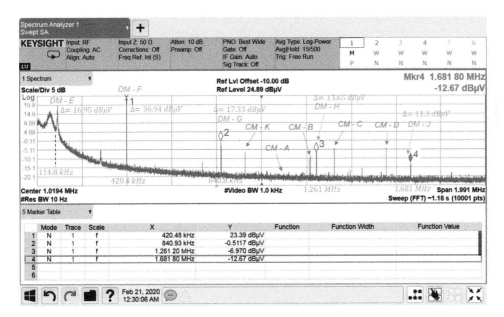

Figure 9.29 Total and DM-current measurement results – power wire.

Differential-mode current	Frequency	Magnitude (dBµV) ground wire	Magnitude (dBµV) power wire
DM-E	114.8 kHz	14.94	16.95
DM-E	420.4 kHz	37.83	36.94
DM-F	840.9 kHz	19.09	17.33
DM-G	1.261 MHz	13.65	12.51
DM-J	1.681 MHz	10.1	11.3

Figure 9.30 Tabulated DM-current measurement results.

References

Bogdan Adamczyk. *Foundations of Electromagnetic Compatibility*. John Wiley & Sons, Ltd, Chichester, UK, 2017. ISBN 9781119120810. doi: https://doi.org/10.1002/9781119120810.

Bogdan Adamczyk. Common-Mode Current Creation and Suppression. *In Compliance Magazine*, August 2019. URL https://incompliancemag.com/article/common-mode-current-creation-and-suppression/.

Bogdan Adamczyk. Measuring Differential- and Common-Mode Current Radiation from Cables. *In Compliance Magazine*, October 2020. URL https://incompliancemag.com/article/measuring-differential-and-common-mode-current-radiation-from-cables/.

Murata. CM Choke, 2023. URL https://www.murata.com/en-us/products/emc/emifil/products-search/selectionguide/highspeed.

Henry W. Ott. *Electromagnetic Compatibility Engineering*. John Wiley & Sons, Inc., Hoboken, New Jersey, 2009. ISBN 978-0-470-18930-6.

Clayton R. Paul. *Introduction to Electromagnetic Compatibility*. John Wiley & Sons, Inc., Hoboken, New Jersey, 2nd edition, 2006. ISBN 978-0-471-75500-5.

Chapter 10: Return-Current Path, Flow, and Distribution

10.1 Return-Current Path

Consider a two-layer PCB with a single trace on top and solid copper reference (ground) plane on the bottom, shown in Figure 10.1.

At points A and B, vias connect the top trace to the ground plane [Adamczyk, 2017]. The forward current flows on the top trace as shown in Figure 10.2.

Upon reaching the point B the current travels through a via to the ground plane and returns to the source. But how does the return current flow back to the source? The return current has a few options: the direct path from B to A or an alternative path underneath the top trace, or a combination of both, as shown in Figure 10.3.

Current returns to the source through the path of least impedance. Using the lumped-parameter circuit theory, we can associate this impedance, Z_g, with the ground path and model it as

$$Z_g = R_g + j\omega L_g \tag{10.1}$$

where R_g and L_g are the ground path resistance and inductance, respectively.

At low frequencies the ground current will take the path of least resistance, the direct path from B to A (which corresponds to the path of the lowest impedance). This is shown in Figure 10.4.

At high frequencies the return current will take the path of least inductance, which is directly underneath the trace, because this represents the smallest loop area (smallest inductance). This is shown in Figure 10.5.

Figure 10.6 shows the CST Studio model of a two-sided PCB with a solid copper return plane and a microstrip trace on the top layer.

Figure 10.7 shows the return-current path on the reference plane at different frequencies.

Observations: At 10 Hz the return current spreads wide over the reference plane, flowing both under the top trace and directly from the source port to the termination port. As the frequency increases to 100 Hz, more of the return current flows under the trace (with a narrower spread), and less of it flows directly from the source port to the termination port. This trend continues as the frequency increases to 1 kHz.

The results for 10 kHz and beyond show something very interesting. As the frequency increases, the return-current path remains virtually unchanged. In other words, the return-current path and current density no longer depend on frequency. The frequency is high enough that the resistance of the return plane is negligible compared to its inductive reactance [Ott, 2009].

Johnson and Graham [1993], explain this phenomenon as follows: "The current distribution, … , balances two opposing forces. Were the current more tightly drawn together, it would have higher inductance (a skinny wire has more inductance than a broad, flat one). Were the current spread farther apart from the signal trace, the total loop area between the outgoing and returning signal paths would increase, raising the inductance."

Principles of Electromagnetic Compatibility: Laboratory Exercises and Lectures, First Edition. Bogdan Adamczyk.
© 2024 John Wiley & Sons Ltd. Published 2024 by John Wiley & Sons Ltd.
Companion website: www.wiley.com/go/principlesofelectromagneticcompatibility

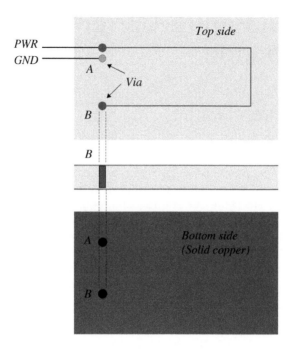

Figure 10.1   Two-sided PCB with a solid ground plane.

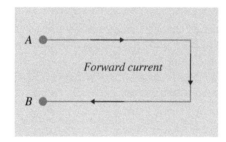

Figure 10.2   Forward current flow.

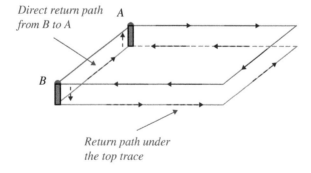

Figure 10.3   Return-current alternative paths.

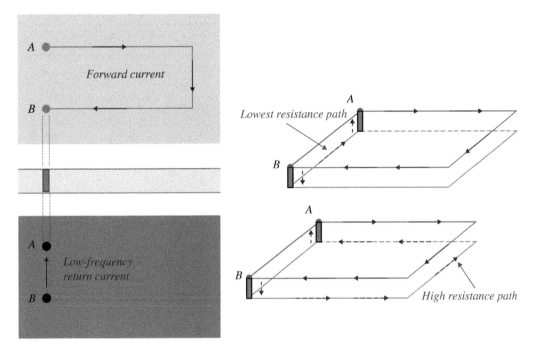

Figure 10.4   Low-frequency return-current path.

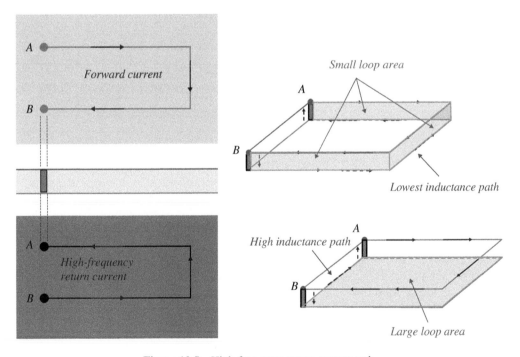

Figure 10.5   High-frequency return-current path.

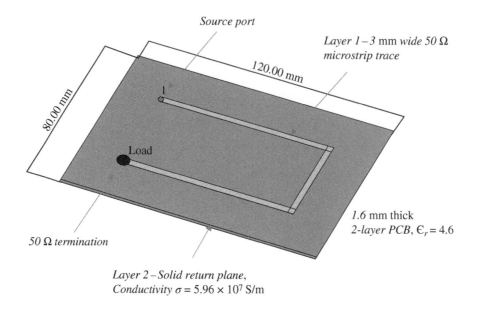

Figure 10.6   Two-sided PCB used in CST Studio simulations.

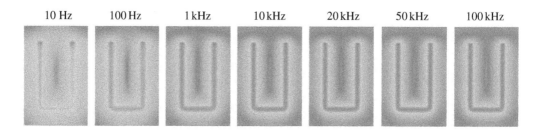

Figure 10.7   Return-current path at different frequencies.

The just-presented PCB return-current path discussion can be directly extended to a coaxial cable configuration shown in Figure 10.8 [Van Doren, 2007].

The forward current flows in the center conductor. The return current can either flow through a direct path (low-frequency return path) or the shield (high-frequency return path). We will return to this configuration in the Laboratory Exercises section of this chapter.

The preceding discussion considered the return-current flow when the return path was through a solid (uninterrupted) ground plane. Let us now investigate the effect of a discontinuity in the ground plane, as shown in Figure 10.9.

When the return plane contains a slot or a cutout, the return current flows around it creating a larger current loop area. This is shown in Figure 10.9. Larger current loop area results in the increase in both the radiation and the loop inductance (both detrimental effects).

The clearance holes impede the current flow to a lesser degree. If the holes do not overlap and the return current can flow between them, there is no significant increase in

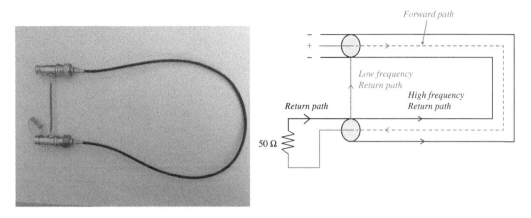

Figure 10.8 Coaxial-cable return-current path.

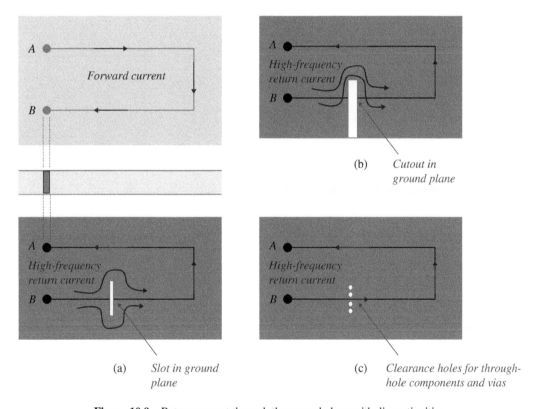

Figure 10.9 Return current through the ground plane with discontinuities.

the radiation. We will return to this configuration in the Laboratory Exercises section of this chapter.

Figure 10.10 shows the CST Studio model of a two-sided PCB with cutouts added in the return (ground) plane.

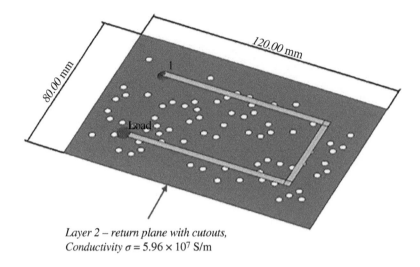

Layer 2 – return plane with cutouts,
Conductivity $\sigma = 5.96 \times 10^7$ S/m

Figure 10.10   Two-sided PCB with cutouts in the return plane.

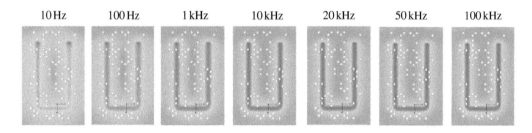

Figure 10.11   Return-current path with cutouts in the reference plane.

Figure 10.11 shows the return current-path on the reference plane at different frequencies.

Observations: The results look very similar to the ones from the solid return plane. We will look closer at these results in Section 10.3.1.1.

10.2 Return-Current Flow

Before we discuss the return-current distribution for the microstrip and stripline configurations, let's address the topic of the return-current flow.

The return currents will flow on the adjacent reference plane underneath the trace where the electric field lines terminate, as shown in Figure 10.12, for a microstrip line.

The reference plane could be a ground or a power plane, as shown in Figure 10.13, depicting one of the possible four-layer PCB configurations.

Let's stop here for a minute and answer this important question: How is it possible for the return current to flow on the power plane?

This seems to contradict what we have learned in a basic circuit course. In a "classical" circuit course we always assumed that the current flows out of the positive terminal of the

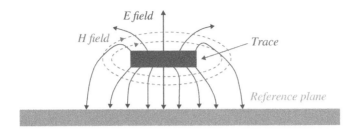

Figure 10.12 Microstrip line and the surrounding fields.

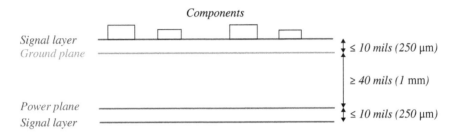

Figure 10.13 Microstrip configuration in a four-layer PCB.

source, flows along the forward path, through the load, and returns to the source on the return (ground) path (conductor or plane).

This is true for pure DC currents where we model the current flow as moving charges along the conductor, flowing with a drift velocity proportional to the voltage of the source. We often ignore the drift velocity and assume that there is no time delay in the current flow and the voltages and currents appear instantaneously everywhere in the circuit when the source is connected to it.

Pure AC current (regardless of its frequency) does not involve the charges moving along a conductor; it is modeled as the charges oscillating back and forth (with respect to their original position) as the polarity of the source changes. This ac model is valid for the functional, i.e. intentional ac currents. When analyzing such circuits, we still assume that the current flows from the positive terminal of the source, along the forward conductor to the load, and returns to the source along the return (ground) path (conductor or a plane). And again, we often ignore the time delay.

Now, the situation is quite different for high-frequency noise currents. Actually, for any-frequency noise currents, we usually ignore the low-frequency noise and focus on high-frequency noise currents. These high-frequency noise currents (created for instance during the switching the DC voltage levels), are superimposed on the existing functional DC or AC currents.

In understanding their impact, it helps to think of them as the electromagnetic waves traveling on the surface of the conductor. These waves disturb the functional behavior of the charges moving in a conductor, regardless of what DC potential the conductor is at! These high-frequency noise waves (current) will "flow" on any conducting surface, regardless whether it is a ground or power plane!

Figure 10.14   High-frequency return currents.

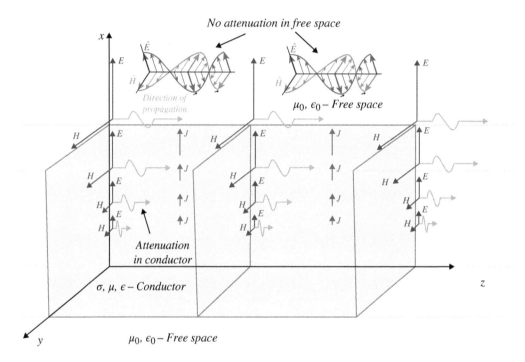

Figure 10.15   Fields and current density inside a conductor.

Now, having established the fact that the high-frequency noise current can flow on either the ground or power plane, let's look into more details of such a flow. The high-frequency current flows within a few skin depths of the reference plane surface as shown in Figure 10.14.

Figure 10.15 shows the fields and the current density inside the current-carrying conductor [Adamczyk, 2020a].

Since the current density decays to virtually zero within a few skin depths in a conductor, the high-frequency currents are often considered to be the surface currents. Consequently, the reference plane can be considered as two different conductors. The high-frequency current flowing on the top surface is different from the high-frequency current flowing on the bottom surface. This is illustrated in Figure 10.16.

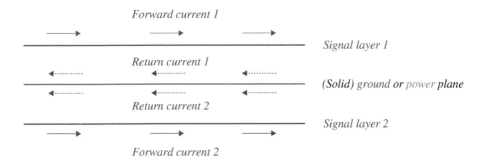

Figure 10.16   Currents on both sides of the reference plane.

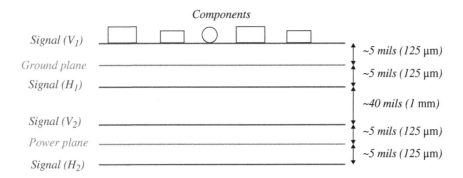

Figure 10.17   Six-layer PCB with shared reference planes.

Figure 10.17 shows one of the possible six-layer PCB configurations, where such a scenario takes place.

10.3 Return-Current Distribution

10.3.1 Microstrip Line PCB

Figure 10.18 shows a microstrip geometry where the trace of width w is at a height h above a reference plane; x is the distance from the center of the trace.

Figure 10.19 shows a microstrip configuration in a four-layer PCB.

The current distribution on the reference plane underneath a microstrip is described by its current density $J(x)$ [Ott, 2009],

$$J(x) = \frac{I}{w\pi}\left[\tan^{-1}\left(\frac{2x-w}{2h}\right) - \tan^{-1}\left(\frac{2x+w}{2h}\right)\right] \qquad (10.2)$$

where I is the total current flowing in the loop. The current density underneath the center of the trace is

$$J(0) = \frac{I}{w\pi}\left[\tan^{-1}\left(\frac{-w}{2h}\right) - \tan^{-1}\left(\frac{w}{2h}\right)\right] \qquad (10.3)$$

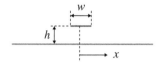

Figure 10.18   Microstrip line geometry.

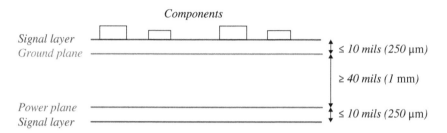

Figure 10.19   Four-layer microstrip PCB.

Figure 10.20 shows the (normalized) current density underneath a microstrip line as a function of x/h. Matlab script is shown in Figure 10.21.

Observations: (i) Majority of the current tends to remain close to the area underneath the trace. (ii) As the distance from the center underneath the trace increases the current density becomes smaller.

Figure 10.22 shows the % of the total microstrip return current contained in the portion of the plane between $\pm x/h$ of the centerline underneath the trace. Matlab script is shown in Figure 10.23.

Observations: (i) 50% of the current is contained within a distance $\pm h$. (ii) 80% of the current is contained within a distance $\pm 3\,h$. (iii) 97% of the current is contained within a distance $\pm 20\,h$.

Figure 10.24 shows more detailed results for other distances from the centerline underneath the trace [Ott, 2009].

10.3.1.1 CST Simulation Results

Let's verify the analytical results of Section 10.3.1 through simulation.

Figure 10.25 shows the dimensions of a two-layer microstrip PCB with a solid return plane, used in the CST Studio simulations.

Figure 10.26 shows the details of the board geometry and the centerline location.

Normalized current distributions at different frequencies along the z axis are shown in Figure 10.27.

Observation: When the current frequency is above 10 kHz, the current distribution is virtually unaltered, i.e. does not depend on frequency. This is consistent with the result presented earlier in Figure 10.7 and repeated here, as Figure 10.28, for the ease of comparison.

Next, let's look at the current distribution at different frequencies for the board with the return plane cutouts, shown in Figure 10.29. Except for the cutouts, board geometry is the same as the one with solid return plane.

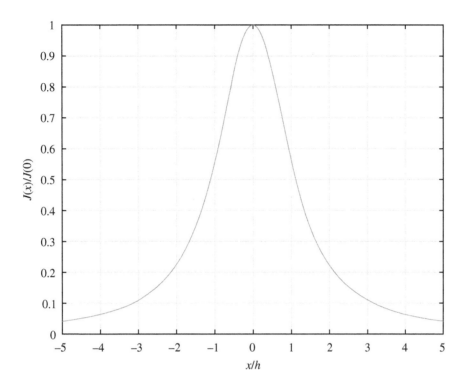

Figure 10.20 Normalized reference plane current density underneath a microstrip line.

```
% Microstrip - Normalized current density
clc;
clear all;
x=-5:0.01:5; % distance from center of the trace
w=1; % width of the trace
h=1; % height of the trace from reference plane
I=1; % I is the total current flowing in the loop
% Current density at distance x
J=-(I/(w*pi))*(atan(((2*x)-w)/(2*h))-atan(((2*x)+w)/(2*h)));
% current density at x=0
J0=-(I/(w*pi))*(atan(((2*0)-w)/(2*h))-atan(((2*0)+w)/(2*h)));
Jx=J/J0; % Normalized current density
plot(x/h,Jx)
grid on;
xlabel('x/h');
ylabel('J(x)/J(0)');
```

Figure 10.21 Normalized reference plane current density underneath a microstrip line – Matlab script.

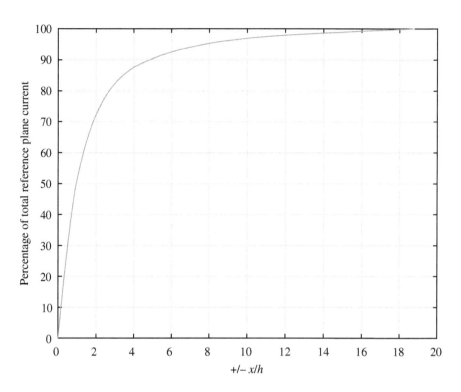

Figure 10.22   Integral of the normalized reference plane current density underneath a microstrip line.

```
% Microstrip - cumulative current density
clc;
clear all;
x=0:0.01:20; % Distance from the center of the trace
h=1; % height of the trace from reference plane
w=1; % width of the trace
I=1; % I total current flowing in the loop
% Current density at distance x
J=-(I./(w.*pi))*(atan(((2.*x)-w)/(2.*h))-
atan(((2.*x)+w)/(2.*h)));
% Current density at x = 0
J0=-(I./(w.*pi))*(atan(((2.*0)-w)/(2.*h))-
atan(((2.*0)+w)/(2.*h)));
Jx=cumtrapz(x,J);% integral of current density
normalized_Jx=Jx./max(Jx); %normalization
plot(x,(normalized_Jx.*100));
xlim ([0 20]);
grid on;
xlabel('+/- x/h');
ylabel('Percentage of total reference plane current');
```

Figure 10.23   Integral of the normalized reference plane current density underneath a microstrip line – Matlab script.

x/h	% of current
1	50
2	70
3	80
5	87
10	94
20	97
50	99
100	99.4
500	99.9

Figure 10.24   % of the microstrip return current contained in the portion of the plane between $\pm h$ of the trace centerline.

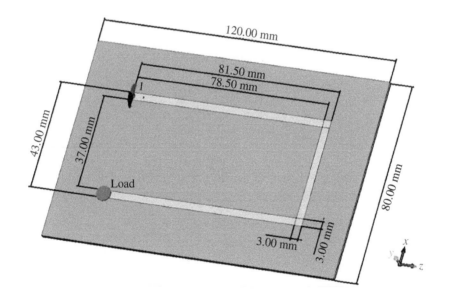

Figure 10.25   Dimensions of the PCB used in simulations.

Normalized current distributions at different frequencies along the z axis are shown in Figure 10.30.

Observations: At lower frequencies (below 10 kHz), the impact of the cutouts is more pronounced than at higher frequencies. Beyond 10 kHz the impact of cutouts remains virtually unchanged.

Finally, let's compare the current distributions between the boards with solid return plane and the return plane with cutouts. We will do this for a few selected frequencies.

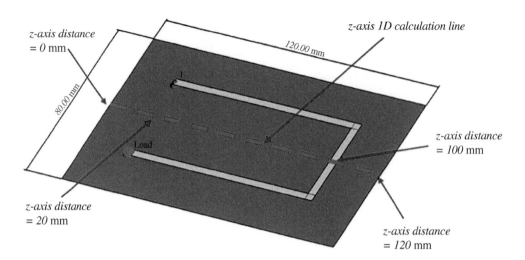

Figure 10.26   Board geometry and the centerline location.

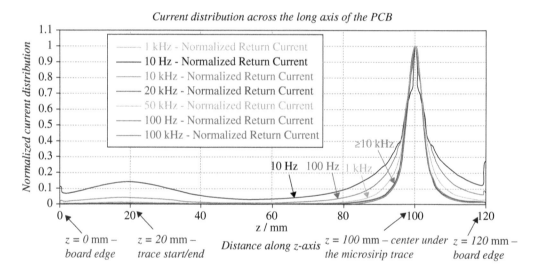

Figure 10.27   Normalized current distributions at different frequencies.

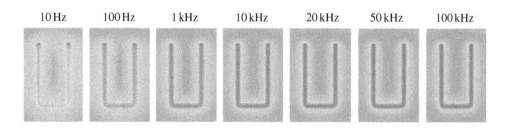

Figure 10.28   Return-current path for different frequencies.

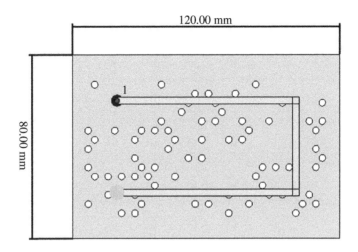

Figure 10.29   PCB with the return plane cutouts.

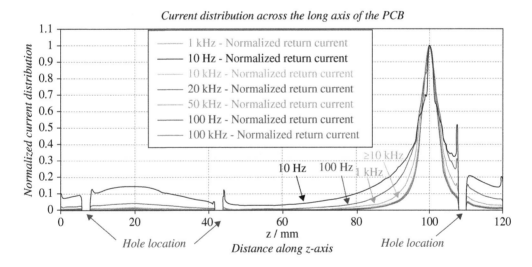

Figure 10.30   Normalized current distribution for a PCB with cutouts.

Let's start with the lower frequencies. Figure 10.31 compares the distributions for at 100 Hz, while Figure 10.32 compares the distributions at 1 kHz.

Observation: At lower frequencies the impact of the cutouts is more noticeable when compared to the solid return plane. The impact, however, is not significant, except at and adjacent to the cutouts.

Lastly, let's look at the impact of the cutouts at higher frequencies. Figure 10.33 compares the distributions for at 10 kHz, while Figure 10.34 compares the distributions at 100 kHz.

Observations: At higher frequencies the impact of the cutouts is minimal, except at and adjacent to the cutouts. Additionally, this minimal impact does not change with frequency.

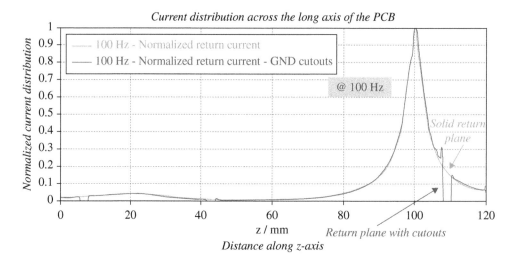

Figure 10.31   Distributions at 100 Hz – solid return plane vs. plane with cutouts.

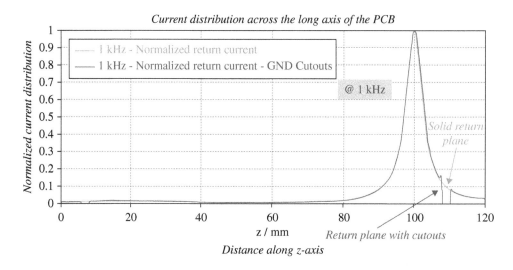

Figure 10.32   Distributions at 1 kHz – solid return plane vs. plane with cutouts.

10.3.2 Stripline PCB

10.3.2.1 Symmetric Stripline PCB

Next, let's turn our attention to a symmetric stripline configuration shown in Figure 10.35, where the trace of width *w* is located in between two planes, at a distance *h* from each reference plane; *x* is the distance from the center of the trace [Adamczyk, 2020b].

The possible plane combinations are shown in Figure 10.36.

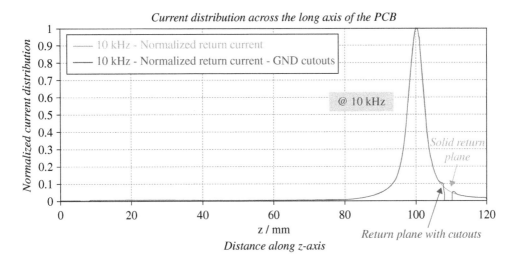

Figure 10.33 Distributions at 10 kHz – solid return plane vs. plane with cutouts.

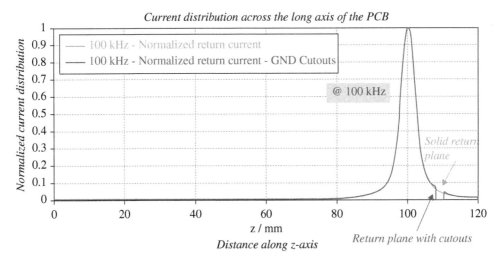

Figure 10.34 Distributions at 100 kHz – solid return plane vs. plane with cutouts.

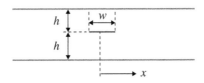

Figure 10.35 Symmetric stripline geometry.

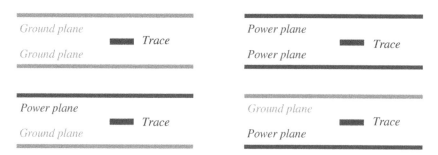

Figure 10.36   Plane combinations for a symmetric stripline.

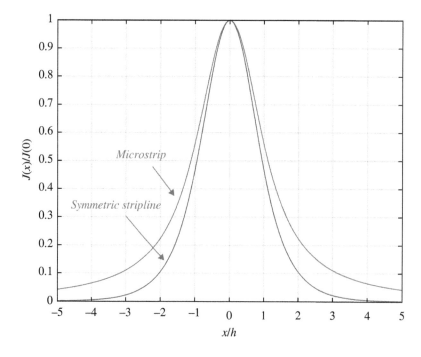

Figure 10.37   Normalized reference plane current density – symmetric stripline vs. microstrip.

The reference plane current density $J(x)$ at a distance x from the center is given by [Ott, 2009]:

$$J(x) = \frac{I}{w\pi}\left[\tan^{-1}e^{\left(\frac{\pi(x-w/2)}{2h}\right)} - \tan^{-1}e^{\left(\frac{\pi(x+w/2)}{2h}\right)}\right] \qquad (10.4)$$

where I is the total current flowing in the loop. Eq. (10.4) represents the current density in just one of the two reference planes. The total reference plane current density is twice of that in Eq. (10.4).

```
% Microstrip vs. symmetric stripline
clc; clear all;
x=-5:0.01:5; w=1; h=1; I=1;
% Current density at distance x (microstrip)
J=-(I/(w*pi))*(atan((((2*x)-w)/(2*h))-atan((((2*x)+w)/(2*h))));
% Current density (symmetric stripline)
Js=-(I/(w*pi))*((atan(exp((pi*(x-(w/2)))/(2*h)))-
atan(exp((pi*(x+(w/2)))/(2*h))))));
% current density at x=0 (microstrip)
JO=-(I./(w.*pi))*(atan((((2.*0)-w)/(2.*h))-
atan((((2*0)+w)/(2*h))));
% current density at x = 0 (symmetric stripline)
JOs=-(I/(w*pi))*((atan(exp((pi*(0-(w/2)))/(2*h)))-
atan(exp((pi*(0+(w/2)))/(2*h))))));
Jx=J/JO; % Normalization (microstrip)
Jxs=Js/JOs; % Normalization (symmetric stripline)
plot(x,Jx); hold on
plot(x,Jxs); grid on;
xlabel('x/h'); ylabel('J(x)/J(0)');
```

Figure 10.38 Normalized reference plane current density – symmetric stripline vs. microstrip – Matlab script.

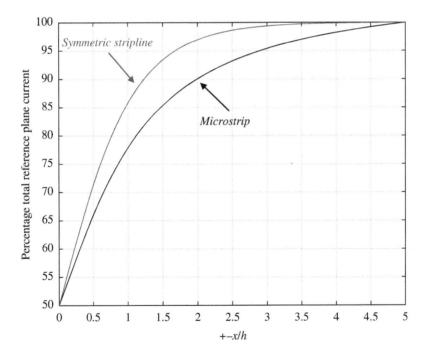

Figure 10.39 % of the total reference plane current symmetric stripline vs. microstrip.

Figure 10.37 shows the (normalized) current density as a function of x/h, for both the symmetric stripline and a microstrip configurations. Matlab script is shown in Figure 10.38.

Note that the stripline current does not spread out as far as in the case of a microstrip line. At a distance $\pm 4\, x/h$ from the center, current density in a stripline rapidly decays toward zero, while in a microstrip there's still a noticeable non-zero current density.

Figure 10.39 shows the % of the total return current for both configurations contained within a distance $\pm x/h$ of the trace centerline. In the symmetric stripline case, half of the total current flows in each reference plane.

Matlab script is shown in Figure 10.40.

Figure 10.41 shows the tabulated results for the symmetric stripline [Ott, 2009].

```
% Microstrip vs. symmetric stripline - cumulative
J_microstrip=-(I./(w.*pi))*(atan(((2.*x)-w)/(2.*h))-
atan(((2.*x)+w)/(2.*h)));
J0_microstrip=-(I./(w.*pi))*(atan(((2.*0)-w)/(2.*h))-
atan(((2.*0)+w)/(2.*h)));
Jx_microstrip=cumtrapz(x,J_microstrip);
normalized_Jx_microstrip= Jx_microstrip./max(Jx_microstrip);
percentage_Jx_microstrip=(Jx_microstrip./max(Jx_microstrip)).*100;
J_stripline=-(I./(w.*pi))*((atan(exp((pi.*(x-(w./2)))/(2.*h)))-
atan(exp((pi.*(x+(w./2)))/(2.*h)))));
J0_stripline=-(I./(w.*pi))*((atan(exp((pi.*(0-(w./2)))/(2.*h)))-
atan(exp((pi.*(0+(w./2)))/(2.*h)))));
Jx_stripline=cumtrapz(x,J_stripline);
normalized_Jx_stripline = Jx_stripline./max(Jx_stripline);
percentage_Jx_stripline = (Jx_stripline./max(Jx_stripline)).*100;
plot(x,percentage_Jx_microstrip,'k',x,percentage_Jx_stripline,'r');
xlim([0 5]); grid on; xlabel('+-x/h');
ylabel('Percentage total reference plane current');
```

Figure 10.40 Cumulative distribution of the return current – symmetric stripline vs. microstrip – Matlab script.

x/h	% of Current (symmetric stripline)	% of Current (microstrip)
1	74	50
2	94	70
3	99	80
5	99.95	87
10	99.9999756	94

Figure 10.41 % of the stripline return current contained in the portion of the plane between $\pm x/h$ of the trace centerline – stripline vs. microstrip.

10.3.2.2 Asymmetric Stripline PCB

Consider an asymmetric stripline configuration, shown in Figure 10.42, where h_1 is the distance between the trace and the closest plane, while h_2 is the distance between the trace and the furthest plane.

The asymmetric configuration with $h_2 = 2h_1$ is commonly used on PCBs where two orthogonally routed signal layers are located between two reference planes, as shown in Figure 10.43.

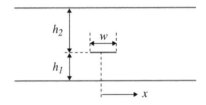

Figure 10.42 Asymmetric stripline configuration.

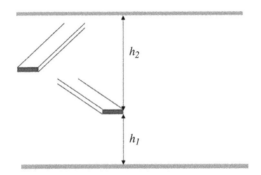

Figure 10.43 Asymmetric stripline with orthogonal traces.

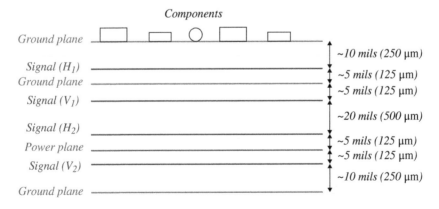

Figure 10.44 Eight-layer PCB with orthogonal traces.

Figure 10.44 shows an eight-layer PCB with such a configuration.

The asymmetric stripline current density for the *close* and *far* reference planes, respectively, is given by [Ott, 2009],

$$J_{close}(x) = \frac{I}{w\pi}\left\{\tan^{-1}\left[\frac{e^{\left(\frac{\pi(x-w/2)}{h_1+h_2}\right)}-\cos\left(\frac{\pi h_1}{h_1+h_2}\right)}{\sin\left(\frac{\pi h_1}{h_1+h_2}\right)}\right] - \tan^{-1}\left[\frac{e^{\left(\frac{\pi(x+w/2)}{h_1+h_2}\right)}-\cos\left(\frac{\pi h_1}{h_1+h_2}\right)}{\sin\left(\frac{\pi h_1}{h_1+h_2}\right)}\right]\right\}$$ (10.5)

$$J_{far}(x) = \frac{I}{w\pi}\left\{\tan^{-1}\left[\frac{e^{\left(\frac{\pi(x-w/2)}{h_1+h_2}\right)}-\cos\left(\frac{\pi h_2}{h_1+h_2}\right)}{\sin\left(\frac{\pi h_2}{h_1+h_2}\right)}\right] - \tan^{-1}\left[\frac{e^{\left(\frac{\pi(x+w/2)}{h_1+h_2}\right)}-\cos\left(\frac{\pi h_2}{h_1+h_2}\right)}{\sin\left(\frac{\pi h_2}{h_1+h_2}\right)}\right]\right\}$$ (10.6)

where I is the total current flowing in the loop.

Figure 10.45 shows the current densities for the *close* and *far* reference planes with $h_2 = 2h_1$, as a function of x/h_1.

Matlab script is shown in Figure 10.46.

Directly under the trace, approximately 74% of the current is on the close plane and 26% of the current is on the far plane. At distances greater than approximately $3 \times x/h_1$, the currents in both planes become the same.

Figure 10.47 shows an eight-layer asymmetric stripline PCB configuration with $h_2 = 3h_1$.

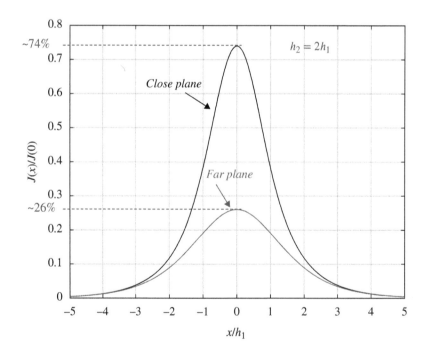

Figure 10.45   Normalized current densities for the close and far reference planes, $h_2 = 2h_1$.

```
% Asymmetric stripline
clear all;
x=-5:0.01:5; w=1; I = 1;
h1=1; % height of the trace from the close plane
h2_1=2; % height of the trace from the far plane
% current density near plane
Jx1_close=-(I/(w*pi))*((atan(((exp((pi*(x-(w/2)))/(h1+h2_1))-
cos((pi*h1)/(h1+h2_1)))/(sin((pi*h1)/(h1+h2_1))))))-
(atan(((exp((pi*(x+(w/2)))/(h1+h2_1)))-
cos((pi*h1)/(h1+h2_1)))/(sin((pi*h1)/(h1+h2_1)))))));
% current density for the far plane
Jx1_far=-(I/(w*pi))*((atan(((exp((pi*(x-(w/2)))/(h1+h2_1)))-
cos((pi*h2_1)/(h1+h2_1)))/(sin((pi*h2_1)/(h1+h2_1))))))-
(atan(((exp((pi*(x+(w/2)))/(h1+h2_1)))-
cos((pi*h2_1)/(h1+h2_1)))/(sin((pi*h2_1)/(h1+h2_1)))))));
normalized_Jx1_close=Jx1_close/max(Jx1_close+Jx1_far);
normalized_Jx1_far=Jx1_far/max(Jx1_far+Jx1_close);
plot(x,normalized_Jx1_close,'k',x,normalized_Jx1_far,'r');
grid on; xlabel('x/h1');ylabel('J(x)/J(0)');
```

Figure 10.46   Asymmetric stripline – Matlab script.

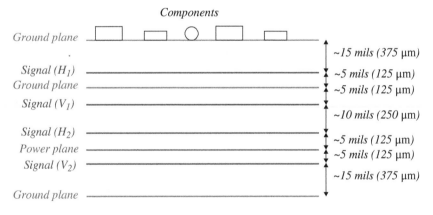

Figure 10.47   Eight-layer PCB – asymmetric stripline with $h_2 = 3h_1$.

Figure 10.48 shows the current densities for the *close* and *far* reference planes with $h_2 = 3h_1$, as a function of x/h_1.

Directly under the trace, approximately 85% of the current is on the close plane and 15% of the current is on the far plane. At distances greater than approximately $4 \times x/h_1$, the currents in both planes become the same.

Figure 10.49 shows the normalized current densities for the *close* and *far* reference planes, at $x = 0$, and total current in each plane, for asymmetric stripline with $h_2 = 2h_1$ and $h_2 = 3h_1$ [Ott, 2009].

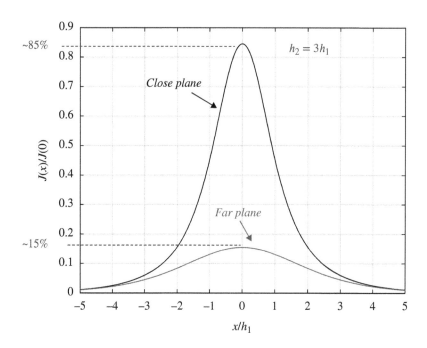

Figure 10.48 Current density for the close and far reference planes, $h_2 = 3h_1$.

h_2/h_1	Current density at $x = 0$		Total current	
	Close plane	Far plane	Close plane	Far plane
2	74%	26%	67%	33%
3	85%	15%	75%	25%

Figure 10.49 Current density at $x = 0$, and total current.

10.4 Laboratory Exercises

Note: This laboratory requires a custom PCB [French, 2023], for which the Altium files and BOM are provided.

10.4.1 Path of the Return Current

Objective: This laboratory examines the path of the return current and the impact of the discontinuities on the return-current flow.

10.4.1.1 Laboratory Equipment and Supplies

Function Generator: Tektronix AFG3252 (or equivalent), shown in Figure 10.50.

RG58C Coaxial Cable, shown in Figure 10.51.
BNC Male Plug to BNC Male Plug, length 4 feet (2 sets),
Supplier: Pomona Electronics, Part Number: 2249-C-48
Digi-Key Part Number: 501-1019-ND

RG58C Coaxial Cable
BNC Male Plug to BNC Male Plug, length 2 feet
Supplier: Pomona Electronics, Part Number: 2249-C-24
Digi-Key Part Number: 501-1016-ND

Current Probe Amplifier: Tektronix TCPA A300 (or equivalent), shown in Figure 10.52.
Current Probe: Tektronix TCP312 (or equivalent), shown in Figure 10.53.
Oscilloscope: Tektronix MDO3104 (or equivalent), shown in Figure 10.54.

Adapter Coaxial Connector, (set of 2), shown in Figure 10.55.
Supplier: Amphenol RF, 31-219-RFX,
Digi-Key Part Number: ARFX1069-ND

Copper Wire, 12-gauge, length 6 inches
Resistor, wire-wound, 50Ω, 0.25-2W rating
Custom PCB, shown in Figure 10.56

Loop Antenna: This antenna, shown in Figure 10.57, was designed in Chapter 8.

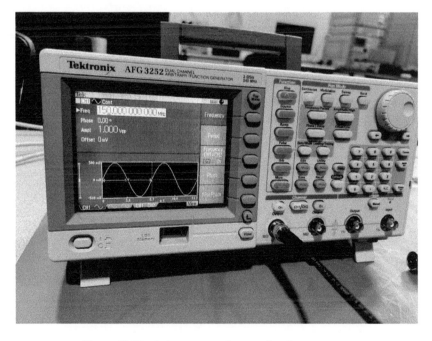

Figure 10.50 Laboratory equipment: function generator.

Figure 10.51   Laboratory equipment: coaxial cable.

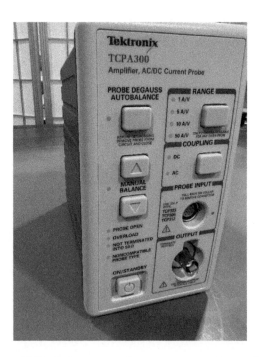

Figure 10.52   Laboratory equipment: current probe amplifier.

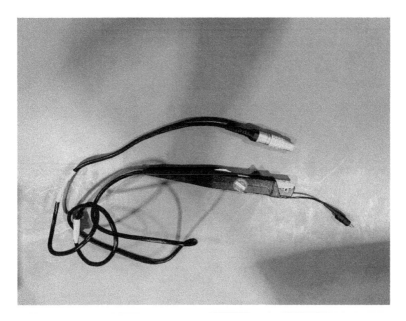

Figure 10.53   Laboratory equipment: current probe.

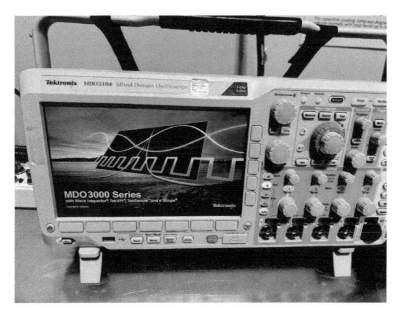

Figure 10.54   Laboratory equipment: oscilloscope.

Figure 10.55   Adapter coaxial connector.

Figure 10.56   Custom PCB.

10.4.1.2 Uninterrupted Current Flow

10.4.1.3 Laboratory Procedure

Solder the copper wire to the adapter coaxial connectors, as shown in Figure 10.58. Solder the load resistor as shown.

Figure 10.59 shows the experimental setup used for the return-current measurements.

This section describes the return-current measurements, described in Adamczyk [2017].

Laboratory Exercise: After reading this section, replicate the measurement with your own custom-made coaxial cable assembly and the custom current probe built in the Laboratory Exercise of Chapter 8.

Circuit Operation: The signal from the function generator travels along the center conductor of the coaxial cable and through the 50Ω load resistor. The return current has two different paths to return to the source: the direct path over the copper wire or the path

Figure 10.57 Loop antenna.

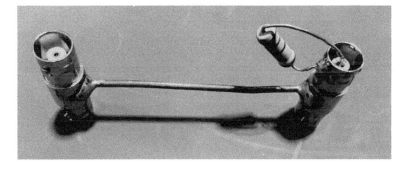

Figure 10.58 Adapter, wire, resistor assembly.

through the shield of the coaxial cable. A current probe is placed over the copper wire and a sinusoidal signal is generated by the function generator. The frequency of this signal is varied and current through the copper wire is measured.

Figure 10.60 shows the measurement results.

As expected, as the frequency increases more current returns through the shield as it provides the lower impedance path than the direct copper wire.

Laboratory Exercise: Recreate this experiment with your own custom coaxial cable/wire assembly. Vary the frequency of the sinusoidal signal and measure the current flowing through the copper wire. Comment on the results.

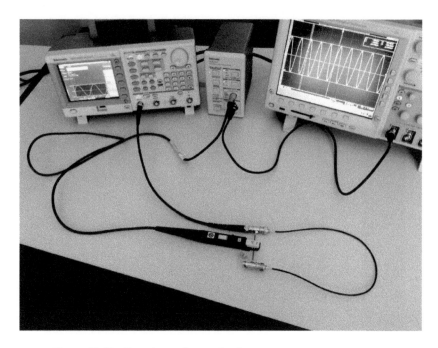

Figure 10.59 Experimental setup for the return-current measurements.

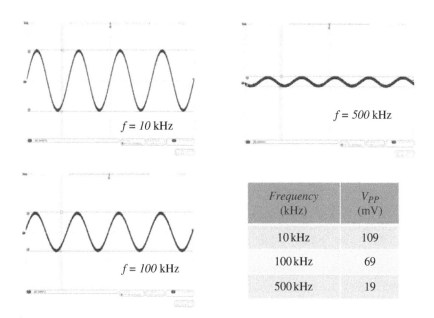

Frequency (kHz)	V_{PP} (mV)
10 kHz	109
100 kHz	69
500 kHz	19

Figure 10.60 Coaxial cable measurement results.

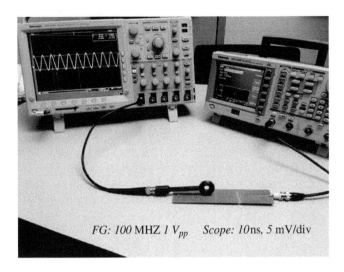

FG: 100 MHZ 1 V_{pp} Scope: 10ns, 5 mV/div

Figure 10.61   Measurement setup - PCB with the return path discontinuities.

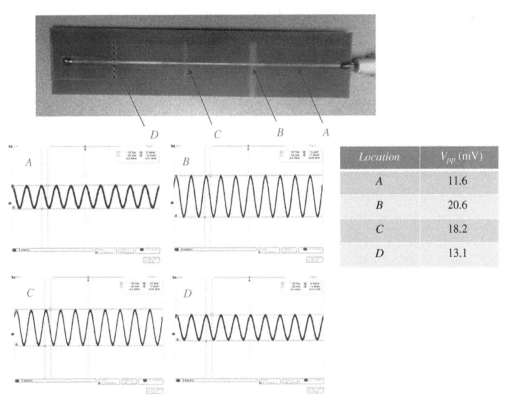

Location	V_{pp} (mV)
A	11.6
B	20.6
C	18.2
D	13.1

Figure 10.62   H-field measurement results.

10.4.1.4 Discontinuities in the Return-Current Path

10.4.1.5 Laboratory Procedure

Figure 10.61 shows the measurement setup used for investigating the effects of discontinuities on the flow of return current.

The forward current flows in the solid copper trace on the top surface and returns on the copper ground plane which contains several discontinuities. An *H*-field probe placed above the trace at several locations measures the radiated field caused by the current flowing along different loops. The larger the loop, the larger the emissions. The measurement results shown in Figure 10.62 verify these conclusions and are consistent with the discussion earlier in this chapter.

Laboratory Exercise: Recreate this experiment with your custom PCB and the custom **H** field probe, built as a part of Chapter 8 laboratory exercise. Place the probe at the locations indicated in Figure 10.62, and measure the **H**-field. Tabulate and comment on the results.

References

Bogdan Adamczyk. Alternative Paths of the Return Current. *In Compliance Magazine*, May 2017. URL https://incompliancemag.com/article/alternative-paths-of-the-return-current/.

Bogdan Adamczyk. Skin Depth in Good Conductors. *In Compliance Magazine*, February 2020a. URL https://incompliancemag.com/article/skin-depth-in-good-conductors/.

Bogdan Adamczyk. PCB Return-Current Distribution in the Stripline Configurations. *In Compliance Magazine*, December 2020b. URL https://incompliancemag.com/article/pcb-return-current-distribution-in-the-stripline-configurations/.

Matt French. Return Current PCB Designer, 2023.

Howard W. Johnson and Martin Graham. *High-Speed Digital Design, a Handbook of Black Magic*. Prentice Hall, Upper Saddle River, New Jersey, 1993. ISBN 0-13-395724-1.

Henry W. Ott. *Electromagnetic Compatibility Engineering*. John Wiley & Sons, Inc., Hoboken, New Jersey, 2009. ISBN 978-0-470-18930-6.

Tom Van Doren. Grousnding and Shielding of Electronic Systems - EMC Course, 2007.

Chapter 11: Shielding to Prevent Radiation

Shielding theory is based on three fundamental concepts:

- reflection and transmission of electromagnetic waves at the boundaries of two media,
- radiated fields of the electric and magnetic dipole antennas,
- wave impedance of an electromagnetic wave.

The first concept leads to the analytical formulas for the far-field shielding effectiveness of a metallic shield. When combined with the concepts of the fundamental dipole antennas and wave impedance, the far-field formulas lead to the expressions for the near-field shielding effectiveness.

We will begin our shielding discussion with a concept of a *uniform plane wave*.

11.1 Uniform Plane Wave

Uniform plane wave is a simple and a very useful type of wave that serves as a building block in the study of electromagnetic waves. Both descriptors in the name: *uniform* and *plane* are very important. The term *plane* means that the *E* and *H* vectors associated with the wave lie in a plane and as the wave propagates, the planes defined by these vectors are parallel [Paul, 2006]. The term *uniform* means that *E* and *H* vectors do not depend on the location within each plane, i.e. they have the same amplitudes and directions over the entire plane. Such a wave, propagating in the $+z$ direction, is shown in Figure 11.1.

It is customary to have the *E* field point in the positive *x* direction, as shown in Figure 11.1. That is,

$$E = \left[E_x\left(z,t\right), 0, 0 \right] \tag{11.1}$$

Since the uniform plane wave is an electromagnetic wave, it must satisfy Maxwell's curl equations, which for the source-free media, in time domain, are given by

$$\nabla \times E = -\mu \frac{\partial H}{\partial t} \tag{11.2}$$

$$\nabla \times H = \sigma E + \epsilon \frac{\partial E}{\partial t} \tag{11.3}$$

Uniformity assumption, combined with equations (11.1)–(11.3), reveals the fact that the *H* field is perpendicular to the *E* field in the plane and is pointing in the $+y$ direction [Adamczyk, 2017], as shown in Figure 11.1. Thus,

$$H = \left[0, H_y\left(z,t\right), 0 \right] \tag{11.4}$$

Principles of Electromagnetic Compatibility: Laboratory Exercises and Lectures, First Edition. Bogdan Adamczyk.
© 2024 John Wiley & Sons Ltd. Published 2024 by John Wiley & Sons Ltd.
Companion website: www.wiley.com/go/principlesofelectromagneticcompatibility

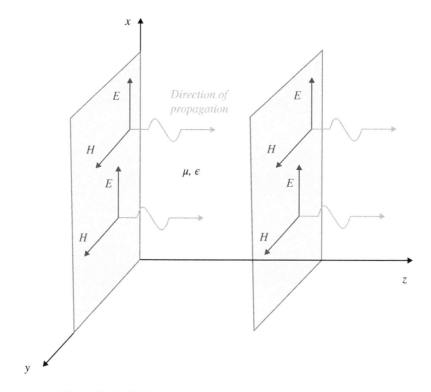

Figure 11.1   Uniform plane wave propagating in the $+z$ direction.

For arbitrary time variations, in any medium, equations (11.2) and (11.3) lead to the (uniform plane) wave equations

$$\frac{\partial^2 E_x(z,t)}{\partial z^2} = \mu\sigma\frac{\partial E_x(z,t)}{\partial t} + \mu\epsilon\frac{\partial^2 E_x(z,t)}{\partial t^2} \tag{11.5}$$

$$\frac{\partial^2 H_y(z,t)}{\partial z^2} = \mu\sigma\frac{\partial H_y(z,t)}{\partial t} + \mu\epsilon\frac{\partial^2 H_y(z,t)}{\partial t^2} \tag{11.6}$$

In sinusoidal steady-state equations (11.5) and (11.6) become

$$\frac{d^2\hat{E}_x}{dz^2} = \hat{\gamma}^2\hat{E}_x(z) \tag{11.7}$$

$$\frac{d^2\hat{H}_y}{dz^2} = \hat{\gamma}^2\hat{H}_y(z) \tag{11.8}$$

where

$$\hat{\gamma} = \sqrt{j\omega\mu(\sigma + j\omega\epsilon)} \tag{11.9}$$

is the *propagation constant*. The solution of the wave equations (11.7) and (11.8) is well known and is given by [Paul, 2006],

$$\hat{E}_x = \hat{E}_m^+ e^{-\hat{\gamma}z} + \hat{E}_m^- e^{\hat{\gamma}z} \tag{11.10}$$

$$\hat{H}_y = \frac{\hat{E}_m^+}{\hat{\eta}} e^{-\hat{\gamma}z} - \frac{\hat{E}_m^-}{\hat{\eta}} e^{\hat{\gamma}z} \tag{11.11}$$

where

$$\hat{\eta} = \sqrt{\frac{j\omega\mu}{\sigma + j\omega\epsilon}} \tag{11.12}$$

is the *intrinsic impedance* of the medium.

The solutions in equations (11.10) and (11.11) consist of the superposition of two waves

$$\hat{E}_x = \hat{E}_x^f + \hat{E}_x^b \tag{11.13}$$

$$\hat{H}_y = \hat{H}_y^f + \hat{H}_y^b \tag{11.14}$$

where the forward propagating waves (in +z direction) are

$$\hat{E}_x^f = \hat{E}_m^+ e^{-\hat{\gamma}z} \tag{11.15}$$

$$\hat{H}_y^f = \frac{\hat{E}_m^+ e^{-\hat{\gamma}z}}{\hat{\eta}} \tag{11.16}$$

and the backward propagating waves (in −z direction) are

$$\hat{E}_x^b = \hat{E}_m^- e^{\hat{\gamma}z} \tag{11.17}$$

$$\hat{H}_y^b = -\frac{\hat{E}_m^- e^{\hat{\gamma}z}}{\hat{\eta}} \tag{11.18}$$

Expressing the propagation constant $\hat{\gamma}$ in terms of its real and imaginary parts

$$\hat{\gamma} = \alpha + j\beta \tag{11.19}$$

and the intrinsic impedance $\hat{\eta}$ in terms of its magnitude and angle

$$\hat{\eta} = \eta \angle \theta_\eta = \eta e^{j\theta_\eta} \tag{11.20}$$

(α is the *attenuation constant* in Np/m and β is the *phase constant* in rad/m) we can write the solution in equations (11.10) and (11.11) as

$$\hat{E}_x = \hat{E}_m^+ e^{-\alpha z} e^{-j\beta z} + \hat{E}_m^- e^{\alpha z} e^{j\beta z} \tag{11.21}$$

$$\hat{H}_y = \frac{\hat{E}_m^+}{\eta} e^{-\alpha z} e^{-j\beta z} e^{-j\theta_\eta} - \frac{\hat{E}_m^-}{\eta} e^{\alpha z} e^{j\beta z} e^{-j\theta_\eta} \tag{11.22}$$

In a lossless medium $\alpha = 0$ and $\hat{\eta} = \eta \angle 0$, and the phasor solutions in equations (11.21) and (11.22) become

$$\hat{E}_x = \hat{E}_m^+ e^{-j\beta z} + \hat{E}_m^- e^{j\beta z} \tag{11.23}$$

$$\hat{H}_y = \frac{\hat{E}_m^+}{\eta} e^{-j\beta z} - \frac{\hat{E}_m^-}{\eta} e^{j\beta z} \tag{11.24}$$

Thus, in a perfect dielectric, the amplitudes of the forward and backward propagating waves are *not* attenuated.

If the medium is considered a *good conductor* of conductivity σ, then equations (11.21) and (11.22) apply, and the attenuation constant, at a frequency f is given by

$$\alpha = \sqrt{\pi f \mu \sigma} \tag{11.25}$$

11.1.1 Skin Depth

Let's consider a forward propagating wave, in a positive z direction, in a lossy medium. This wave is described by

$$\hat{E}_x^f = \hat{E}_m^+ e^{-\hat{\gamma} z} = \hat{E}_m^+ e^{-\alpha z} e^{-j\beta z} \tag{11.26}$$

$$\hat{H}_y^f = \frac{\hat{E}_m^+ e^{-\hat{\gamma} z}}{\hat{\eta}} = \frac{\hat{E}_m^+}{\eta} e^{-\alpha z} e^{-j\beta z} e^{-j\theta_\eta} \tag{11.27}$$

The magnitudes (amplitudes) of the **E** and **H** fields associated with this wave are given by

$$|\hat{E}_x^f| = E_f^x = E_m^+ e^{-\alpha z} \tag{11.28}$$

$$|\hat{H}_y^f| = H_y^f = \frac{E_m^+}{\eta} e^{-\alpha z} \tag{11.29}$$

Therefore, as the wave travels in a lossy medium, the amplitudes of the **E** and **H** fields associated with it are attenuated by a factor $e^{-\alpha z}$. The distance δ through which the wave amplitude decreases by a factor of e is called *skin depth* of the medium [Adamczyk, 2020].

Thus, we have

$$E_m^+ e^{-\alpha(z=\delta)} = E_m^+ e^{-1} \tag{11.30}$$

or

$$\delta = \frac{1}{\alpha} \tag{11.31}$$

The expression in equation (11.31) is valid in any lossy medium. Usually, the concept of skin depth is applied to good conductors. Using equation (11.25) we obtain the well-known formula for skin depth in good conductors

$$\delta = \frac{1}{\sqrt{\pi f \mu \sigma}} \tag{11.32}$$

Note: The skin depth is inversely proportional to the square root of frequency. The higher the frequency, the lower the skin depth. Equation (11.32) also reveals that the skin depth of the ideal conductor is zero, since its conductivity is infinite. Practical conductors carry currents up to several skin depths as discussed next.

11.1.2 Current Density in Conductors

Consider a typical circuit model shown in Figure 11.2.

Both the forward and the return differential-mode currents have the differential-mode EM waves, traveling in free space, associated with them [Adamczyk, 2019], as shown in Figure 11.3.

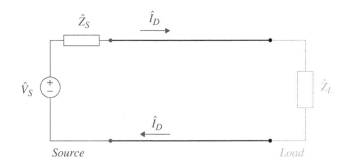

Figure 11.2   A typical circuit model.

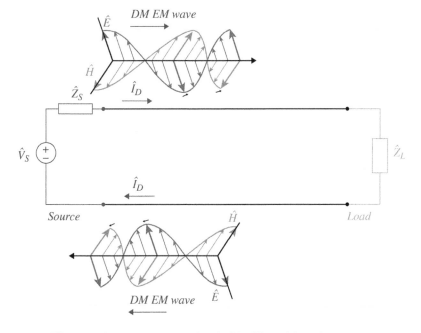

Figure 11.3   EM waves associated with differential-mode currents.

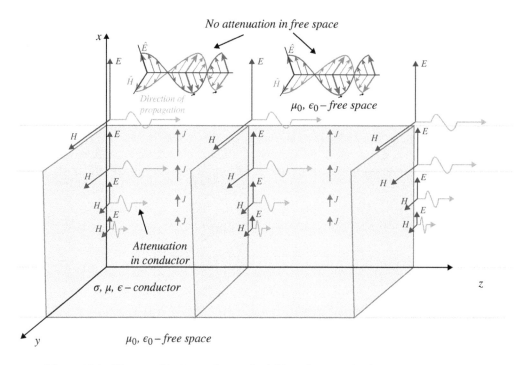

Figure 11.4   Wave outside the conductor, the fields and current density inside the conductor.

Let's focus on the forward conductor. As the wave propagates in free space outside the conductor it suffers no attenuation, since free space is an ideal dielectric. The **E** and **H** fields associated with this wave penetrate the conductor and are attenuated as shown in Figure 11.4.

The current density inside the conductor is related to the electric field by

$$\boldsymbol{J} = \sigma\boldsymbol{E} \tag{11.33}$$

Therefore, as the fields penetrate the surface of the conductor, the current density decays exponentially, as shown in Figure 11.4.

11.1.3 Reflection and Transmission at a Normal Boundary

In Section 11.2 we will discuss the electromagnetic wave shielding in far field. In order to derive the equations describing this phenomenon we need to understand the reflection and transmission of electromagnetic waves at the boundaries of two media. We will consider a normal incidence of a uniform plane wave on the boundary between two media, as shown in Figure 11.5.

When the wave encounters the boundary between two media, a reflected and transmitted wave is created [Paul, 2006].

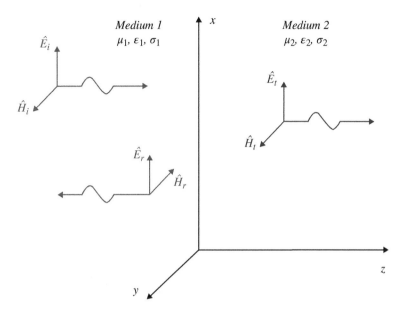

Figure 11.5   Reflection and transmission of a uniform wave at the boundary.

The incident wave is described by

$$\hat{\mathbf{E}}_i = \hat{E}_i e^{-\hat{\gamma}_1 z}\mathbf{a}_x = \hat{E}_i e^{-\alpha_1 z}e^{-j\beta_1 z}\mathbf{a}_x \tag{11.34}$$

$$\hat{\mathbf{H}}_i = \frac{\hat{E}_i}{\hat{\eta}_1}e^{-\hat{\gamma}_1 z}\mathbf{a}_y = \frac{\hat{E}_i}{\eta_1}e^{-\alpha_1 z}e^{-j\beta_1 z}e^{-j\theta_{\eta 1}}\mathbf{a}_y \tag{11.35}$$

while the reflected wave is expressed as

$$\hat{\mathbf{E}}_r = \hat{E}_r e^{\hat{\gamma}_1 z}\mathbf{a}_x = \hat{E}_r e^{\alpha_1 z}e^{j\beta_1 z}\mathbf{a}_x \tag{11.36}$$

$$\hat{\mathbf{H}}_r = -\frac{\hat{E}_r}{\hat{\eta}_1}e^{\hat{\gamma}_1 z}\mathbf{a}_y = -\frac{\hat{E}_r}{\eta_1}e^{\alpha_1 z}e^{j\beta_1 z}e^{-j\theta_{\eta 1}}\mathbf{a}_y \tag{11.37}$$

where the propagation constant and the intrinsic impedance in medium 1 are given by

$$\hat{\gamma}_1 = \sqrt{j\omega\mu_1(\sigma_1 + j\omega\epsilon_1)} = \alpha_1 + j\beta_1 \tag{11.38}$$

$$\hat{\eta}_1 = \sqrt{\frac{j\omega\mu_1}{\sigma_1 + j\omega\epsilon_1}} = \eta_1 e^{j\theta_{\eta 1}} \tag{11.39}$$

The transmitted wave is described by

$$\hat{\mathbf{E}}_t = \hat{E}_t e^{-\hat{\gamma}_2 z}\mathbf{a}_x = \hat{E}_t e^{-\alpha_2 z}e^{-j\beta_2 z}\mathbf{a}_x \tag{11.40}$$

$$\hat{\mathbf{H}}_t = \frac{\hat{E}_t}{\hat{\eta}_t}e^{-\hat{\gamma}_2 z}\mathbf{a}_y = \frac{\hat{E}_t}{\eta_2}e^{-\alpha_2 z}e^{-j\beta_2 z}e^{-j\theta_{\eta 2}}\mathbf{a}_y \tag{11.41}$$

where the propagation constant and the intrinsic impedance in medium 2 are given by

$$\hat{\gamma}_2 = \sqrt{j\omega\mu_2(\sigma_2 + j\omega\epsilon_2)} = \alpha_2 + j\beta_2 \qquad (11.42)$$

$$\hat{\eta}_2 = \sqrt{\frac{j\omega\mu_2}{\sigma_2 + j\omega\epsilon_2}} = \eta_2 e^{j\theta_{\eta_2}} \qquad (11.43)$$

At the boundary of two media, the tangential component of the electric field intensity is continuous (Adamczyk 2017).
Thus,

$$\hat{\mathbf{E}}_i + \hat{\mathbf{E}}_r = \hat{\mathbf{E}}_t, \quad z = 0 \qquad (11.44)$$

or

$$\hat{E}_i e^{-\hat{\gamma}_1 z} + \hat{E}_r e^{\hat{\gamma}_1 z} = \hat{E}_t e^{-\hat{\gamma}_2 z}, \quad z = 0 \qquad (11.45)$$

leading to

$$\hat{E}_i + \hat{E}_r = \hat{E}_t \qquad (11.46)$$

The boundary condition imposed on the magnetic field requires that tangential component of the magnetic field intensity must be continuous [Paul, 2006]. Thus,

$$\hat{\mathbf{H}}_i + \hat{\mathbf{H}}_r = \hat{\mathbf{H}}_t, \quad z = 0 \qquad (11.47)$$

$$\frac{\hat{E}_i}{\hat{\eta}_1} e^{-\hat{\gamma}_1 z} - \frac{\hat{E}_r}{\hat{\eta}_1} e^{\hat{\gamma}_1 z} = \frac{\hat{E}_t}{\hat{\eta}_2} e^{-\hat{\gamma}_2 z}, \quad z = 0 \qquad (11.48)$$

leading to

$$\frac{\hat{E}_i}{\hat{\eta}_1} - \frac{\hat{E}_r}{\hat{\eta}_1} = \frac{\hat{E}_t}{\hat{\eta}_2} \qquad (11.49)$$

Substituting equation (11.46) into equation (11.49) results in

$$\frac{\hat{E}_i}{\hat{\eta}_1} - \frac{\hat{E}_r}{\hat{\eta}_1} = \frac{\hat{E}_i + \hat{E}_r}{\hat{\eta}_2} \qquad (11.50)$$

or

$$\frac{\hat{E}_i}{\hat{\eta}_1} - \frac{\hat{E}_i}{\hat{\eta}_2} = \frac{\hat{E}_r}{\hat{\eta}_1} + \frac{\hat{E}_r}{\hat{\eta}_2} \qquad (11.51)$$

or

$$\hat{E}_i \left(\frac{1}{\hat{\eta}_1} - \frac{1}{\hat{\eta}_2} \right) = \hat{E}_r \left(\frac{1}{\hat{\eta}_1} + \frac{1}{\hat{\eta}_2} \right) \qquad (11.52)$$

or

$$\hat{E}_i \left(\frac{\hat{\eta}_2 - \hat{\eta}_1}{\hat{\eta}_1 \hat{\eta}_2} \right) = \hat{E}_r \left(\frac{\hat{\eta}_2 + \hat{\eta}_1}{\hat{\eta}_1 \hat{\eta}_2} \right) \qquad (11.53)$$

leading to the definition of the *reflection coefficient* at the boundary as

$$\hat{\Gamma} = \Gamma \angle \theta_\Gamma = \frac{\hat{E}_r}{\hat{E}_i} = \frac{\hat{\eta}_2 - \hat{\eta}_1}{\hat{\eta}_2 + \hat{\eta}_1} \tag{11.54}$$

Thus the reflected wave is related to the incident wave by

$$\hat{E}_r = \hat{\Gamma}\hat{E}_i \tag{11.55}$$

From equation (11.46) we get

$$\hat{E}_r = \hat{E}_t - \hat{E}_i \tag{11.56}$$

Substituting equation (11.56) into equation (11.49) results in

$$\frac{\hat{E}_i}{\hat{\eta}_1} - \frac{\hat{E}_t - \hat{E}_i}{\hat{\eta}_1} = \frac{\hat{E}_i + \hat{E}_r}{\hat{\eta}_2} \tag{11.57}$$

or

$$\frac{\hat{E}_i}{\hat{\eta}_1} + \frac{\hat{E}_i}{\hat{\eta}_1} = \frac{\hat{E}_t}{\hat{\eta}_2} + \frac{\hat{E}_t}{\hat{\eta}_1} \tag{11.58}$$

or

$$\hat{E}_i \left(\frac{2}{\hat{\eta}_1} \right) = \hat{E}_t \left(\frac{\hat{\eta}_1 + \hat{\eta}_2}{\hat{\eta}_1 \hat{\eta}_2} \right) \tag{11.59}$$

leading to the definition of the *transmission coefficient* at the boundary as

$$\hat{T} = T \angle \theta_T = \frac{\hat{E}_t}{\hat{E}_i} = \frac{2\hat{\eta}_2}{\hat{\eta}_2 + \hat{\eta}_1} \tag{11.60}$$

Thus the transmitted wave is related to the incident wave by

$$\hat{E}_t = \hat{T}\hat{E}_i \tag{11.61}$$

11.2 Far-Field Shielding

Consider a conducting shield of thickness t, conductivity σ, permittivity ϵ, and permeability μ, surrounded on both sides by air (free space), as shown in Figure 11.6.

A uniform plane wave is normally incident on its left interface. Since the wave is a uniform wave, the shield is assumed to be in the far field of the radiation source.

The incident wave, upon arrival at the leftmost boundary, $(\hat{E}_i, \hat{H}_i)$, will be partially reflected, $(\hat{E}_r, \hat{H}_r)$, and partially transmitted, $(\hat{E}_1, \hat{H}_1)$, through the shield.

The transmitted wave, $(\hat{E}_1, \hat{H}_1)$, upon arrival at the rightmost boundary will be partially reflected, $(\hat{E}_2, \hat{H}_2)$, and partially transmitted, $(\hat{E}_t, \hat{H}_t)$, through the shield.

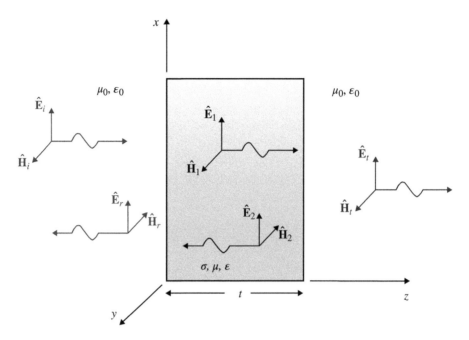

Figure 11.6   Uniform plane wave incident on a metallic shield.

The reflected wave, $(\hat{\mathbf{E}}_2, \hat{\mathbf{H}}_2)$, propagates back through the shield and strikes the first interface, incident from the right.

Once again, a portion of this wave is transmitted through the left interface and adds to the total reflected field in the left medium, and a portion is reflected and proceeds to the right.

The process continues in the like fashion, but the additional reflected and transmitted fields are progressively attenuated. If a shield has a thickness that is much greater than the skin depth of the material at the frequency of the incident field, these multiple reflections and transmissions can be disregarded, and only the initial reflection and transmission at the left and right interfaces need be considered.

The incident wave, $(\hat{\mathbf{E}}_i, \hat{\mathbf{H}}_i)$, is described by

$$\hat{\mathbf{E}}_i = \hat{E}_i e^{-j\beta_0 z} \mathbf{a}_x \tag{11.62}$$

$$\hat{\mathbf{H}}_i = \frac{\hat{E}_i}{\eta_0} e^{-j\beta_0 z} \mathbf{a}_y \tag{11.63}$$

where

$$\beta_0 = \omega \sqrt{\mu_0 \epsilon_0} \tag{11.64}$$

$$\eta_0 = \sqrt{\frac{\mu_0}{\epsilon_0}} \tag{11.65}$$

The reflected wave, $(\hat{\mathbf{E}}_r, \hat{\mathbf{H}}_r)$, is described by

$$\hat{\mathbf{E}}_r = \hat{E}_r e^{j\beta_0 z}\mathbf{a}_x \tag{11.66}$$

$$\hat{\mathbf{H}}_r = -\frac{\hat{E}_r}{\eta_0}e^{j\beta_0 z}\mathbf{a}_y \tag{11.67}$$

The wave transmitted through the left interface is described by

$$\hat{\mathbf{E}}_1 = \hat{E}_1 e^{-\hat{\gamma}z}\mathbf{a}_x \tag{11.68}$$

$$\hat{\mathbf{H}}_1 = \frac{\hat{E}_1}{\hat{\eta}}e^{-\hat{\gamma}z}\mathbf{a}_y \tag{11.69}$$

where

$$\hat{\gamma} = \sqrt{j\omega\mu(\sigma + j\omega\epsilon)} = \alpha + j\beta \tag{11.70}$$

$$\hat{\eta} = \sqrt{\frac{j\omega\mu}{\sigma + j\omega\epsilon}} = \eta e^{j\theta_\eta} \tag{11.71}$$

The wave reflected at the right interface is described by

$$\hat{\mathbf{E}}_2 = \hat{E}_2 e^{\hat{\gamma}z}\mathbf{a}_x \tag{11.72}$$

$$\hat{\mathbf{H}}_2 = -\frac{\hat{E}_2}{\hat{\eta}}e^{\hat{\gamma}z}\mathbf{a}_y \tag{11.73}$$

Finally, the wave transmitted through the right interface is described by

$$\hat{\mathbf{E}}_t = \hat{E}_t e^{-j\beta_0 z}\mathbf{a}_x \tag{11.74}$$

$$\hat{\mathbf{H}}_t = \frac{\hat{E}_t}{\eta_0}e^{-j\beta_0 z}\mathbf{a}_y \tag{11.75}$$

The effectiveness of the shield, or shielding effectiveness, SE, can be determined by evaluating the ratio of the incident field magnitude to the transmitted field magnitude.

$$(SE)_E = \frac{|\hat{E}_i|}{|\hat{E}_t|} \tag{11.76}$$

$$(SE)_H = \frac{|\hat{H}_i|}{|\hat{H}_t|} \tag{11.77}$$

Since the wave is a uniform wave, the two definitions are identical, since the electric and magnetic field magnitudes are related by the intrinsic impedance of the medium.

Usually, the shielding effectiveness is expressed in decibels. Then, the definition in equation (11.76) becomes

$$(SE)_{\mathrm{dB}} = 20 \log_{10}\frac{|\hat{E}_i|}{|\hat{E}_t|} \tag{11.78}$$

11.2.1 Shielding Effectiveness – Exact Solution

The magnitude of incident field, $\hat{E}_i$, is assumed to be known. In order to determine the magnitude of the transmitted field, $\hat{E}_t$, we need to determine the magnitudes of the remaining waves, i.e. $\hat{E}_r, \hat{E}_1, \hat{E}_2$. Thus, we need four equations in four unknowns. These are generated by enforcing the boundary conditions on the field vectors at the two boundaries, $z = 0$ and $z = t$.

Continuity condition of the tangential components of the electric fields at the left interface produces

$$\hat{\mathbf{E}}_i\Big|_{z=0} + \hat{\mathbf{E}}_r\Big|_{z=0} = \hat{\mathbf{E}}_1\Big|_{z=0} + \hat{\mathbf{E}}_2\Big|_{z=0} \tag{11.79}$$

or

$$\hat{E}_i e^{-j\beta_0 z}\mathbf{a}_x\Big|_{z=0} + \hat{E}_r e^{j\beta_0 z}\mathbf{a}_x\Big|_{z=0} = \hat{E}_1 e^{-\hat{\gamma}z}\mathbf{a}_x\Big|_{z=0} + \hat{E}_2 e^{\hat{\gamma}z}\mathbf{a}_x\Big|_{z=0} \tag{11.80}$$

leading to

$$\hat{E}_i + \hat{E}_r = \hat{E}_1 + \hat{E}_2 \tag{11.81}$$

Continuity condition of the tangential components of the magnetic fields at the left interface produces

$$\hat{\mathbf{H}}_i\Big|_{z=0} + \hat{\mathbf{H}}_r\Big|_{z=0} = \hat{\mathbf{H}}_1\Big|_{z=0} + \hat{\mathbf{H}}_2\Big|_{z=0} \tag{11.82}$$

or

$$\frac{\hat{E}_i}{\eta_0} e^{-j\beta_0 z}\mathbf{a}_y\Big|_{z=0} - \frac{\hat{E}_r}{\eta_0} e^{j\beta_0 z}\mathbf{a}_y\Big|_{z=0} = \frac{\hat{E}_1}{\hat{\eta}} e^{-\hat{\gamma}z}\mathbf{a}_y\Big|_{z=0} - \frac{\hat{E}_2}{\hat{\eta}} e^{\hat{\gamma}z}\mathbf{a}_y\Big|_{z=0} \tag{11.83}$$

leading to

$$\frac{\hat{E}_i}{\eta_0} - \frac{\hat{E}_r}{\eta_0} = \frac{\hat{E}_1}{\hat{\eta}} - \frac{\hat{E}_2}{\hat{\eta}} \tag{11.84}$$

Continuity condition of the tangential components of the electric fields at the right interface produces

$$\hat{\mathbf{E}}_1\Big|_{z=t} + \hat{\mathbf{E}}_2\Big|_{z=t} = \hat{\mathbf{E}}_t\Big|_{z=t} \tag{11.85}$$

or

$$\hat{E}_1 e^{-\hat{\gamma}t}\mathbf{a}_x\Big|_{z=t} + \hat{E}_2 e^{\hat{\gamma}z}\mathbf{a}_x\Big|_{z=t} = \hat{E}_t e^{-j\beta_0 z}\mathbf{a}_x\Big|_{z=t} \tag{11.86}$$

leading to

$$\hat{E}_1 e^{-\hat{\gamma}t} + \hat{E}_2 e^{\hat{\gamma}t} = \hat{E}_t e^{-j\beta_0 t} \tag{11.87}$$

Continuity condition of the tangential components of the magnetic fields at the right interface produces

$$\hat{\mathbf{H}}_1\Big|_{z=t} + \hat{\mathbf{H}}_2\Big|_{z=t} = \hat{\mathbf{H}}_t\Big|_{z=t} \tag{11.88}$$

or

$$\frac{\hat{E}_1}{\hat{\eta}} e^{-\hat{\gamma}z}\mathbf{a}_y\Big|_{z=t} - \frac{\hat{E}_2}{\hat{\eta}} e^{\hat{\gamma}z}\mathbf{a}_y\Big|_{z=t} = \frac{\hat{E}_t}{\eta_0} e^{-j\beta_0 z}\mathbf{a}_y\Big|_{z=t} \tag{11.89}$$

leading to

$$\frac{\hat{E}_1}{\hat{\eta}}e^{-\hat{\gamma}t} - \frac{\hat{E}_2}{\hat{\eta}}e^{\hat{\gamma}t} = \frac{\hat{E}_t}{\eta_0}e^{-j\beta_0 t} \tag{11.90}$$

Thus, we need to solve four equations: (11.81), (11.84), (11.87), and (11.90), repeated here

$$\hat{E}_i + \hat{E}_r = \hat{E}_1 + \hat{E}_2 \tag{11.91}$$

$$\frac{\hat{E}_i}{\eta_0} - \frac{\hat{E}_r}{\eta_0} = \frac{\hat{E}_1}{\hat{\eta}} - \frac{\hat{E}_2}{\hat{\eta}} \tag{11.92}$$

$$\hat{E}_1 e^{-\hat{\gamma}t} + \hat{E}_2 e^{\hat{\gamma}t} = \hat{E}_t e^{-j\beta_0 t} \tag{11.93}$$

$$\frac{\hat{E}_1}{\hat{\eta}}e^{-\hat{\gamma}t} - \frac{\hat{E}_2}{\hat{\eta}}e^{\hat{\gamma}t} = \frac{\hat{E}_t}{\eta_0}e^{-j\beta_0 t} \tag{11.94}$$

Toward this end, let us divide equation (11.93) by $\hat{\eta}$ to obtain

$$\frac{\hat{E}_1}{\hat{\eta}}e^{-\hat{\gamma}t} + \frac{\hat{E}_2}{\hat{\eta}}e^{\hat{\gamma}t} = \frac{\hat{E}_t}{\hat{\eta}}e^{-j\beta_0 t} \tag{11.95}$$

Adding equations (11.94) and (11.95) gives

$$2\frac{\hat{E}_1}{\hat{\eta}}e^{-\hat{\gamma}t} = \hat{E}_t e^{-j\beta_0 t}\left(\frac{1}{\eta_0} + \frac{1}{\hat{\eta}}\right) \tag{11.96}$$

or

$$2\frac{\hat{E}_1}{\hat{\eta}}e^{-\hat{\gamma}t} = \hat{E}_t e^{-j\beta_0 t}\left(\frac{\hat{\eta} + \eta_0}{\eta_0\hat{\eta}}\right) \tag{11.97}$$

From which, we obtain $\hat{E}_1$ as

$$\hat{E}_1 = \hat{E}_t\left(\frac{\eta_0 + \hat{\eta}}{2\eta_0}\right)e^{-j\beta_0 t}e^{\hat{\gamma}t} \tag{11.98}$$

Subtracting equation (11.94) from equation (11.95) gives

$$2\frac{\hat{E}_2}{\hat{\eta}}e^{\hat{\gamma}t} = \hat{E}_t e^{-j\beta_0 t}\left(\frac{1}{\hat{\eta}} - \frac{1}{\eta_0}\right) \tag{11.99}$$

or

$$2\frac{\hat{E}_2}{\hat{\eta}}e^{\hat{\gamma}t} = \hat{E}_t e^{-j\beta_0 t}\left(\frac{\eta_0 - \hat{\eta}}{\eta_0\hat{\eta}}\right) \tag{11.100}$$

From which, we obtain $\hat{E}_2$ as

$$\hat{E}_2 = \hat{E}_t\left(\frac{\eta_0 - \hat{\eta}}{2\eta_0}\right)e^{-j\beta_0 t}e^{-\hat{\gamma}t} \tag{11.101}$$

Next, toward this end, let us divide equation (11.91) by η_0 to obtain

$$\frac{\hat{E}_i}{\eta_0} + \frac{\hat{E}_r}{\eta_0} = \frac{\hat{E}_1}{\eta_0} + \frac{\hat{E}_2}{\eta_0} \tag{11.102}$$

Adding equations (11.92) and (11.102) gives

$$2\frac{\hat{E}_i}{\eta_0} = \hat{E}_1\left(\frac{1}{\eta_0} + \frac{1}{\hat{\eta}}\right) + \hat{E}_2\left(\frac{1}{\eta_0} - \frac{1}{\hat{\eta}}\right) \tag{11.103}$$

or

$$2\frac{\hat{E}_i}{\eta_0} = \hat{E}_1\left(\frac{\eta_0 + \hat{\eta}}{\eta_0\hat{\eta}}\right) - \hat{E}_2\left(\frac{\eta_0 - \hat{\eta}}{\eta_0\hat{\eta}}\right) \tag{11.104}$$

From which, we obtain $\hat{E}_i$ as

$$\hat{E}_i = \hat{E}_1\left(\frac{\eta_0 + \hat{\eta}}{2\hat{\eta}}\right) - \hat{E}_2\left(\frac{\eta_0 - \hat{\eta}}{2\hat{\eta}}\right) \tag{11.105}$$

Substituting for $\hat{E}_1$ from equation (11.98), and for $\hat{E}_2$ from equation (11.101) we obtain

$$\hat{E}_i = \hat{E}_t\left(\frac{\eta_0 + \hat{\eta}}{2\eta_0}\right)e^{-j\beta_0 t}e^{\hat{\gamma}t}\left(\frac{\eta_0 + \hat{\eta}}{2\hat{\eta}}\right) - \hat{E}_t\left(\frac{\eta_0 - \hat{\eta}}{2\eta_0}\right)e^{-j\beta_0 t}e^{-\hat{\gamma}t}\left(\frac{\eta_0 - \hat{\eta}}{2\hat{\eta}}\right) \tag{11.106}$$

or

$$\frac{\hat{E}_i}{\hat{E}_t} = \left(\frac{\eta_0 + \hat{\eta}}{2\eta_0}\right)e^{-j\beta_0 t}e^{\hat{\gamma}t}\left(\frac{\eta_0 + \hat{\eta}}{2\hat{\eta}}\right) - \hat{E}_t\left(\frac{\eta_0 - \hat{\eta}}{2\eta_0}\right)e^{-j\beta_0 t}e^{-\hat{\gamma}t}\left(\frac{\eta_0 - \hat{\eta}}{2\hat{\eta}}\right) \tag{11.107}$$

or

$$\frac{\hat{E}_i}{\hat{E}_t} = \frac{(\eta_0 + \hat{\eta})^2}{4\eta_0\hat{\eta}}e^{-j\beta_0 t}e^{\hat{\gamma}t} - \frac{(\eta_0 - \hat{\eta})^2}{4\eta_0\hat{\eta}}e^{-j\beta_0 t}e^{-\hat{\gamma}t} \tag{11.108}$$

or

$$\frac{\hat{E}_i}{\hat{E}_t} = \frac{(\eta_0 + \hat{\eta})^2}{4\eta_0\hat{\eta}}\left[1 - \frac{(\eta_0 - \hat{\eta})^2}{(\eta_0 + \hat{\eta})^2}e^{-2\hat{\gamma}t}\right]e^{-j\beta_0 t}e^{\hat{\gamma}t} \tag{11.109}$$

and thus

$$\frac{\hat{E}_i}{\hat{E}_t} = \frac{(\eta_0 + \hat{\eta})^2}{4\eta_0\hat{\eta}}\left[1 - \left(\frac{\eta_0 - \hat{\eta}}{\eta_0 + \hat{\eta}}\right)^2 e^{-2\hat{\gamma}t}\right]e^{-j\beta_0 t}e^{\hat{\gamma}t} \tag{11.110}$$

Now, let us express the propagation constant as

$$\hat{\gamma} = \alpha + j\beta \tag{11.111}$$

For a good conductor the attenuation constant is related to the skin depth by

$$\alpha = \frac{1}{\delta} \tag{11.112}$$

and thus the propagation constant becomes

$$\hat{\gamma} = \frac{1}{\delta} + j\beta \tag{11.113}$$

Using equation (11.113) in equation (11.110) gives

$$\frac{\hat{E}_i}{\hat{E}_t} = \frac{(\eta_0 + \hat{\eta})^2}{4\eta_0\hat{\eta}} \left[1 - \left(\frac{\eta_0 - \hat{\eta}}{\eta_0 + \hat{\eta}} \right)^2 e^{-2\left(\frac{1}{\delta} + j\beta\right)t} \right] e^{-j\beta_0 t} e^{\left(\frac{1}{\delta} + j\beta\right)t} \tag{11.114}$$

and finally, the *shielding effectiveness* becomes

$$SE = \frac{\hat{E}_i}{\hat{E}_t} = \frac{(\eta_0 + \hat{\eta})^2}{4\eta_0\hat{\eta}} \left[1 - \left(\frac{\eta_0 - \hat{\eta}}{\eta_0 + \hat{\eta}} \right)^2 e^{\frac{-2t}{\delta}} e^{-j2\beta t} \right] e^{\frac{t}{\delta}} e^{j\beta t} e^{-j\beta_0 t} \tag{11.115}$$

The magnitude of the shielding effectiveness is

$$\left| \frac{\hat{E}_i}{\hat{E}_t} \right| = \left| \frac{(\eta_0 + \hat{\eta})^2}{4\eta_0\hat{\eta}} \left[1 - \left(\frac{\eta_0 - \hat{\eta}}{\eta_0 + \hat{\eta}} \right)^2 e^{\frac{-2t}{\delta}} e^{-j2\beta t} \right] e^{\frac{t}{\delta}} e^{j\beta t} e^{-j\beta_0 t} \right| \tag{11.116}$$

or

$$\left| \frac{\hat{E}_i}{\hat{E}_t} \right| = \left| \frac{(\eta_0 + \hat{\eta})^2}{4\eta_0\hat{\eta}} \right| \left| \left[1 - \left(\frac{\eta_0 - \hat{\eta}}{\eta_0 + \hat{\eta}} \right)^2 e^{\frac{-2t}{\delta}} e^{-j2\beta t} \right] e^{\frac{t}{\delta}} \right| \tag{11.117}$$

or

$$\left| \frac{\hat{E}_i}{\hat{E}_t} \right| = \left| \frac{(\eta_0 + \hat{\eta})^2}{4\eta_0\hat{\eta}} \right| \left| e^{\frac{t}{\delta}} \right| \left| \left[1 - \left(\frac{\eta_0 - \hat{\eta}}{\eta_0 + \hat{\eta}} \right)^2 e^{\frac{-2t}{\delta}} e^{-j2\beta t} \right] \right| \tag{11.118}$$

It is convenient to express the shielding effectiveness in decibels

$$SE_{dB} = 20 \log_{10} \left| \frac{\hat{E}_i}{\hat{E}_t} \right| \tag{11.119}$$

Then, utilizing equation (11.118) the shielding effectiveness in dB becomes

$$SE_{dB} = 20 \log_{10} \left| \frac{\hat{E}_i}{\hat{E}_t} \right| = \underbrace{20 \log_{10} \left| \frac{(\eta_0 + \hat{\eta})^2}{4\eta_0\hat{\eta}} \right|}_{R_{dB}} + \underbrace{20 \log_{10} e^{\frac{t}{\delta}}}_{A_{dB}}$$

$$+ \underbrace{20 \log_{10} \left| \left[1 - \left(\frac{\eta_0 - \hat{\eta}}{\eta_0 + \hat{\eta}} \right)^2 e^{\frac{-2t}{\delta}} e^{-j2\beta t} \right] \right|}_{M_{dB}} \tag{11.120}$$

or

$$SE_{\mathrm{dB}} = R_{\mathrm{dB}} + A_{\mathrm{dB}} + M_{\mathrm{dB}} \tag{11.121}$$

R_{dB} is called the *reflection loss* and represents the portion of the incident field that is reflected at the shield interface. It is given by

$$R_{\mathrm{dB}} = 20 \log_{10} \left| \frac{(\eta_0 + \hat{\eta})^2}{4\eta_0\hat{\eta}} \right| \tag{11.122}$$

A_{dB} is called the *absorption loss* and represents the portion of the incident field that crosses the shield surface, and is attenuated as it travels through the shield. It is given by

$$A_{\mathrm{dB}} = 20 \log_{10} e^{\frac{t}{\delta}} \tag{11.123}$$

M_{dB} is called the *multiple-reflection loss* and represents the portion of the incident field that undergoes multiple reflections within the shield. It is given by

$$M_{\mathrm{dB}} = 20 \log_{10} \left| \left[1 - \left(\frac{\eta_0 - \hat{\eta}}{\eta_0 + \hat{\eta}} \right)^2 e^{\frac{-2t}{\delta}} e^{-j2\beta t} \right] \right| \tag{11.124}$$

The reflection and absorption losses are positive numbers (in dB), while the multiple reflection loss is a negative number (in dB). It, therefore, reduces the shielding effectiveness.

The solution in equation (11.120) was obtained for the shield made of a good conductor under the assumption of normal incidence of the uniform wave, i.e. when the shield is in the far field of the radiation source. The solution in equation (11.120) is often referred to as the exact solution.

Next, we will make some reasonable approximation that will greatly simplify this solution without any significant loss of accuracy.

11.2.2 Shielding Effectiveness – Approximate Solution – Version 1

The approximate solution for the shielding effectiveness is obtained from the exact solution of Section 11.2.1. To the requirement that shield, made of a good conductor is in the far field of the source, we add the requirement that shield is much thicker than the skin depth, at the frequency of interest, $t \gg \delta$.

Recall the exact solution given by equation (11.115), and repeated here

$$SE = \frac{\hat{E}_i}{\hat{E}_t} = \frac{(\eta_0 + \hat{\eta})^2}{4\eta_0\hat{\eta}} \left[1 - \left(\frac{\eta_0 - \hat{\eta}}{\eta_0 + \hat{\eta}} \right)^2 e^{\frac{-2t}{\delta}} e^{-j2\beta t} \right] e^{\frac{t}{\delta}} e^{j\beta t} e^{-j\beta_0 t} \tag{11.125}$$

The magnitude of this solution was given by equation (11.116) as

$$\left| \frac{\hat{E}_i}{\hat{E}_t} \right| = \left| \frac{(\eta_0 + \hat{\eta})^2}{4\eta_0 \hat{\eta}} \left[1 - \left(\frac{\eta_0 - \hat{\eta}}{\eta_0 + \hat{\eta}} \right)^2 e^{\frac{-2t}{\delta}} e^{-j2\beta t} \right] e^{\frac{t}{\delta}} e^{j\beta t} e^{-j\beta_0 t} \right| \tag{11.126}$$

which can be written as

$$\left| \frac{\hat{E}_i}{\hat{E}_t} \right| = \left| \frac{(\eta_0 + \hat{\eta})^2}{4\eta_0 \hat{\eta}} \left[e^{\frac{t}{\delta}} e^{j\beta t} e^{-j\beta_0 t} - \left(\frac{\eta_0 - \hat{\eta}}{\eta_0 + \hat{\eta}} \right)^2 e^{\frac{-t}{\delta}} e^{-j\beta t} e^{-j\beta_0 t} \right] \right| \tag{11.127}$$

Let's investigate the consequence of the assumption that shield is made of good conductor. Intrinsic impedance of a good conductor, at the frequencies of interest in shielding, is much smaller than the intrinsic impedance of free space. That is, $\hat{\eta} \ll \eta_0$ (for instance, the magnitude of the intrinsic impedance of copper at 1 MHz is $3.69 \times 10^{-4}\,\Omega \ll 377\,\Omega$).
It follows,

$$\hat{\eta} \ll \eta_0 \Rightarrow \frac{\eta_0 - \hat{\eta}}{\eta_0 + \hat{\eta}} \approx 1 \tag{11.128}$$

If the shield is thick, $t \gg \delta$, then we have

$$t \gg \delta \Rightarrow e^{-\frac{t}{\delta}} \ll 1 \tag{11.129}$$

and the right-hand side of the equation (11.127) can be approximated by

$$\left| \frac{\hat{E}_i}{\hat{E}_t} \right| \approx \left| \frac{(\eta_0 + \hat{\eta})^2}{4\eta_0 \hat{\eta}} \left[e^{\frac{t}{\delta}} e^{j\beta t} e^{-j\beta_0 t} \right] \right| \tag{11.130}$$

or

$$\left| \frac{\hat{E}_i}{\hat{E}_t} \right| \approx \left| \frac{(\eta_0 + \hat{\eta})^2}{4\eta_0 \hat{\eta}} \right| e^{\frac{t}{\delta}} \tag{11.131}$$

Furthermore, for a good conductor, we have

$$\hat{\eta} \ll \eta_0 \Rightarrow |(\eta_0 + \hat{\eta})^2| \approx |\eta_0^2| \tag{11.132}$$

and equation (11.131) simplifies to

$$\left| \frac{\hat{E}_i}{\hat{E}_t} \right| \approx \left| \frac{\eta_0}{4\hat{\eta}} \right| e^{\frac{t}{\delta}} \tag{11.133}$$

This is the approximate solution for a *good and thick conductor in far field*. In dB, this solution becomes

$$SE_{\mathrm{dB}} = 20 \log_{10} \left| \frac{\hat{E}_i}{\hat{E}_t} \right| \approx \underbrace{20 \log_{10} \left| \frac{\eta_0}{4\hat{\eta}} \right|}_{R_{\mathrm{dB}}} + \underbrace{20 \log_{10} e^{\frac{t}{\delta}}}_{A_{\mathrm{dB}}} \tag{11.134}$$

or

$$SE_{\mathrm{dB}} = R_{\mathrm{dB}} + A_{\mathrm{dB}} \tag{11.135}$$

where

$$R_{dB} \approx 20 \log_{10} \left| \frac{\eta_0}{4\hat{\eta}} \right| \tag{11.136}$$

$$A_{dB} = 20 \log_{10} e^{\frac{t}{\delta}} \tag{11.137}$$

Note that the approximate reflection loss is different from the exact reflection loss, given by equation (11.122), while the absorption loss is the same as in the exact solution. Also note that the multiple-reflection loss is not present in equation (11.135), which means, that for a good and thick conductor in far field, it can be ignored.

11.2.3 Shielding Effectiveness – Approximate Solution – Version 2

The approximate solution for the reflection loss given by equation (11.136), and the exact solution for the absorption loss given by equation (11.137) can be expressed in more practical forms.

In order to derive these alternative forms we need some parameter relationships. Recall the expressions defining the propagation constant and the intrinsic impedance.

$$\hat{\gamma} = \sqrt{j\omega\mu(\sigma + j\omega\epsilon)} \tag{11.138}$$

$$\hat{\eta} = \sqrt{\frac{j\omega\mu}{\sigma + j\omega\epsilon}} \tag{11.139}$$

Thus,

$$\hat{\gamma}\hat{\eta} = \sqrt{j\omega\mu(\sigma + j\omega\epsilon)} \sqrt{\frac{j\omega\mu}{\sigma + j\omega\epsilon}} = j\omega\mu \tag{11.140}$$

or

$$\hat{\eta} = \frac{j\omega\mu}{\hat{\gamma}} \tag{11.141}$$

We will return to this equation shortly. The propagation constant in equation (11.138) can be expressed as

$$\hat{\gamma} = \sqrt{j\omega\mu\sigma \left(1 + j\frac{\omega\epsilon}{\sigma} \right)} \tag{11.142}$$

For good conductors

$$\frac{\sigma}{\omega\epsilon} \gg 1 \Rightarrow \frac{\omega\epsilon}{\sigma} \ll 1 \tag{11.143}$$

Thus, the propagation constant in equation (11.142) can be approximated by

$$\hat{\gamma} \approx \sqrt{j\omega\mu\sigma} = \sqrt{\omega\mu\sigma} \angle 45° \tag{11.144}$$

Using this result in equation (11.141) we get

$$\hat{\eta} = \frac{\omega\mu\angle 90°}{\sqrt{\omega\mu\sigma}\angle 45°} \tag{11.145}$$

or

$$\hat{\eta} = \sqrt{\frac{\omega\mu}{\sigma}}\angle 45° \tag{11.146}$$

and thus

$$|\hat{\eta}| = \sqrt{\frac{\omega\mu}{\sigma}} = \sqrt{\frac{2\pi f\mu}{\sigma}} \tag{11.147}$$

The absolute permeability can be expressed in terms of the relative permeability (with respect to free space) as

$$\mu = \mu_r\mu_0 = \mu_r \times 4\pi \times 10^{-7} \tag{11.148}$$

The absolute conductivity can be expressed in terms of the relative conductivity (with respect to copper) as

$$\sigma = \sigma_r\sigma_{Cu} = \sigma_r \times 5.8 \times 10^7 \tag{11.149}$$

Using equations (11.148) and (11.149) in equation (11.147) we have

$$|\hat{\eta}| = \sqrt{\frac{2\pi f\mu_r \times 4\pi \times 10^{-7}}{\sigma_r \times 5.8 \times 10^7}} \tag{11.150}$$

or

$$|\hat{\eta}| = \sqrt{\frac{8\pi^2 f\mu_r}{\sigma_r \times 5.8 \times 10^{14}}} \tag{11.151}$$

Now we are ready to derive an alternative form of the reflection loss given by equation (11.136), repeated here

$$R_{dB} \approx 20\log_{10}\left|\frac{\eta_0}{4\hat{\eta}}\right| \tag{11.152}$$

Substituting $\eta_0 = 120\pi$ and using equation (11.151) in equation (11.152) we get

$$R_{dB} \approx 20\log_{10}\frac{30\pi}{\sqrt{\frac{8\pi^2 f\mu_r}{\sigma_r \times 5.8 \times 10^{14}}}} \tag{11.153}$$

or

$$R_{dB} \approx 20\log_{10}\sqrt{\frac{30^2\pi^2\sigma_r \times 5.8 \times 10^{14}}{8\pi^2 f\mu_r}} = 10\log_{10}\frac{30^2\sigma_r \times 5.8 \times 10^{14}}{8f\mu_r} \tag{11.154}$$

or

$$R_{dB} \approx 10\log_{10}\frac{30^2 \times 5.8 \times 10^{14}}{8} + 10\log_{10}\frac{\sigma_r}{f\mu_r} \tag{11.155}$$

Thus, the alternative expression for the reflection loss for good and thick conductors in far field is

$$R_{dB} \approx 168.15 + 10 \log_{10} \frac{\sigma_r}{f \mu_r} \tag{11.156}$$

Note that the reflection loss is greatest for high-conductivity, low-permeability materials, and at low frequencies.

Now, let's derive an alternative expression for the absorption loss. The exact formula is given by equation (11.137), repeated here

$$A_{dB} = 20 \log_{10} e^{\frac{t}{\delta}} = 20 \left(\frac{t}{\delta} \right) \log_{10} e = 20 (\log_{10} 2.71828) \left(\frac{t}{\delta} \right) \tag{11.157}$$

or

$$A_{dB} = 8.68588 \left(\frac{t}{\delta} \right) \tag{11.158}$$

Skin depth was given by equation (11.32)

$$\delta = \frac{1}{\sqrt{\pi f \mu \sigma}} \tag{11.159}$$

Using equations (11.148) and (11.149) in equation (11.159) we get

$$\delta = \frac{1}{\sqrt{\pi f \mu_r \times 4\pi \times 10^{-7} \sigma_r \times 5.8 \times 10^7}} = \frac{1}{15.132 \sqrt{f \mu_r \sigma_r}} \tag{11.160}$$

Substituting this result into equation (11.158) we get the alternative formula for the absorption loss of good conductors in far field as (with the conductor thickness t in meters)

$$A_{dB} = 131.434 t \sqrt{f \mu_r \sigma_r} \tag{11.161}$$

When the conductor thickness is expressed in inches this formula becomes

$$A_{dB} = 3.338 t \sqrt{f \mu_r \sigma_r} \tag{11.162}$$

Note that the absorption loss is greatest for high-conductivity, high-permeability materials, and at high frequencies.

Figure 11.7 shows the relative conductivity and relative permeability for various shield materials.

11.2.4 Shielding Effectiveness – Simulations

In this section we compare far-field shielding effectiveness of copper and steel.

Let's begin with the reflection loss, computed from

$$R_{dB} \approx 168.15 + 10 \log_{10} \frac{\sigma_r}{f \mu_r} \tag{11.163}$$

Figure 11.8 shows the reflection loss in the frequency range 100 Hz–1 GHz

Material	σ_r	μ_r
Copper	1	1
Aluminum	0.61	1
Brass	0.26	1
Bronze	0.18	1
Zinc	0.32	1
Nickel	0.2	100–600
Steel (SAE 1045)	0.1	1000

Figure 11.7 Relative conductivity and permeability for various shields.

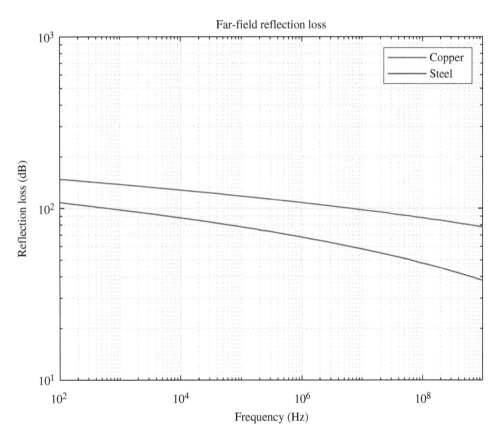

Figure 11.8 Reflection loss – copper vs. steel.

```
%% Far-Field Reflection Loss
%% Material Constants
copper_cond = 1; % copper conductivity
copper_perm = 1; % copper permittivity
steel_cond = 0.1; % steel conductivity
steel_perm = 1000; % steel permittivity
f = 1e2:1e3:3e10; % frequency vector
% reflection loss
R_dB_copper = 168+10*log10(copper_cond./(copper_perm.*f));
R_dB_steel = 168+10*log10(steel_cond./(steel_perm.*f));
figure; % create figure
loglog(f,R_dB_copper,'-r',f,R_dB_steel,'-b'); % plot equations
% add labels to plots
title('Far-Field Reflection Loss');
xlabel('Frequency (Hz)');
ylabel('Reflection Loss (dB)');
legend('Copper','Steel'); %create plot legend
xlim([1e2 1e9]); % create x axis limits
ylim([10 10e2]); % create y axis limits
grid; % add grid to plot
```

Figure 11.9 Matlab code – reflection loss – copper vs. steel.

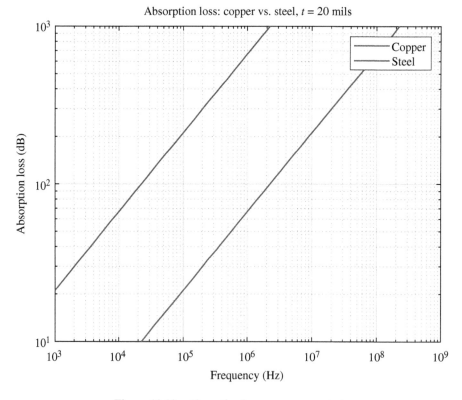

Figure 11.10 Absorption loss – copper vs. steel.

```
%% Absorption Loss - Copper vs. Steel t = 20 mils
%% Material Constants
copper_cond = 1; % copper conductivity
copper_perm = 1; % copper permittivity
steel_cond = 0.1; % steel conductivity
steel_perm = 1000; % steel permittivity
%% Absorption Loss with Changing Frequency
t = 0.02; % fixed thickness
f = 1e2:1e3:1e9; % frequency vector
% absorption loss equations for copper and steel
A_dB_copper = 3.34*t*sqrt(f*copper_perm*copper_cond);
A_dB_steel = 3.34*t*sqrt(f*steel_perm*steel_cond);
figure; % create figure
loglog(f,A_dB_copper,'-r',f,A_dB_steel,'-b'); %plot equations
% add labels to plots
title('Absorption Loss: Copper vs. Steel, t = 20 mils');
xlabel('Frequency (Hz)');
ylabel('Absorption Loss (dB)');
legend('Copper','Steel'); %create plot legend
xlim([10e2 1e9]); % create x axis limits
ylim([10 10e2]); % create y axis limits
grid; % add grid to plot
```

Figure 11.11   Matlab code – absorption loss – copper vs. steel.

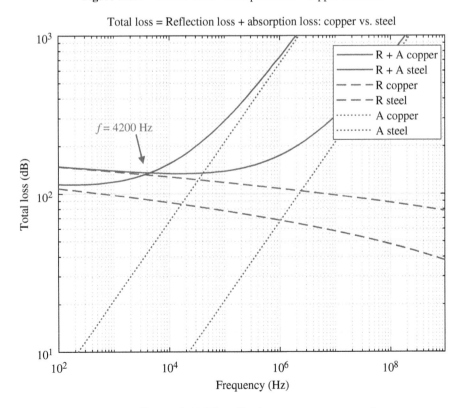

Figure 11.12   Total shielding effectiveness – copper vs. steel.

Note that the reflection loss of copper is higher over the entire frequency range. Matlab code used in the simulation is shown in Figure 11.9.

The absorption loss, for 20-mil thick shields, is calculated from

$$A_{\mathrm{dB}} = 3.338t\sqrt{f\mu_r\sigma_r} \tag{11.164}$$

and is shown in Figure 11.10.

Note that the absorption loss of steel is higher over the entire frequency range. Matlab code used in the simulation is shown in Figure 11.11.

The total shielding effectiveness is calculated from

$$SE_{\mathrm{dB}} = R_{\mathrm{dB}} + A_{\mathrm{dB}} = 168.15 + 10\log_{10}\frac{\sigma_r}{f\mu_r} + 3.338t\sqrt{f\mu_r\sigma_r} \tag{11.165}$$

and is shown in Figure 11.12.

Note that up to the frequency of about 4200 Hz, the shielding effectiveness of copper is higher than that of steel. Beyond that frequency the opposite is true. Matlab code used in the simulation is shown in Figure 11.13.

```matlab
%% Far-Field Total Loss
%% Material Constants
copper_cond = 1; % copper conductivity
copper_perm = 1; % copper permittivity
steel_cond = 0.1; % steel conductivity
steel_perm = 1000; % steel permittivity
f = 1e2:1e2:1e9; % frequency vector
% Absorption loss
A_dB_copper = 3.34*t*sqrt(f*copper_perm*copper_cond);
A_dB_steel = 3.34*t*sqrt(f*steel_perm*steel_cond);
% reflection loss
R_dB_copper = 168+10*log10(copper_cond./(copper_perm.*f));
R_dB_steel = 168+10*log10(steel_cond./(steel_perm.*f));
% Total loss = Reflection loss + absorption loss
T_dB_copper = R_dB_copper + A_dB_copper;
T_dB_steel = R_dB_steel + A_dB_steel;
figure; % create figure
% plot equations
loglog(f,T_dB_copper,'-r',f,T_dB_steel,'-b',f,R_dB_copper,'--r',f,R_dB_steel,'--
b',f,A_dB_copper,':r',f,A_dB_steel,':b');
% add labels to plots
title('Total Loss = Reflection Loss + Absorption Loss: Copper vs. Steel');
xlabel('Frequency (Hz)');
ylabel('Total Loss (dB)');
legend('R+A copper','R+A steel','R copper','R steel','A copper','A steel');
xlim([1e2 1e9]); % create x axis limits
ylim([10 10e2]); % create y axis limits
grid; % add grid to plot
```

Figure 11.13   Matlab code – total shielding effectiveness – copper vs. steel.

11.3 Near-Field Shielding

Note: The following derivations are valid under the assumptions of a good conductor and thick shield.

The total shielding effectiveness in the near field is

$$SE_{dB} = R_{dB} + A_{dB} \tag{11.166}$$

just like it was in the far field, but the reflection loss for electric sources is different from the reflection loss for the magnetic sources (in the far field the reflection loss for the two sources was the same).

The absorption loss in the near field is the same for the electric and magnetic sources, and is the same as it was in the far field. That is,

$$A_{dB} = 20 \log_{10} e^{\frac{t}{\delta}} \tag{11.167}$$

Thus, the shielding effectiveness in the near field for electric sources is

$$SE_{dB,e} = R_{dB,e} + A_{dB} \tag{11.168}$$

while the shielding effectiveness in the near field for magnetic sources is

$$SE_{dB,m} = R_{dB,m} + A_{dB} \tag{11.169}$$

Near-field shielding formulas for the reflection loss can be derived using the far-field shielding results for the reflection loss, and the concept of the near-field wave impedance for the electric and magnetic sources.

11.3.1 Electric Field Sources

Recall equation (11.136) for the reflection loss of a good, thick conductor in the far field, repeated here,

$$R_{dB,e} \approx 20 \log_{10} \left| \frac{\eta_0}{4\hat{\eta}} \right| \tag{11.170}$$

Recall the near-field wave impedance for electric sources given by equation (8.30), repeated here,

$$|\hat{Z}_{w,e}| \approx 60 \frac{\lambda_0}{r} \tag{11.171}$$

The reflection loss for the near-field electric sources is obtained by substituting this wave impedance for the intrinsic impedance in equation (11.170).

Thus,

$$R_{dB,e} \approx 20 \log_{10} \left| \frac{\hat{Z}_{w,e}}{4\hat{\eta}} \right| \tag{11.172}$$

or

$$R_{dB,e} \approx 20 \log_{10} \left| \frac{60 \frac{\lambda_0}{r}}{4\hat{\eta}} \right| \tag{11.173}$$

Wavelength can be expressed as

$$\lambda_0 = \frac{v_0}{f} = \frac{3 \times 10^8}{f} \tag{11.174}$$

Recall equation (11.151), repeated here,

$$|\hat{\eta}| = \sqrt{\frac{8\pi^2 f \mu_r}{\sigma_r \times 5.8 \times 10^{14}}} \tag{11.175}$$

Substituting equations (11.174) and (11.175) into equation (11.173) we get

$$R_{dB,e} \approx 20 \log_{10} \frac{\frac{180 \times 10^8}{fr}}{4\sqrt{\frac{8\pi^2 f \mu_r}{\sigma_r \times 5.8 \times 10^{14}}}} \tag{11.176}$$

or

$$R_{dB,e} \approx 20 \log_{10} \frac{45 \times 10^8}{fr} \sqrt{\frac{\sigma_r \times 5.8 \times 10^{14}}{8\pi^2 f \mu_r}} \tag{11.177}$$

or

$$R_{dB,e} \approx 20 \log_{10} \sqrt{\frac{\sigma_r \times 5.8 \times 10^{14} \times 45^2 \times 10^{16}}{8\pi^2 f^3 r^2 \mu_r}} \tag{11.178}$$

or

$$R_{dB,e} \approx 10 \log_{10} \frac{\sigma_r \times 5.8 \times 10^{14} \times 45^2 \times 10^{16}}{8\pi^2 f^3 r^2 \mu_r} \tag{11.179}$$

or

$$R_{dB,e} \approx 10 \log_{10} \frac{5.8 \times 10^{30} \times 45^2}{8\pi^2} + 10 \log_{10} \frac{\sigma_r}{f^3 r^2 \mu_r} \tag{11.180}$$

leading to the final result

$$R_{dB,e} \approx 321.72 + 10 \log_{10} \frac{\sigma_r}{f^3 r^2 \mu_r} \tag{11.181}$$

Thus, the near-field shielding effectiveness for electric field sources is

$$SE_{dB,e} = R_{dB,e} + A_{dB} \approx 321.72 + 10 \log_{10} \frac{\sigma_r}{f^3 r^2 \mu_r} + 20 \log_{10} e^{\frac{t}{\delta}} \tag{11.182}$$

11.3.2 Magnetic Field Sources

Again, recall equation (11.136) for the reflection loss of a good, thick conductor in the far field, repeated here,

$$R_{dB,m} \approx 20 \log_{10} \left| \frac{\eta_0}{4\hat{\eta}} \right| \tag{11.183}$$

Recall the near-field wave impedance for magnetic sources given by equation (8.57), repeated here,

$$|\hat{Z}_{w,m}| \approx 2369 \frac{r}{\lambda_0} \tag{11.184}$$

The reflection loss for the near-field magnetic sources is obtained by substituting this wave impedance for the intrinsic impedance in equation (11.183).
Thus,

$$R_{dB,m} \approx 20 \log_{10} \left| \frac{\hat{Z}_{w,m}}{4\hat{\eta}} \right| \tag{11.185}$$

or

$$R_{dB,m} \approx 20 \log_{10} \left| \frac{2369 \frac{r}{\lambda_0}}{4\hat{\eta}} \right| \tag{11.186}$$

where

$$\lambda_0 = \frac{v_0}{f} = \frac{3 \times 10^8}{f} \tag{11.187}$$

$$|\hat{\eta}| = \sqrt{\frac{8\pi^2 f \mu_r}{\sigma_r \times 5.8 \times 10^{14}}} \tag{11.188}$$

Thus,

$$R_{dB,m} \approx 20 \log_{10} \left| \frac{2369 \frac{r}{\frac{3 \times 10^8}{f}}}{4\sqrt{\frac{8\pi^2 f \mu_r}{\sigma_r \times 5.8 \times 10^{14}}}} \right| \tag{11.189}$$

or

$$R_{dB,m} \approx 20 \log_{10} \frac{197.417 fr}{10^8} \sqrt{\frac{\sigma_r \times 5.8 \times 10^{14}}{8\pi^2 f \mu_r}} \tag{11.190}$$

or

$$R_{dB,m} \approx 20 \log_{10} \sqrt{\frac{\sigma_r \times 5.8 \times 10^{14} \times (197.417)^2 f^2 r^2}{8\pi^2 f \mu_r \times 10^{16}}} \tag{11.191}$$

or

$$R_{dB,m} \approx 10 \log_{10} \frac{\sigma_r \times 5.8 \times 10^{14} \times (197.417)^2 f^2 r^2}{8\pi^2 f \mu_r \times 10^{16}} \tag{11.192}$$

or

$$R_{\text{dB},m} \approx 10 \log_{10} \frac{5.8 \times (197.417)^2}{8\pi^2 \times 10^2} + 10 \log_{10} \frac{fr^2\sigma_r}{\mu_r} \tag{11.193}$$

leading to

$$R_{\text{dB},m} \approx 14.57 + 10 \log_{10} \frac{fr^2\sigma_r}{\mu_r} \tag{11.194}$$

or equivalently

$$R_{\text{dB},m} \approx 14.57 + 10 \log_{10} fr^2 + 10 \log_{10} \frac{\sigma_r}{\mu_r} \tag{11.195}$$

Thus, the near-field shielding effectiveness for magnetic field sources is

$$SE_{\text{dB},m} = R_{\text{dB},m} + A_{\text{dB}} \approx 14.57 + 10 \log_{10} \frac{fr^2\sigma_r}{\mu_r} + 20 \log_{10} e^{\frac{t}{\delta}} \tag{11.196}$$

11.3.3 Shielding Effectiveness – Simulations

In this section we compare near-field shielding effectiveness of copper and steel.

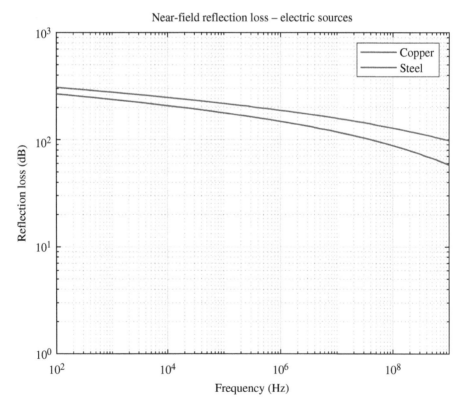

Figure 11.14   Reflection loss – electric field – copper vs. steel.

```
%% Near-Field Reflection Loss - Electric Source
%% Material Constants
copper_cond = 1; % copper conductivity
copper_perm = 1; % copper permittivity
steel_cond = 0.1; % steel conductivity
steel_perm = 1000; % steel permittivity
f = 1e2:1e3:1e9; % frequency vector
r = 0.005;
% reflection loss
R_e_dB_copper = 322+10*log10(copper_cond./(copper_perm.*f.*f.*f.*r.*r));
R_e_dB_steel = 322+10*log10(steel_cond./(steel_perm.*f.*f.*f.*r.*r));
figure; % create figure
loglog(f,R_e_dB_copper,'-r',f,R_e_dB_steel,'-b'); % plot equations
% add labels to plots
title('Near-Field Reflection Loss - Electric Sources');
xlabel('Frequency (Hz)');
ylabel('Reflection Loss (dB)');
legend('Copper','Steel'); %create plot legend
xlim([1e2 1e9]); % create x axis limits
ylim([1 10e2]); % create y axis limits
grid; % add grid to plot
```

Figure 11.15   Matlab code – reflection loss – electric field – copper vs. steel.

Let's begin with the reflection loss for electric field sources, at a distance of 5 mm, computed from

$$R_{dB,e} \approx 321.72 + 10 \log_{10} \frac{\sigma_r}{f^3 r^2 \mu_r} \tag{11.197}$$

Figure 11.14 shows the electric field reflection loss in the frequency range 100 Hz–1 GHz.

Note that the reflection loss of copper is higher over the entire frequency range. Matlab code used in simulation is shown in Figure 11.15.

Next, we compare the reflection loss for magnetic field sources. It is calculated at a distance of 5 mm, from the source using

$$R_{dB,m} \approx 14.57 + 10 \log_{10} \frac{f r^2 \sigma_r}{\mu_r} \tag{11.198}$$

and is shown in Figure 11.16.

Note that the reflection loss of copper is higher over the entire frequency range. Both reflection losses are quite small. Matlab code used in simulation is shown in Figure 11.17.

The absorption loss, for 20-mil thick shields, is calculated from (same as in far field)

$$A_{dB} = 3.338t \sqrt{f \mu_r \sigma_r} \tag{11.199}$$

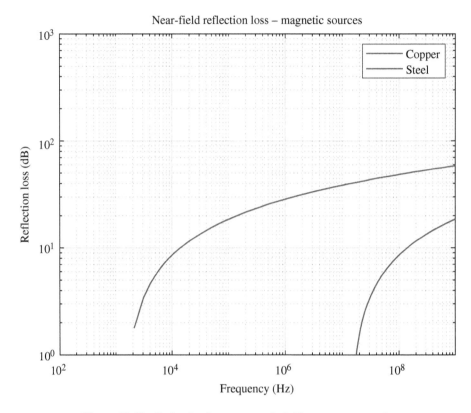

Figure 11.16   Reflection loss – magnetic field – copper vs. steel.

```
%% Near-Field Reflection Loss - Electric Source
%% Material Constants
copper_cond = 1; % copper conductivity
copper_perm = 1; % copper permittivity
steel_cond = 0.1; % steel conductivity
steel_perm = 1000; % steel permittivity
f = 1e2:1e3:1e9; % frequency vector
r = 0.005;
% reflection loss
R_e_dB_copper = 322+10*log10(copper_cond./(copper_perm.*f.*f.*f.*r.*r));
R_e_dB_steel = 322+10*log10(steel_cond./(steel_perm.*f.*f.*f.*r.*r));
figure; % create figure
loglog(f,R_e_dB_copper,'-r',f,R_e_dB_steel,'-b'); % plot equations
% add labels to plots
title('Near-Field Reflection Loss - Electric Sources');
xlabel('Frequency (Hz)');
ylabel('Reflection Loss (dB)');
legend('Copper','Steel'); %create plot legend
xlim([1e2 1e9]); % create x axis limits
ylim([1 10e2]); % create y axis limits
grid; % add grid to plot
```

Figure 11.17   Matlab code – reflection loss – magnetic field – copper vs. steel.

and was shown earlier in Figure 11.10, repeated here for reference,

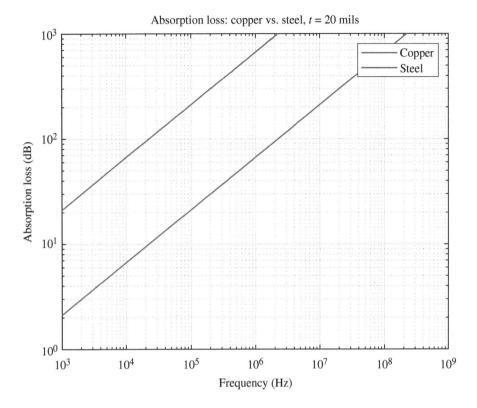

Note that the absorption loss of steel is higher over the entire frequency range.
The total shielding effectiveness for electric field sources is calculated from

$$SE_{dB} = R_{dB} + A_{dB} = 168.15 + 10 \log_{10} \frac{\sigma_r}{f \mu_r} + 3.338t\sqrt{f \mu_r \sigma_r} \qquad (11.200)$$

and is shown in Figure 11.18.

Note that up to the frequency of about 4400 Hz, the shielding effectiveness of copper is higher than that of steel. Beyond that frequency the opposite is true. Matlab code used in simulation is shown in Figure 11.19.

The total shielding effectiveness for magnetic field sources is calculated from

$$SE_{dB} = R_{dB} + A_{dB} = 14.57 + 10 \log_{10} \frac{f r^2 \sigma_r}{\mu_r} + 3.338t\sqrt{f \mu_r \sigma_r} \qquad (11.201)$$

and is shown in Figure 11.20.

Note that up to the frequency of about 4400 Hz, the shielding effectiveness of copper is higher than that of steel. Beyond that frequency the opposite is true. Matlab code used in simulation is shown in Figure 11.21.

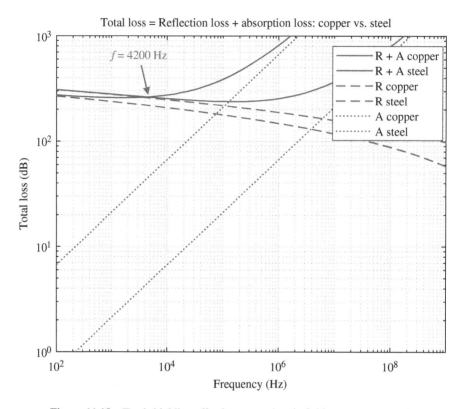

Figure 11.18   Total shielding effectiveness – electric field – copper vs. steel.

11.3.4 Shielding Effectiveness – Measurements

In this section, we compare the shielding effectiveness of various solid metallic shields by direct **H**-field measurements in the near field of the source [Adamczyk and Spencer, 2018]. We also show the detrimental effect of the shield apertures on the shielding effectiveness. The source of the **H** field is a step-down (buck) SMPS operating at a fundamental frequency of 118 kHz. The shields evaluated are: phosphor-bronze, nickel-silver, and cold-rolled steel. The measurements are taken 5 mm from the source of the field, in the frequency range of 100 kHz–1 MHz. The shields thickness is either 8 or 15 mils, and thus the shields are considered to be thick in the range of the frequency of interest.

The reflection loss and the absorption loss are given by

$$R_{dB,m} \approx 14.57 + 10 \log_{10} f r^2 + 10 \log_{10} \frac{\sigma_r}{\mu_r} \tag{11.202}$$

$$A_{dB} = 3.338 t \sqrt{f} \sqrt{\mu_r \sigma_r} \tag{11.203}$$

In the above equations, f is the frequency in Hz, r is the distance in meters, t represents the thickness of the shield in inches, and μ_r and σ_r are the relative permeability (with respect to free space) and relative conductivity (with respect to copper), respectively.

```
%% Near-Field Total Loss - Electric Source
%% Material Constants
copper_cond = 1; % copper conductivity
copper_perm = 1; % copper permittivity
steel_cond = 0.1; % steel conductivity
steel_perm = 1000; % steel permittivity
f = 1e2:1e2:1e9; % frequency vector
r = 0.005; % distance from the source
% Absorption loss
A_dB_copper = 3.34*t*sqrt(f*copper_perm*copper_cond);
A_dB_steel = 3.34*t*sqrt(f*steel_perm*steel_cond);
% reflection loss
R_e_dB_copper = 322+10*log10(copper_cond./(copper_perm.*f.*f.*f.*r.*r));
R_e_dB_steel = 322+10*log10(steel_cond./(steel_perm.*f.*f.*f.*r.*r));
% Total loss = Reflection loss + absorption loss
T_dB_copper = R_e_dB_copper + A_dB_copper;
T_dB_steel = R_e_dB_steel + A_dB_steel;
figure; % create figure
% plot equations
loglog(f,T_dB_copper,'-r',f,T_dB_steel,'-b',f,R_e_dB_copper,'--r',f,R_e_dB_steel,'--
b',f,A_dB_copper,':r',f,A_dB_steel,':b');
% add labels to plots
title('Total Loss = Reflection Loss + Absorption Loss: Copper vs. Steel');
xlabel('Frequency (Hz)');
ylabel('Total Loss (dB)');
legend('R+A copper','R+A steel','R copper','R steel','A copper','A steel');
xlim([1e2 1e9]); % create x axis limits
ylim([1 10e2]); % create y axis limits
grid; % add grid to plot
```

Figure 11.19 Matlab code – total shielding effectiveness – electric field – copper vs. steel.

The total shielding effectiveness is the sum of the reflection and absorption losses

$$SE_{dB} = R_{dB} + A_{dB} \qquad (11.204)$$

To validate the shielding formulas we used a step-down SMPS operating at a fundamental frequency of 118 kHz and performed **H**-field probe measurements with the probe placed directly above the source as shown in Figure 11.22.

The shield configurations are shown in Figure 11.23.

First, we compared the phosphor-bronze shields of thicknesses, 8 and 15 mils, respectively. The measurement results are shown in Figure 11.24.

Since the shields are made of the same material, the reflection losses for both shields are the same. The absorption loss depends on the shield thickness, and thus the absorption loss for the 15 mil shield is larger than that for the 8 mil shield. It follows that the total shielding effectiveness for the 15 mil shield is larger than that for the 8 mil shield. Figure 11.24 confirms these conclusions. It should be noted that Figure 11.24 shows the **H**-field measurements (and not the shielding effectiveness); these near-field measurements, however, follow the trend specified by the shielding equations (11.202) and (11.203).

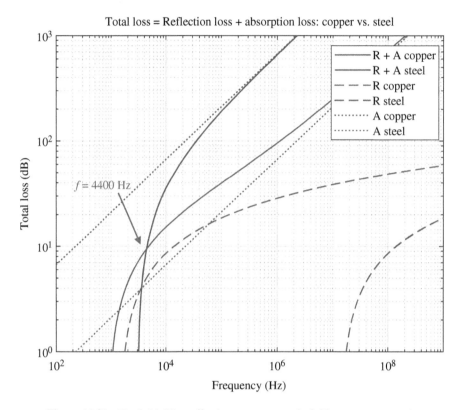

Figure 11.20   Total shielding effectiveness – magnetic field – copper vs. steel.

Next, let's compare phosphor-bronze and nickel-silver shields, both 8 mil thick. The measurement results for these shields are shown in Figure 11.25.

Both shields have the same thickness, same permeability, but different conductivities. The conductivity of the phosphor-bronze is higher than that of nickel silver. Therefore, its reflection and absorption losses are also higher; so is the total shielding effectiveness. This is confirmed in Figure 11.25.

Now, let's compare phosphor-bronze and cold-rolled steel shields, both 15 mils thick. The measurement results for these shields are shown in Figure 11.26.

The measurement results clearly show that the cold-rolled steel shield dramatically outperforms the phosphor-bronze shield, especially above 500 kHz. The explanation why this happens, is a bit more involved, since the shields have different permeabilities and different conductivities.

Evaluating the expression in equation (11.202) for phosphor-bronze we get

$$R_{\mathrm{dB},m} \approx 14.57 + 10 \log_{10} fr^2 + 10 \log_{10} \frac{\sigma_r}{\mu_r} \tag{11.205}$$

```
%% Near-Field Total Loss - Magnetic Source
%% Material Constants
copper_cond = 1; % copper conductivity
copper_perm = 1; % copper permittivity
steel_cond = 0.1; % steel conductivity
steel_perm = 1000; % steel permittivity
f = 1e2:1e2:1e9; % frequency vector
r = 0.005; % distance from the source
% Absorption loss
A_dB_copper = 3.34*t*sqrt(f*copper_perm*copper_cond);
A_dB_steel = 3.34*t*sqrt(f*steel_perm*steel_cond);
% reflection loss
R_m_dB_copper = 14.57+10*log10(f.*r.*r.*copper_cond/copper_perm);
R_m_dB_steel = 14.57+10*log10(f.*r.*r.*steel_cond/steel_perm);
% Total loss = Reflection loss + absorption loss
T_dB_copper = R_m_dB_copper + A_dB_copper;
T_dB_steel = R_m_dB_steel + A_dB_steel;
figure; % create figure
% plot equations
loglog(f,T_dB_copper,'-r',f,T_dB_steel,'-b',f,R_m_dB_copper,'--r',f,R_m_dB_steel,'--
b',f,A_dB_copper,':r',f,A_dB_steel,':b');
% add labels to plots
title('Total Loss = Reflection Loss + Absorption Loss: Copper vs. Steel');
xlabel('Frequency (Hz)');
ylabel('Total Loss (dB)');
legend('R+A copper','R+A steel','R copper','R steel','A copper','A steel');
xlim([1e2 1e9]); % create x axis limits
ylim([1 10e2]); % create y axis limits
grid; % add grid to plot
```

Figure 11.21 Matlab code – total shielding effectiveness – magnetic field – copper vs. steel.

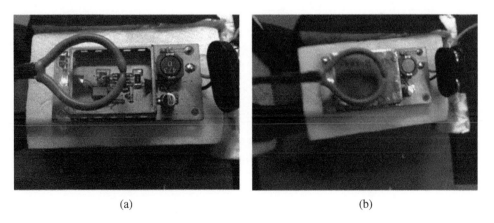

(a)   (b)

Figure 11.22 SMPS – (a) with no shield and (b) with a shield.

Material	Thickness	Relative conductivity σ_r	Relative permeability μ_r
Phosphor bronze	0.008	1	0.15
Phosphor bronze	0.015	1	0.15
Nickel silver	0.008	1	0.058
Cold-rolled steel	0.015	100	0.106

Figure 11.23   Shield configurations.

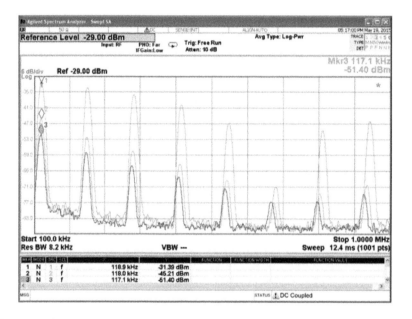

Figure 11.24   1 – no shield; 2 – phosphor-bronze 8 mils; 3 – phosphor-bronze 15 mils.

or

$$R_{\mathrm{dB},m} \approx 14.57 + 10\log_{10} f r^2 + 10\log_{10}\frac{0.15}{1} \tag{11.206}$$

or

$$R_{\mathrm{dB},m} \approx 14.57 + 10\log_{10} f r^2 - 8.24 \tag{11.207}$$

While for cold-rolled steel we have

$$R_{\mathrm{dB},m} \approx 14.57 + 10\log_{10} f r^2 + 10\log_{10}\frac{\sigma_r}{\mu_r} \tag{11.208}$$

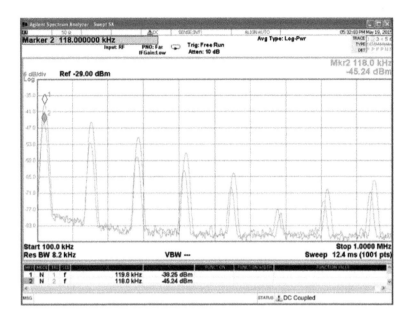

Figure 11.25   1 – Nickel-silver 8 mils; 2 – phosphor-bronze 8 mils.

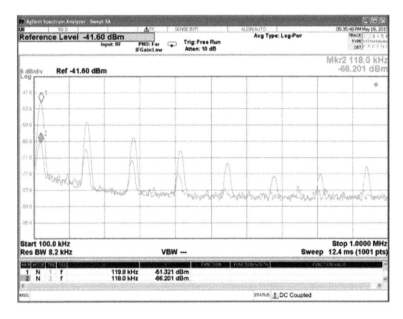

Figure 11.26   1 – phosphor-bronze 15 mils; 2 – cold-rolled steel 15 mils.

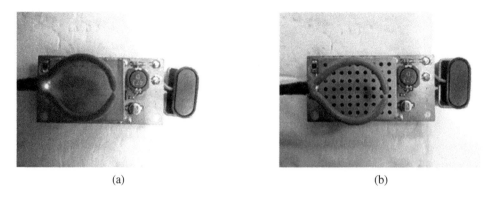

(a)   (b)

Figure 11.27   Cold-rolled steel 15 mils (a) solid shield and (b) shield with apertures.

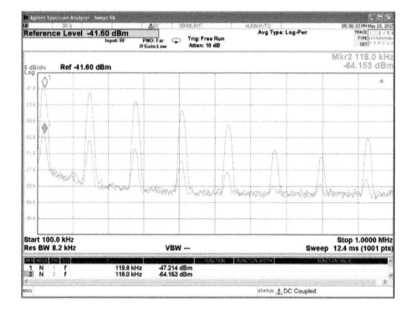

Figure 11.28   1 – cold-rolled steel w/holes 15 mils; 2 – cold-rolled steel solid 15 mils.

or

$$R_{\mathrm{dB},m} \approx 14.57 + 10\log_{10} fr^2 + 10\log_{10}\frac{0.106}{100} \tag{11.209}$$

$$R_{\mathrm{dB},m} \approx 14.57 + 10\log_{10} fr^2 - 29.75 \tag{11.210}$$

Thus, the reflection loss for phosphor-bronze is about 21.5 dB higher than that for cold-rolled steel, at all frequencies.

Now, let's look at the absorption loss. Evaluating the expression in equation (11.203) for phosphor-bronze we get

$$A_{\mathrm{dB}} = 3.338t\sqrt{f}\sqrt{\mu_r\sigma_r} \tag{11.211}$$

or

$$A_{dB} = 3.338t\sqrt{f}\sqrt{(1)(0.15)} = 0.39(3.338t\sqrt{f}) \tag{11.212}$$

While for cold-rolled steel we have

$$A_{dB} = 3.338t\sqrt{f}\sqrt{(100)(0.106)} = 3.26(3.338t\sqrt{f}) \tag{11.213}$$

Thus, the absorption loss for cold-rolled steel is about 8.36 times that of the phosphor-bronze. Note that this difference increases with frequency. Thus, as the frequency increases, the total loss, i.e. the total shielding effectiveness of the cold-rolled steel becomes higher.

At 100 kHz the absorption loss of the cold-rolled steel is about 15 dB higher than that of the phosphor-bronze. At 500 kHz the difference is almost 38 dB. This explains why the cold-rolled steel outperforms phosphor-bronze especially at higher frequencies.

Finally, we show detrimental effect of the shield apertures on the shielding effectiveness. Figure 11.27 shows the 15-mil thick cold-rolled steel solid shield, as well as a shield with apertures (holes) in it.

H-field measurement results for these shields are shown in Figure 11.28.

It is apparent from the measurement results shown in Figure 11.28 that the presence of apertures reduces the shielding effectiveness of a solid shield.

11.4 Laboratory Exercises

Objective: This laboratory exercise is devoted to the far-field and near-field shielding, and consists of two parts: simulation and measurement.

11.4.1 Shielding Effectiveness – Simulations

Repeat the far-field simulations of Section 11.2.4 and near-field simulations of Section 11.3.3. Comment on the results.

11.4.2 Shielding Effectiveness – Measurements

11.4.2.1 Laboratory Equipment and Supplies

Spectrum analyzer: SIGLENT SVA 1032X Spectrum Analyzer (or equivalent), shown in Figure 11.29.

DC Power Supply: BK PRECISION 1685B Switching Mode Power Supply (or equivalent), shown in Figure 11.30.

SMPS: Designed and build in Laboratory Exercise of Chapter 12; shown in Figure 11.31.

Loop antenna: This antenna, shown in Figure 11.32, was designed in the Laboratory Exercise of Chapter 8.

RG58C coaxial cable, shown in Figure 11.33
BNC Male Plug to BNC Male Plug, Length 2 feet

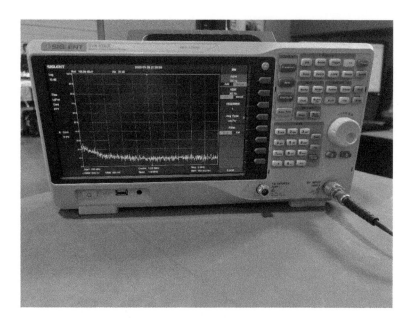

Figure 11.29 Laboratory equipment: spectrum analyzer.

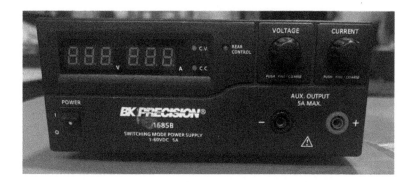

Figure 11.30 Laboratory equipment: DC power supply.

Supplier: Pomona Electronics, Part Number: 2249-C-24
Digi-Key Part Number: 501-1016-ND

RF Shield Frame, shown in Figure 11.34
Supplier: Laird Technologies EMI, Part Number:BMI-S-209-F
Digi-Key Part Number: 903-1212-1-ND

RF Shield, shown in Figure 11.35
Supplier: Laird Technologies EMI, Part Number:BMI-S-209-C
Digi-Key Part Number: 903-1185-ND

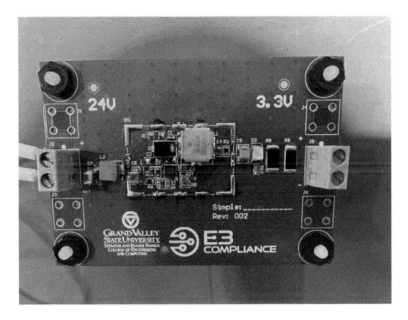

Figure 11.31 Laboratory equipment: SMPS.

Figure 11.32 Laboratory equipment: loop antenna.

Figure 11.33   Laboratory equipment: coax cable.

Figure 11.34   Laboratory equipment: RF shield frame.

Figure 11.35   Laboratory equipment: RF shield.

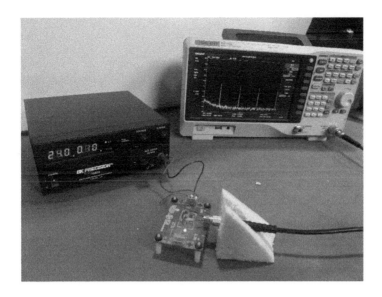

Figure 11.36 Measurement setup – near-field measurements of *H* field.

11.4.2.2 Laboratory Procedure

Figure 11.36 shows the measurement setup used for the near-field measurements of *H* field radiating from the SMPS.

Step 1: Place the *H*-field probe over the SMPS and capture the emissions in the frequency range spanning the first 20 harmonics.
Step 2: Attach (or solder) the RF shield frame to the SMPS.
Step 3: Place the shield over the frame and repeat the measurements taken in Step 1.
Step 4: Tabulate and comment on the results

References

Bogdan Adamczyk. *Foundations of Electromagnetic Compatibility*. John Wiley & Sons, Ltd, Chichester, UK, 2017. ISBN 9781119120810. doi: https://doi.org/10.1002/9781119120810.

Bogdan Adamczyk. Common-Mode Current Creation and Suppression. *In Compliance Magazine*, August 2019. URL https://incompliancemag.com/article/common-mode-current-creation-and-suppression/.

Bogdan Adamczyk. Skin Depth in Good Conductors. *In Compliance Magazine*, February 2020. URL https://incompliancemag.com/article/skin-depth-in-good-conductors/.

Bogdan Adamczyk and Bill Spencer. Near-Field Shield Effectiveness Evaluation Using H-Field Probe. *In Compliance Magazine*, April 2018. URL https://incompliancemag.com/article/near-field-shield-effectiveness-evaluation-using-h-field-probe/.

Clayton R. Paul. *Introduction to Electromagnetic Compatibility*. John Wiley & Sons, Inc., Hoboken, New Jersey, 2nd edition, 2006. ISBN 978-0-471-75500-5.

Chapter 12: SMPS Design for EMC

12.1 Basics of SMPS Operation

In this section the basics of a step-down (buck) DC Switched-Mode Power Supply (SMPS) are presented. The content of this section is based on the material presented in Hart [2011]. This section should serve as an entry-level tutorial and a building step toward the more advanced design in Section 12.2.

12.1.1 Basic SMPS Topology

The main functional objective of the buck SMPS is to step down a DC signal, V_{IN}, to a lower DC value, V_{OUT}, as shown in Figure 12.1.

The first step in this process consists of creating a Pulse-Width Modulated (PWM) version of the DC input signal, as shown in Figure 12.2.

The output signal shown in Figure 12.2 is far from the desired output signal described in our objective. Namely: (i) it is a constant signal only when the transistor is ON, (ii) its level, when the transistor is ON, is not lower than the input signal, and (iii) it contains high harmonic content during the transition times.

Let's address the third aspect by placing a low-pass LC filter on the output side of the circuit, as shown in Figure 12.3.

To reduce the unwanted power dissipation in the circuit, RC and RL filters are avoided, and the basic designs utilize a simple LC filter.

Let's assume that the transistor is OFF, there is no energy stored in the LC filter, and the output voltage is zero. When the transistor turns ON we have the circuit shown in Figure 12.4a (assuming an ideal transistor with no voltage drop).

The output voltage gradually increases. Assuming the ON time is long enough, this voltage eventually reaches the steady-state value $V_{OUT} = V_{IN}$. In steady state the voltage across the inductor, v_L, is zero, and a dc current, I_L, flows through the inductor, as shown in Figure 12.4b. The magnetic energy is stored in the inductor.

When the switch opens, a large negative voltage develops across the inductor and subsequently the switch. The magnetic energy stored in the inductor is dissipated in the arc across the switch contacts or is radiated, as shown in Figure 12.5a.

This behavior is often destructive to the switch and some sort of a protective circuitry is required [Ott, 2009]. The simplest solution is to provide a path for the inductor current during this switching event by inserting a diode in the circuit, as shown in Figure 12.5b. We have arrived at one of the simplest step-down (buck) SMPS.

The basic design of this SMPS amounts to the proper choice of the components, L and C, to satisfy the imposed design requirements. The component values are determined through the circuit analysis when the transistor is ON (switch closed) and when it is OFF (switch open). The respective circuits and selected circuit variables are shown in Figure 12.6.

Note that in both cases, when the switch closed and open, the inductor current is positive and flows in the same direction. If the switch remains open long enough the inductor

Principles of Electromagnetic Compatibility: Laboratory Exercises and Lectures, First Edition. Bogdan Adamczyk.
© 2024 John Wiley & Sons Ltd. Published 2024 by John Wiley & Sons Ltd.
Companion website: www.wiley.com/go/principlesofelectromagneticcompatibility

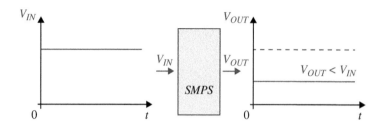

Figure 12.1 Objective of the buck SMPS.

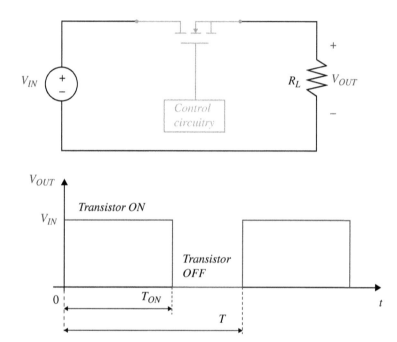

Figure 12.2 PWM signal.

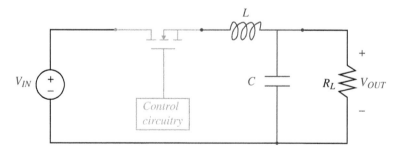

Figure 12.3 SMPS circuity with a low-pass filter.

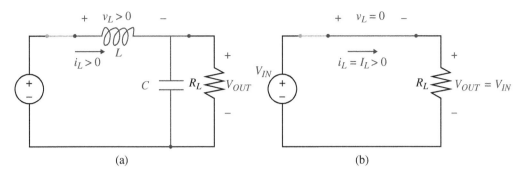

Figure 12.4 Transistor ON for the first time: (a) transient behavior, and (b) steady state.

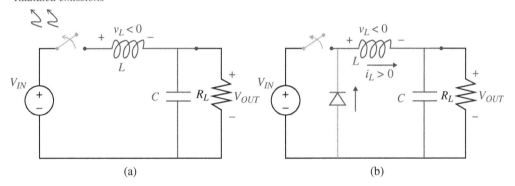

Figure 12.5 Transistor switches OFF: (a) unwanted behavior and (b) protective diode.

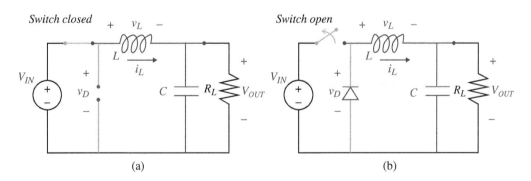

Figure 12.6 Step-down SMPS circuit: (a) transistor ON and (b) transistor OFF.

current decays to zero and subsequently the output voltage goes to zero. If the switch closes before the inductor current (and output voltage) goes to zero in the switching cycle, the SMPS will operate in the so-called *continuous conduction mode*. This is the preferred mode in EMC – it results in a smaller output ripple, smaller load-current variations, and lower EMC emissions.

When the switch subsequently opens, the output voltage rises. When it reaches the desired value, $V_{OUT} < V_{IN}$, the switch opens again. In a continuous mode and a steady-state operation, the inductor current and output voltage always stay positive and never go to zero.

12.1.2 Basic SMPS Design

The following SMPS design assumes that the components are ideal – transistor and diode voltage drops are zero, inductor and capacitor are ideal (no parasitics). There are no losses in the circuitry – power supplied by the source equals the power delivered to the load. The approach discussed here is based on the material presented in Hart [2011]. The SMPS operates in a steady state, in a continuous conduction mode with the duty cycle, D, of the PWM signal equal to

$$D = \frac{T_{ON}}{T} = f_{SW}T_{ON} \tag{12.1}$$

where the switching frequency f_{SW} is constant. Switch is closed for time

$$T_{ON} = DT \tag{12.2}$$

and is open for time

$$T_{OFF} = T - T_{ON} = T - DT = (1 - D)T \tag{12.3}$$

When the switch is closed, the diode is reverse biased and we have the circuit shown in Figure 12.7.

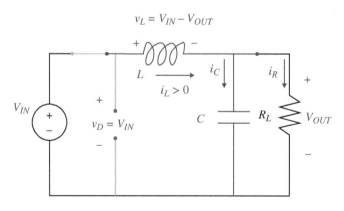

Figure 12.7   Circuit with the switch closed.

The voltage across the inductor is

$$V_L = L\frac{di_L}{dt} = V_{IN} - V_{OUT} \tag{12.4}$$

The voltage across the diode is equal to the input voltage, $v_D = V_{IN}$. From Eq. (12.4) we get

$$\frac{di_L}{dt} = \frac{V_{IN} - V_{OUT}}{L} > 0 \tag{12.5}$$

Since this derivative is positive, the inductor current increases linearly during the time when the switch is closed. To determine the (approximate) change in the inductor current during that time we approximate the derivative in Eq. (12.5) by

$$\frac{di_L}{dt} \approx \frac{\Delta i_L}{\Delta t} = \frac{\Delta i_L}{DT} = \frac{V_{IN} - V_{OUT}}{L} \tag{12.6}$$

Thus, the change in the inductor current is

$$(\Delta i_L)_{closed} = \left(\frac{V_{IN} - V_{OUT}}{L} \right) DT \tag{12.7}$$

When the switch is open, the diode is forward biased, $v_D = 0$, and we have the circuit shown in Figure 12.8.

The voltage across the inductor is

$$V_L = L\frac{di_L}{dt} = -V_{OUT} \tag{12.8}$$

From Eq. (12.8) we get

$$\frac{di_L}{dt} = \frac{-V_{OUT}}{L} > 0 \tag{12.9}$$

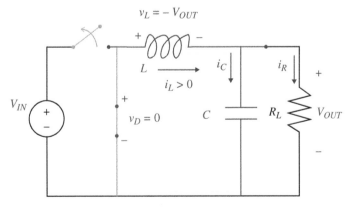

Figure 12.8   Circuit with the switch open.

Since this derivative is negative, the inductor current decreases linearly during the time when the switch is open. The (approximate) change in the inductor current during that time is obtained from

$$\frac{di_L}{dt} \approx \frac{\Delta i_L}{\Delta t} = \frac{\Delta i_L}{(1-D)T} = \frac{-V_{OUT}}{L} \tag{12.10}$$

Thus, the change in the inductor current is

$$(\Delta i_L)_{open} = -\left(\frac{V_{OUT}}{L}\right)(1-D)T \tag{12.11}$$

The variations in the inductor voltage and current are shown in Figure 12.9. Obviously, $(\Delta i_L)_{closed} = (\Delta i_L)_{open}$. From Eqs. (12.7) and (12.11) we get

$$\left(\frac{V_{IN} - V_{OUT}}{L}\right)DT = -\left(\frac{V_{OUT}}{L}\right)(1-D)T \tag{12.12}$$

or

$$(V_{IN} - V_{OUT})D = V_{OUT}(1-D) \tag{12.13}$$

resulting in an input–output relationship for a buck converter

$$V_{OUT} = DV_{IN} \tag{12.14}$$

Since the duty cycle is less than 1, the output voltage is lower than the input voltage. We can control the level of the output voltage by simply changing the duty cycle.

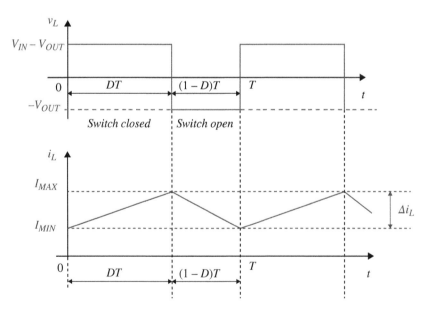

Figure 12.9 Variations in the inductor voltage and current.

Next, let's calculate the average, maximum, and minimum inductor currents. In a steady-state operation the average capacitor current, $I_C = 0$ [Hart, 2011]. It follows that the average inductor current, I_L, must be the same as the average load current, I_R. That is,

$$I_L = I_R = \frac{V_{OUT}}{R_L} \tag{12.15}$$

From Figure 12.9, the minimum and maximum values of the inductor current are

$$I_{MIN} = I_L - \frac{|\Delta i_L|}{2} \tag{12.16}$$

$$I_{MAX} = I_L + \frac{|\Delta i_L|}{2} \tag{12.17}$$

Using Eqs. (12.5) and (12.15) in Eqs. (12.16) and (12.17) we get

$$I_{MIN} = \frac{V_{OUT}}{R_L} - \frac{1}{2}\left(\frac{V_{OUT}}{L}\right)(1-D)T \tag{12.18}$$

$$I_{MAX} = \frac{V_{OUT}}{R_L} + \frac{1}{2}\left(\frac{V_{OUT}}{L}\right)(1-D)T \tag{12.19}$$

or

$$I_{MIN} = V_{OUT}\left(\frac{1}{R_L} - \frac{1-D}{2Lf_{SW}}\right) \tag{12.20}$$

$$I_{MAX} = V_{OUT}\left(\frac{1}{R_L} + \frac{1-D}{2Lf_{SW}}\right) \tag{12.21}$$

where $f_{SW} = 1/T$.

Now, we can calculate the minimum value of the inductance, L_{MIN}, for the continuous mode of operation. At the boundary between the continuous and discontinuous mode $I_{MIN} = 0$. Thus, from Eq. (12.20) we get

$$0 = V_{OUT}\left(\frac{1}{R_L} - \frac{1-D}{2L_{MIN}f_{SW}}\right) \tag{12.22}$$

or

$$L_{MIN} = \frac{(1-D)R_L}{2f_{SW}} \tag{12.23}$$

or, utilizing (12.13)

$$L_{MIN} = \frac{(V_{IN} - V_{OUT})R_L}{2f_{SW}V_{IN}} \tag{12.24}$$

The actual value of the inductance should, of course, be larger than the minimum value given by Eq. (12.24). A reasonable choice is

$$L = 1.25L_{MIN} \tag{12.25}$$

Finally, the output voltage ripple, ΔV_{OUT}, can be obtained by analyzing Figure 12.10, which shows the capacitor current and the output voltage curves [Hart, 2011].

The change in the capacitor charge, ΔQ, is equal to the triangle area under the capacitor current curve when the capacitor is charging. That is,

$$\Delta Q = \frac{1}{2}\left(\frac{T}{2}\right)\left(\frac{\Delta i_L}{2}\right) \tag{12.26}$$

Since

$$C = \frac{Q}{V_{OUT}} \tag{12.27}$$

it follows that

$$Q = CV_{OUT} \tag{12.28}$$

and

$$\Delta Q = C\Delta V_{OUT} \tag{12.29}$$

Using Eq. (12.29) in Eq. (12.26) we get

$$C\Delta V_{OUT} = \frac{1}{2}\left(\frac{T}{2}\right)\left(\frac{\Delta i_L}{2}\right) \tag{12.30}$$

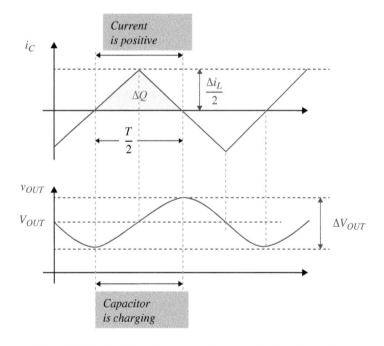

Figure 12.10   Variations in the capacitor current and output voltage.

Now, using Δi_L from Eq. (12.11) in the above equation we obtain

$$C\Delta V_{OUT} = \frac{1}{2}\left(\frac{T}{2}\right)\left(\frac{V_{OUT}}{2L}\right)(1-D)T \tag{12.31}$$

which leads to the output voltage ripple as

$$\Delta V_{OUT} = \frac{V_{OUT}(1-D)}{8LC(f_{SW})^2} \tag{12.32}$$

The relative output voltage ripple is

$$\frac{\Delta V_{OUT}}{V_{OUT}} = \frac{1-D}{8LC(f_{SW})^2} \tag{12.33}$$

which can be used to obtain the required capacitance in terms of the specified voltage ripple as

$$C = \frac{1-D}{8L(\Delta V_{OUT}/V_{OUT})(f_{SW})^2} \tag{12.34}$$

In the design of a SMPS the input and output voltages are usually specified. As are the load and the output voltage ripple. Once the switching frequency is chosen, the minimum inductor value can be calculated from Eq. (12.24) and the capacitor value from Eq. (12.34).

12.2 DC/DC Converter Design with EMC Considerations

This section focuses on the design of the DC/DC converter circuitry using Texas Instrument TPS54360B regulator [Adamczyk et al., 2021]. The converter simplified schematic is shown in Figure 12.11.

The design process for a baseline converter is based on the regulator application note. To begin the design process, the design requirements must be set. These are shown in Figure 12.12.

Additional assumption and design constraints are listed in Figure 12.13.

Next, the details of the design process will be described. We begin with the switching frequency.

12.2.1 Switching Frequency

Selecting the switching frequency is the first step in the design. According to the device specification, the upper limit on the switching frequency is obtained by first determining two frequencies: $f_{SW(maxskip)}$ and $f_{SW(shift)}$:

$$f_{SW(maxskip)} = \frac{1}{t_{ON}} \times \left(\frac{I_{OUT}R_{DC} + V_{OUT} + V_D}{V_{IN(max)} - I_{OUT}R_{DS(ON)} + V_D}\right) \tag{12.35}$$

or

$$f_{SW(maxskip)} = \frac{1}{135 \times 10^{-9}} \times \left(\frac{0.5 \times 0.18 + 3.3 + 0.65}{30 - 0.5 \times 0.092 + 0.65}\right) = 977.84\,\text{kHz} \tag{12.36}$$

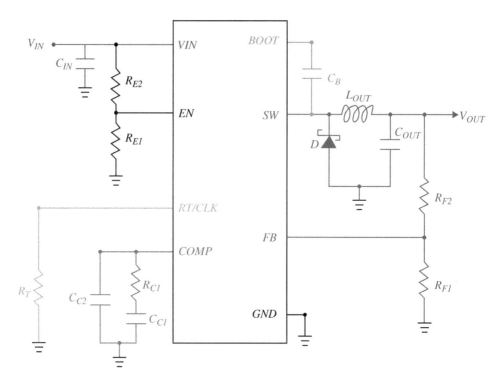

Figure 12.11   DC–DC converter simplified schematic.

Parameter	Value
Input voltage	$V_{IN} = 24$ V
Maximum input voltage	$V_{IN(max)} = 30$ V
Output voltage	$V_{OUT} = 3.3$ V
Maximum output voltage	$V_{OUT(max)} = 3.35$ V
Minimum output voltage	$V_{OUT(min)} = 3.25$ V
Output voltage ripple $(0.5\%V_{OUT})$	$V_{ripple} = 16.5$ mV
Output current	$I_{OUT} = 0.5$ A
Maximum output current $(1.25\,I_{OUT})$	$I_{OUT(max)} = 0.625$ A
Minimum output current $(0.75\,I_{OUT})$	$I_{OUT(min)} = 0.375$ A

Figure 12.12   Design requirements.

Parameter	Value
MOSFET ON time	$t_{ON} = 135$ ns
MOSFET DS ON resistance	$R_{DS(ON)} = 92$ mΩ
Inductor DC resistance	$R_{DC} = 180$ mΩ
Peak inductor current	$I_{CL} = 0.9$ A
Minimum load current (0.75 I_{OUT})	$I_{OUT(min)} = 0.375$ A
Maximum load current (1.75 I_{OUT})	$I_{OUT(max)} = 0.625$ A
Max protection factor for inductor current overload (increases OFF time)	$f_{DIV} = 8$
Diode threshold voltage	$V_D = 0.65$ V
Short-circuit output voltage	$V_{OUT(SC)} = 0.1$ V

Figure 12.13 Design constraints.

and

$$f_{SW(shift)} = \frac{f_{DIV}}{t_{ON}} \times \left(\frac{I_{OUT(max)}R_{DC} + V_{OUT(SC)} + V_D}{V_{IN(max)} - I_{OUT(max)}R_{DS(ON)} + V_D} \right) \tag{12.37}$$

or

$$f_{SW(maxskip)} = \frac{8}{135 \times 10^{-9}} \times \left(\frac{0.625 \times 0.18 + 0.1 + 0.65}{30 - 0.625 \times 0.092 + 0.65} \right) = 1670.71 \, \text{kHz} \tag{12.38}$$

The operating frequency should be lower than the lowest of the two values predicted by Eqs. (12.35) and (12.37). In our design targeted switching frequency $f_{SW} = 600$ kHz.

The switching frequency is adjusted using a (timing) resistor to ground connected to the *RT/CLK* pin. The timing resistance for a given switching frequency is determined from

$$R_T(\text{k}\Omega) = \frac{101\,756}{f_{SW}(\text{kHz})^{1.008}} = \frac{101\,756}{600^{1.008}} = 161.13 \, \text{k}\Omega \approx 160 \, \text{k}\Omega \tag{12.39}$$

The switching frequency corresponding to the nominal value of $R_T = 160$ kΩ is calculated from

$$f_{SW}(\text{kHz}) = \frac{92417}{R_T(\text{k}\Omega)^{0.991}} = \frac{92417}{160^{0.991}} = 604.6 \, \text{kHz} \tag{12.40}$$

12.2.2 Output Inductor

The minimum value of the inductor current is obtained from

$$L_{OUT(min)} = \frac{V_{IN(max)} - V_{OUT}}{I_{OUT} \times K_{IND}} = \frac{30 - 3.3}{0.5 \times 0.3} = 32.39 \, \mu\text{H} \tag{12.41}$$

$K_{IND} = 0.3$ in Eq. (12.41) comes from the regulator data sheet. In our design we chose Würth 744 053 330 inductor with a nominal value of $L_{OUT} = 33\,\mu H$.

The inductor ripple current is calculated from

$$I_{ripple} = \frac{V_{OUT} \times (V_{IN(max)} - V_{OUT})}{V_{IN(max)} \times L_{OUT} \times f_{SW}} = \frac{3.3 \times (30 - 3.3)}{30 \times 33 \times 10^{-6} \times 604.6^3} = 147\,\text{mA} \qquad (12.42)$$

The inductor ripple current is part of the current mode PWM control system and the suggested minimum value of the ripple current for a 5 V regulator is $150\,\text{mA}$. Since we are designing 3.3 V regulator, the value $147\,\text{mA}$ is acceptable.

The inductor *RMS* current is calculated from

$$I_{L(rms)} = \sqrt{I_{OUT} + \frac{1}{12}I_{ripple}^2} = 502\,\text{mA} \qquad (12.43)$$

The peak inductor current is obtained from

$$I_{L(peak)} = I_{OUT} + \frac{I_{ripple}}{2} = 0.5 + \frac{0.147}{2} = 573.5\,\text{mA} \qquad (12.44)$$

The Würth 744 053 330 inductor has a rated current of 900 mA and the saturation current of 750 mA, and thus the calculated values are well within these limits (the maximum output current of 625 mA shown in Figure 12.13 is also within these limits).

12.2.3 Output Capacitor

The output capacitor needs to satisfy several criteria. The first is the allowable change in the output voltage for a maximum change in the load current.

$$C_{OUT} > \frac{2 \times \Delta I_{OUT}}{f_{SW} \times \Delta V_{OUT}} = \frac{2 \times (I_{OUT(max)} - I_{OUT(min)})}{f_{SW} \times (V_{OUT(max)} - V_{OUT(min)})} \qquad (12.45)$$

or

$$C_{OUT} > \frac{2 \times (0.625 - 0.375)}{604.4 \times 10^3 \times (3.35 - 3.25)} = 8.27\,\mu F \qquad (12.46)$$

The output capacitor must also be sized to absorb the energy stored in the inductor when transitioning from a high to a low load current. This value is obtained from

$$C_{OUT} > (L_{OUT})\frac{I_{OUT(max)}^2 - I_{OUT(min)}^2}{V_{OUT(max)}^2 - V_{OUT}^2} \qquad (12.47)$$

or

$$C_{OUT} > (33 \times 10^{-6})\frac{0.625^2 - 0.375^2}{3.35^2 - 3^2} = 24.8\,\mu F \qquad (12.48)$$

The minimum output capacitance needed to satisfy the output voltage ripple is obtained from

$$C_{OUT} > \frac{1}{8f_{SW}} \frac{1}{\frac{V_{ripple}}{I_{ripple}}} = \frac{1}{8 \times 604.6 \times 10^3} \frac{1}{\frac{0.0165}{0.147}} = 1.8\,\mu\text{F} \tag{12.49}$$

The output capacitance should be larger than the largest value calculated in Eqs. (12.47), (12.48), and (12.49). Additionally, we need to account for the derating of the output capacitor; thus, let's double the value given by Eq. (12.48) to arrive at $C_{OUT(min)} = 49.6\,\mu\text{F}$. This value will be later used when designing the compensation network.

To account for the safety margin in our design, we chose two output capacitors in parallel, each of the value $C_{OUT} = 47\,\mu\text{F}$.

Maximum *ESR* of the output capacitor is calculated from

$$R_{ESR} < \frac{V_{ripple}}{I_{ripple}} = \frac{0.0165}{0.147} = 112\,\text{m}\Omega \tag{12.50}$$

In our design we chose Murata capacitor GRM32ER71A476KE15L with an *ESR* of $3\,\text{m}\Omega$ at $100\,\text{kHz}$, which is well below the maximum ESR predicted by Eq. (12.50).

12.2.4 Catch Diode

The regulator requires an external catch diode between the *SW* pin and *GND*. The diode must have a reverse voltage rating equal to or greater than $V_{IN(max)}$. The peak current rating of the diode must be greater than the maximum inductor current. The diode must also have an appropriate power rating. The power dissipation of the diode is calculated from

$$P_D = \frac{(V_{IN(max)} - V_{OUT}) \times I_{OUT} \times V_D}{V_{IN(max)}} + \frac{C_T \times f_{SW} \times (V_{IN(max)} + V_D)^2}{2} \tag{12.51}$$

where C_T is junction capacitance. We chose a Schottky diode B260S1F from Diodes Incorporated. It is rated at the reverse voltage of $60\,\text{V}$, peak current of $2\,\text{A}$, $V_D = 0.65\,\text{V}$, $c_T = 75\,\text{pF}$.

Thus, the power dissipation of the diode is

$$P_D = \frac{(30 - 3.3) \times 0.5 \times 0.65}{30} + \frac{75 \times 10^{-12} \times 604.6 \times 10^3 \times (30 + 0.65)^2}{2} = 310.55\,\text{mW} \tag{12.52}$$

and is well below the power threshold for the rated voltage and current values.

12.2.5 Input Capacitor

The regulator requires a high-quality ceramic type *X5R* or *X7R* input capacitor with at least $3\,\mu\text{F}$ of effective capacitance. The voltage rating of the input capacitor must be greater than the maximum input voltage. In our design we use two Murata GCJ32DR72A225KA01L, $2.2\,\mu\text{F}$, $100\,\text{V}$, ceramic capacitors in parallel.

12.2.6 Bootstrap Capacitor

Bootstrap capacitor between the *BOOT* and *SW* pins is needed to provide the gate voltage to drive the high-side MOSFET. The recommended value is $0.1\,\mu\text{F}$. TI recommends a ceramic capacitor with *X7R* or *X5R* grade dielectric with a voltage rating of $10\,\text{V}$ or higher. In our design we use $0.1\,\mu\text{F}$, $50\,\text{V}$ Murata GCJ188R71H104KA12D capacitor.

12.2.7 Undervoltage Lockout

According to the specifications, the input voltage range of the regulator is $4.5 - 60\,\text{V}$. In our application the input voltage to the regulator is $24\,\text{V}$ *(VIN)*. The regulator is enabled when the input voltage *(VIN)* rises above $4.3\,\text{V}$ and disabled when this voltage drops below $4.3\,\text{V}$. The undervoltage lockout *(UVLO)* of $4.3\,\text{V}$ can be adjusted with two resistors (forming a voltage divider) connected to *EN* pin, as shown in the schematic.

When adjusted to a non-default value of $4.3\,\text{V}$, *UVLO* has two thresholds, one for power up when the input voltage is rising *(UVLO start)*, and one for power down when the input voltage is falling *(UVLO stop)*.

In our application we chose *UVLO start* or $V_{START} = 8\,\text{V}$ and *UVLO stop* or $V_{STOP} = 6.25\,\text{V}$. Then the values of the two resistors are calculated as follows. The value of the resistor between *EV* and *VIN*, R_{UVLO1}, is obtained from

$$R_{UVLO1} = \frac{V_{START} - V_{STOP}}{I_{HYS}} \tag{12.53}$$

I_{HYS} is internally set to $3.4\,\mu\text{A}$. Thus, in our design,

$$R_{UVLO1} = \frac{8 - 6.25}{3.4 \times 10^{-6}} = 514.7 \approx 523\,\text{k}\Omega \tag{12.54}$$

The value of the resistor between *EV* and *GND*, R_{UVLO2}, is obtained from

$$R_{UVLO2} = \frac{V_{ENA}}{\frac{V_{START} - V_{ENA}}{R_{UVLO1}} + I_1} \tag{12.55}$$

I_1 is internally set to $1.2\,\mu\text{A}$. Thus, in our design,

$$R_{UVLO2} = \frac{1.2}{\frac{8 - 1.2}{523 \times 10^3} + 1.2 \times 10^{-6}} \approx 84.5\,\text{k}\Omega \tag{12.56}$$

12.2.8 Feedback Pin

The *FB* pin monitors the output voltage by comparing it to the internal value of $0.8\,\text{V}$. The output voltage is set by a resistor divider from the output node to the *FB* pin. The current flowing through the feedback network should be greater than $1/\mu\text{A}$ to maintain the output voltage accuracy. The resistors comprising the voltage divider should have 1%, or better, tolerance.

If a low-side resistor, $R_{LS} = 10\,\text{k}\Omega$ is used, then the high-side resistor value, R_{HS}, is obtained from

$$R_{HS} = R_{LS}\frac{V_{OUT} - 0.8}{0.8} \qquad (12.57)$$

In our application, $V_{OUT} = 3.3\,V$ resulting in $R_{HS} = 31.25\,k\Omega$. The resulting current in the feedback circuitry is $80\,\mu A$.

12.2.9 Compensation Network

The *COMP* pin is connected to the frequency compensation network. The compensation network internally provides input to the PWM circuitry which controls the switching of the *SW* node and thus controls the current to the load. Specifically, the *COMP* pin voltage controls the peak current on the high-side MOSFET.

To design the compensation network, we follow the procedure outline in the regulator specifications. First, several frequencies are calculated.

$$f_{P(mod)} = \frac{I_{OUT(max)}}{2\pi V_{OUT}C_{OUT(min)}} = \frac{0.625}{2\pi \times 3.3 \times 49.6 \times 10^{-6}} = 607.7\,\text{Hz} \qquad (12.58)$$

$$f_{z(mod)} = \frac{1}{2\pi R_{ESR}C_{OUT(min)}} = \frac{1}{2\pi \times 1.5 \times 10^{-3} \times 49.6 \times 10^{-6}} = 2,139.2\,\text{kHz} \qquad (12.59)$$

$$f_{c01} = \sqrt{f_{P(mod)} \times f_{z(mod)}} = \sqrt{2,139,200 \times 607.7} = 36,055.4\,\text{kHz} \qquad (12.60)$$

$$f_{c02} = \sqrt{f_{P(mod)} \times \frac{f_{SW}}{2}} = \sqrt{607.7 \times 302.3 \times 10^3} = 13,553.9\,\text{Hz} \qquad (12.61)$$

The lower of the two values $f_{co} = f_{co2}$ in Eqs. (12.60) and (12.61) will be used. The resistor R_{C1} in the compensation network is calculated from

$$R_{C1} = \frac{2\pi \times f_{co} \times C_{OUT(min)}}{gmps} \times \frac{V_{OUT}}{V_{REF} \times gmea} \qquad (12.62)$$

From the regulator specifications we use $gmps = 12\,\text{A/V}$, $gmea = 350\,\mu\text{A/V}$, $V_{REF} = 0.8\,\text{V}$. Thus,

$$R_{C1} = \frac{2\pi \times 13,553.9 \times 49.6 \times 10^{-6}}{12} \times \frac{3.3}{0.8 \times 350 \times 10^{-6}} \approx 4,120\,\Omega \qquad (12.63)$$

The capacitor C_{C1} in the compensation network is calculated from

$$C_{C1} = \frac{1}{2\pi \times R_{C1} \times f_{P(mod)}} = \frac{1}{2\pi \times 4120 \times 607.7} = 63.56 \approx 68\,\text{nF} \qquad (12.64)$$

The capacitor to ground, C_{C2}, in the compensation network is calculated from two different equations and the larger value is chosen.

$$C_{C2} = \frac{C_{OUT(min)} \times R_{ESR}}{R_{C1}} = \frac{49.6 \times 10^{-6} \times 1.5 \times 10^{-3}}{4120} = 18.06\,\text{pF} \qquad (12.65)$$

$$C_{C2} = \frac{1}{R_{C1} \times f_{SW} \times \pi} = \frac{1}{4120 \times 604.6 \times 10^{3} \times \pi} = 127.78\,\text{pF} \approx 120\,\text{pF} \qquad (12.66)$$

12.2.10 Complete Regulator Circuitry

Figure 12.14 shows the complete circuitry of the designed regulator.

12.2.11 EMC Considerations

An EMC design review was performed next to identify high, risk areas to meeting emissions requirements. During the review a number of additional components were identified for filtering, damping switching effects, inductor technology, and shielding options. The schematic shown in Figure 12.14 was modified by superimposing the EMC considerations, shown as dashed boxes labeled *A* through *F* in Figure 12.15. The values of these EMC components are chosen based on experience but will likely need to be tuned in the laboratory through experimental measurements.

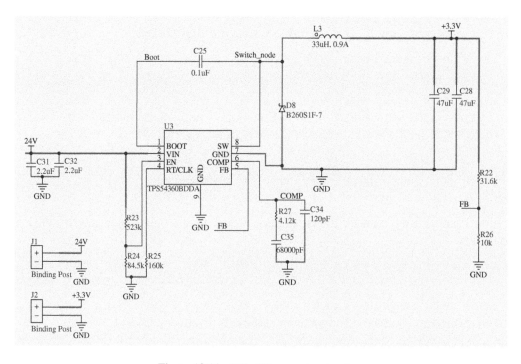

Figure 12.14   DC–DC converter circuitry.

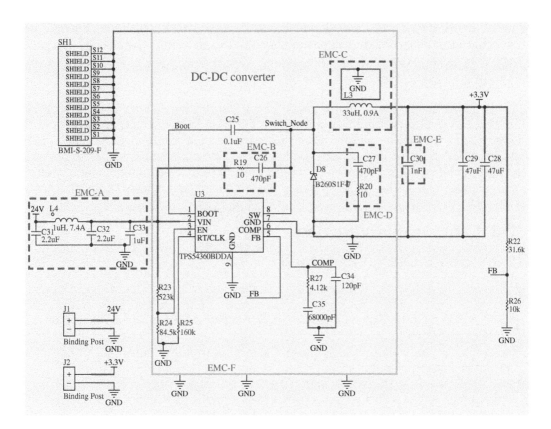

Figure 12.15 DC–DC converter circuitry with EMC considerations.

These considerations are addressed below.

EMC-A: A 1 nF capacitor (C33) is added on the V_{in} pin (U1.2) for high-frequency decoupling to reduce the amount of high-frequency energy generated by the DC–DC converter that can conduct into the 24 V bus. Additionally, a π filter is constructed by adding a 1 μH inductor (L4) between the two input capacitors (C31 and C32).

EMC-B: A snubber circuit (series Resistor R19 and Capacitor C26) is added from SW Node (U1.8) to Vin (U1.2) because the Internal MOSFET is placed between these pins. This snubber is used to control the ringing from the internal MOSFET that results from the step response to the RLC network.

EMC-C: The Würth Elektronik inductor (L3) may be replaced with a Vishay IHLP inductor which is magnetically shielded or an IHLE inductor which is E-Field shielded to reduce radiated and conducted emissions further.

EMC-D: A snubber circuit (series Resistor R20 and Capacitor C27) is added across the catch diode to reduce ringing across the diode junction.

EMC-E: A 1 nF capacitor (C30) is added near the inductor (L3) on the 3.3 V bus to filter high-frequency noise from conducting out onto the 3.3 V bus.

EMC-F: A shield (SH1) may be added if the PCB layout and EMC components are not successful to reduce the emissions from the DC−DC converter. This may be needed in some instances where the EMC requirements are very stringent or the supply may be closely co-located to sensitive receivers.

These EMC components are designed into the first prototype as optional components that will be evaluated during emissions testing. The EMC performance may have tradeoff decisions to make with regard to other requirements such as thermal, reliability, manufacturability, and cost. An example would be selecting the snubber values to optimize the EMC and thermal performance. It is likely that not all of the EMC components will be needed and efforts will be made to remove the unneeded components to optimize cost.

12.3 Laboratory Exercises

12.3.1 SMPS Design and Build

Objective: The objective of this laboratory is to design and build a step-down SMPS.

Note 1: This laboratory requires a custom PCB [Koeller, 2022], for which Altium files and BOM are provided.
Note 2: SMPS designed in this laboratory Exercise is used in Chapter 11 for near-field emissions measurements.
Note 3: SMPS designed in this laboratory Exercise is a baseline design (with minimal EMC consideration). It can be augmented with EMC components to optimize conducted and radiated emissions. This is presented in the Appendix.

12.3.2 Laboratory Equipment and Supplies

DC Power Supply: BK PRECISION 1685B SMPS (or equivalent), shown in Figure 12.16.
  Oscilloscope: Tektronix MDO3104 (or equivalent), shown in Figure 12.17.
  Custom PCB: Shown in Figure 12.18.

Figure 12.16   Laboratory equipment: DC power supply.

Figure 12.17 Laboratory equipment: oscilloscope.

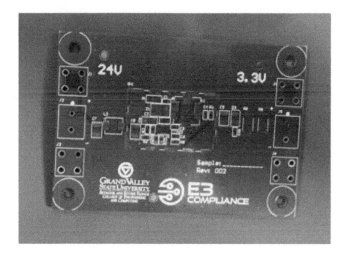

Figure 12.18 Laboratory equipment: custom PCB.

12.3.3 Laboratory Procedure

Design and build a step-down switched-mode power supply (SMPS). The input to the SMPS is 24 V DC and the required output is 3.3 V DC. The output ripple should be less than 50 mV.

Document each step of the design and support the choices with the relevant theory and equations. Capture the input and output waveforms to verify the operation.

Create the project report that includes the project definition, design section, measurement section, and conclusions.

References

Bogdan Adamczyk, Scott Mee, and Nick Koeller. Evaluation of EMC Emissions and Ground Techniques on 1- and 2-layer PCBs with Power Converters - Part 2: DC/DC Converter Design with EMC Considerations. *In Compliance Magazine*, June 2021. URL https://incompliancemag.com/article/evaluation-of-emc-emissions-and-ground-techniques-part2/.

Daniel W. Hart. *Power Electronics*. McGraw Hill, New York, New York, 2011. ISBN 978-0-07-338067-4.

Nick Koeller. SMPS Designer, 2022.

Henry W. Ott. *Electromagnetic Compatibility Engineering*. John Wiley & Sons, Inc., Hoboken, New Jersey, 2009. ISBN 978-0-470-18930-6.

Appendix A: Evaluation of EMC Emissions and Ground Techniques on 1- and 2-Layer PCBs with Power Converters

This appendix is devoted to the design, test, and EMC emissions evaluation of 1- and 2-layer PCBs that contain AC/DC and/or DC/DC converters, and employ different ground techniques. First, a top-level block diagram description of the design problem under research is presented. Subsequently the specific parts of the design are addressed, followed by the evaluation of RF emissions from the PCB assembly. The goal of this study is to evaluate the impact of different grounding strategies and the tradeoff with other design constraints that designers often face.

A.1 Top-Level Description of the Design Problem

Figure A.1 shows the top-level schematic and the functional blocks of the PCB assembly [Adamczyk et al., 2021a].

The board will be capable of accepting either an AC or DC input. The AC to DC conversion will take part in Partition *A* of the board (not drawn to scale). The DC to DC converter in Partition *B* will accept 24 V DC input either from the AC/DC converter in Partition *A* or from an external source.

In Figure A.2 we show the EMC consideration superimposed onto the functionality requirements. These considerations include both conducted and radiated emissions.

The external AC and DC inputs, and I/O circuitry provide noise-coupling paths (for conducted /radiated emissions) from the converters. Additional noise paths exist between the two converters themselves, as well as between the converters and the rest of the circuitry in Partition *B*. The implementation of EMC design controls and PCB layout will affect the EMC performance of the PCB assembly and associated cabling.

A.1.1 Functional Block Details

Figure A.3 shows the block diagram of the AC/DC converter.

The converter stage employs a filtering block, full-wave rectifier, controller, and a transformer which provides isolation between the two partitions.

Figure A.4 shows the block diagram of the DC/DC converter.

24 V DC input to the converter comes either from the AC/DC converter or from an external linear power supply input. The control IC contains the switching transistor and the feedback signal detection.

Figure A.5 shows the block diagram of the I/O circuitry.

Principles of Electromagnetic Compatibility: Laboratory Exercises and Lectures, First Edition. Bogdan Adamczyk.
© 2024 John Wiley & Sons Ltd. Published 2024 by John Wiley & Sons Ltd.
Companion website: www.wiley.com/go/principlesofelectromagneticcompatibility

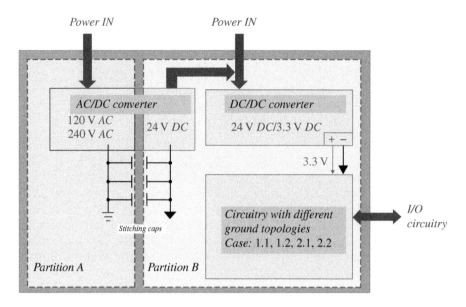

Figure A.1   Top-level schematic – functional blocks.

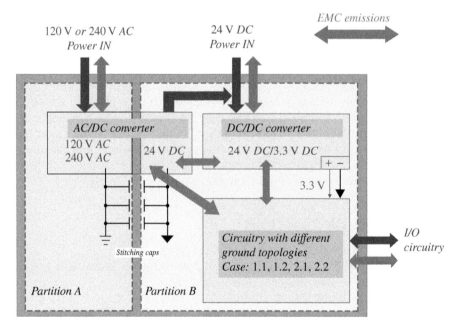

Figure A.2   Functional blocks with EMC considerations.

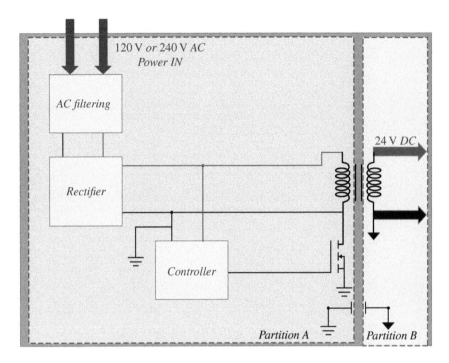

Figure A.3   AC/DC converter – block diagram.

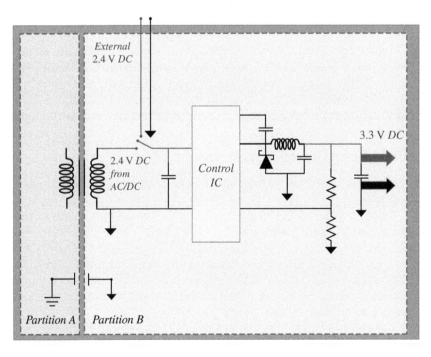

Figure A.4   DC/DC converter – block diagram.

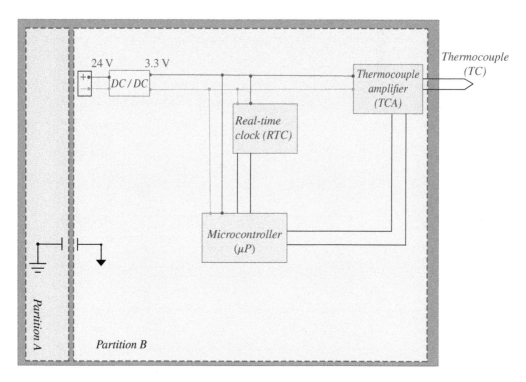

Figure A.5   I/O circuitry – block diagram.

The I/O circuitry contains a microprocessor powered by 3.3 V DC as regulated by the DC/DC converter. A real-time clock is provided so that analog values from the thermocouple can be recorded in memory. An unshielded multi-conductor cable with a length of 1 m will be connected between Partition *B* and a thermocouple. This cable is likely to carry some of the common-mode emissions from the converters and the microcontroller.

A.1.2 One-Layer Board Topologies

This section describes two 1-layer PCB topologies under study, referred to as Cases 1.1 and 1.2, respectively.

Figure A.6 shows grounding scheme for Case 1.1, where the ground is routed exclusively as traces on the top of the board.

This case represents some of the more challenging designs that are subject to significant cost and space constraints. In this scenario the designer has very few options to apply EMC rules-of-thumb, and best design practices. It is likely that this design will have challenges meeting RF emission requirements and may require additional filtering components to address non-compliances.

Figure A.7 shows grounding scheme for Case 1.2, where ground floods are introduced on the top of the board.

Case 1.2 is similar to Case 1.1, but with fewer space constraints in its application. Here the designer has more opportunities to improve grounding and reference areas. Adding

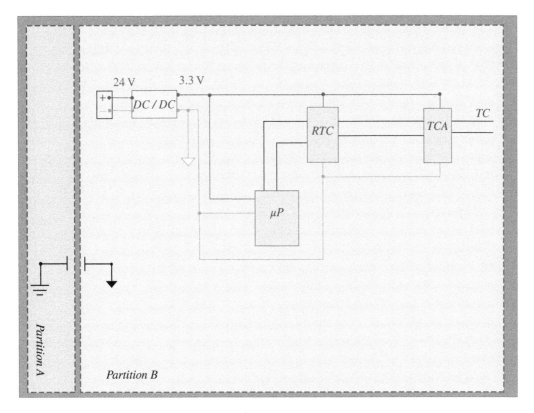

Figure A.6 One-layer board – Case 1.1.

additional ground and/or reference areas improves RF return paths and can reduce RF emissions. The additional copper areas will likely help with thermal power dissipation, as well.

A.1.3 Two-Layer Board Topologies

This section describes two 2-layer PCB topologies under study, referred to as Cases 2.1 and 2.2, respectively.

Figure A.8 shows grounding scheme for Case 2.1, where the bottom layer is a mostly solid reference plane with some slots accounting for the need to route signals on the secondary layer.

This design moves closer to the ideal reference plane implementation on the secondary side of the PCB. It has significantly more reference copper to help reduce RF emissions, but the designer requires some use of the secondary side to route power and signal nets. These nets create cut-outs (slots) in the secondary side of the PCB and can negatively impact RF emissions. Stitching vias are used to connect some copper reference areas on top and bottom layers.

Figure A.9 shows grounding scheme for Case 2.2, where the bottom layer is a complete ground flood with via stitching to the top-layer ground areas.

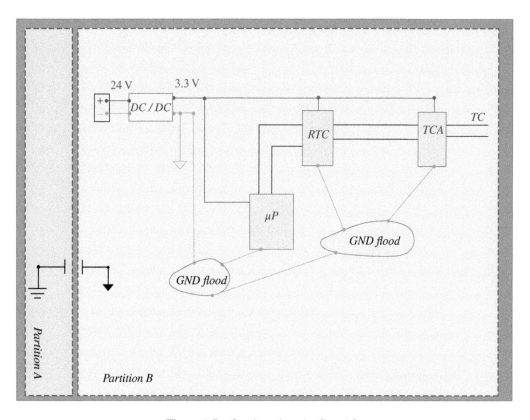

Figure A.7   One-layer board – Case 1.2.

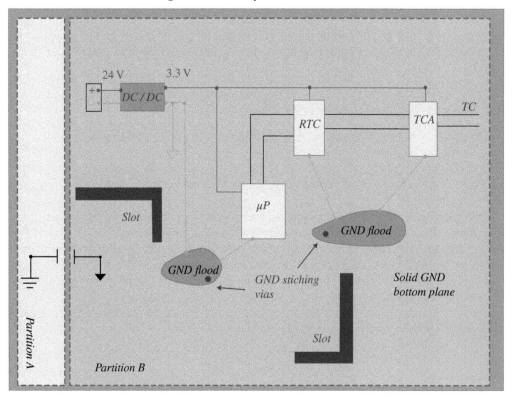

Figure A.8   Two-layer board – Case 2.1.

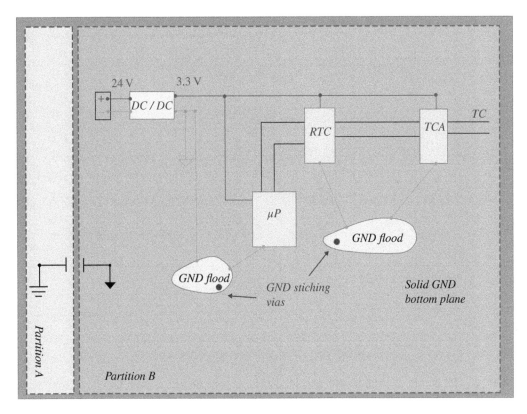

Figure A.9   Two-layer board – Case 2.2.

This design implements a solid reference plane on the secondary side of the PCB. Stitching vias are used to connect the reference planes between top and bottom layers. This design approach can improve RF emissions while potentially reducing the number of filtering components needed for compliance.

A.2 DC/DC Converter – Baseline EMC Emissions Evaluation

In this section we evaluate the performance of the baseline DC/DC converter (e.g., use only IC-vendor-recommended components and no additional EMC countermeasures) [Adamczyk et al., 2021b]. Specifically, we present the test results from the radiated and conducted emissions tests performed according the CISPR 25 Class 5.

Like so many industries at this time, while working on the DC/DC converter we were faced with a semiconductor shortage issue in our design with the main controller-integrated circuit. This forced us to redesign the converter using a different DC/DC IC that is widely available in quantities. After selecting a new integrated circuit, a new design was created and appropriate components were chosen. Then the PCB layout was updated and a "quick turn" PCB fabrication was ordered and received. The

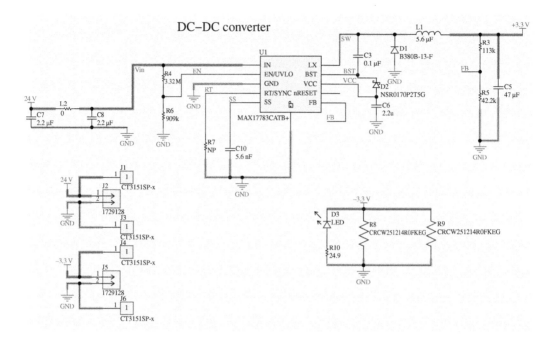

Figure A.10   DC–DC converter baseline schematic.

schematic, PCB layout, and a photograph of the assembly are shown for the new design which was tested and results are discussed in this section.

Due to the aforementioned semiconductor shortage a new design was created using a similar process as described in Section 12.2 and is shown in Figure A.10.

The baseline converter is shown in Figure A.11.

A.2.1 CISPR 25 Radiated Emissions Test Results

The DC–DC switcher was tested according to CISPR 25 4th Edition, Class 5. Radiated emissions test setup is shown in Figure A.12.

A legend for the radiated emissions plot is shown in Figure A.13.

Figure A.14 shows the radiated emissions measurements results in frequency range of 150 kHz–1 GHz. These measurements were made using a monopole antenna from 150 kHz to 30MHz, a biconical antenna from 30 to 300 MHz, and a log-periodic antenna from 300 MHz to 1 GHz.

The monopole range (150 kHz–30 MHz) shows a failure of the Quasi-Peak and Average limits at 978 kHz and 1469 kHz. The emissions in this region are all narrowband and spaced by 487 kHz. This suggests that these emissions are all harmonics of the 487 kHz switching frequency of the power supply.

The biconical range (30–300 MHz) shows failures of the average limit at 36.75 and 182.46MHz, and failure of the peak limit at 182.46 MHz. Measuring the distance between the different peaks that are present in this range shows that these failures are also due to the buck converter, but because of the broadband nature of this noise this would indicate these emissions are likely due to ringing on the switching signal.

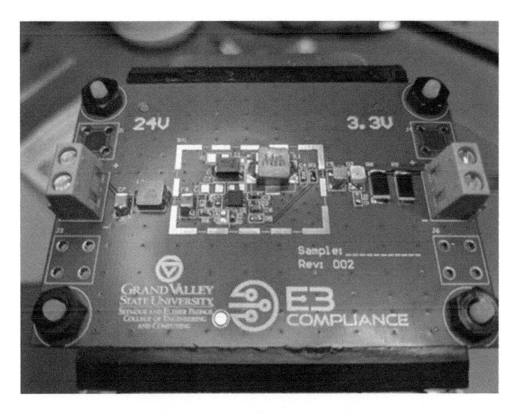

Figure A.11   DC–DC converter assembly.

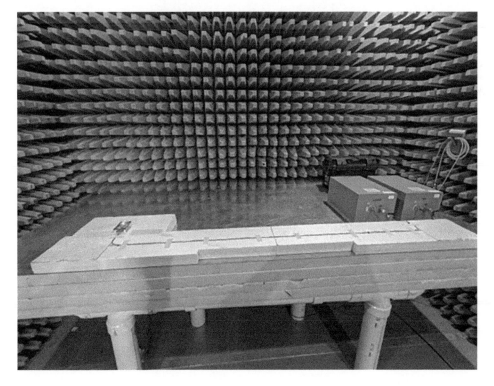

Figure A.12   Radiated emissions test setup.

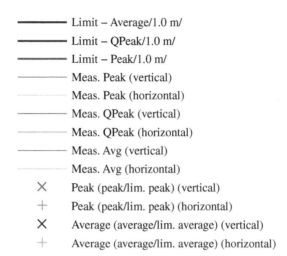

Figure A.13   Radiated emissions legend.

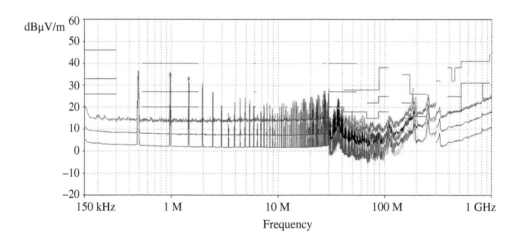

Figure A.14   Radiated emission results in the frequency range 150 kHz–1 GHz.

A.2.2 CISPR 25 Conducted Emissions (Voltage Method) Test Results

Figure A.15 shows the voltage method conducted emissions test setup.

A legend for the conducted emissions plots is shown in Figure A.16.

The test results on the battery line, in the frequency range of 150 kHz–108 MHz, are shown in Figure A.17.

The conducted emissions plot of the battery line, like the monopole region of the RE test data, shows significant emissions at the switching frequency of the buck regulator and the

Figure A.15 Conducted emission test setup (voltage method).

———— Limit – Average/
———— Limit – QPeak/
———— Limit – Peak/
———— Meas. Peak
———— Meas. QPeak
———— Meas. Avg
×   Peak (peak/lim. peak)
×   Average (average/lim. average)
×   QPeak (QPeak/lim. QPeak)

Figure A.16 Conducted emission results legend.

harmonics of this frequency. This measurement was taken with a 9 kHz resolution bandwidth from 150 kHz to 30 MHz, and a 120 kHz resolution bandwidth from 30 to 108 MHz.
This shows failure of all three limits at 978 kHz and 1.4685 MHz. This also shows failures
of the average and quasi-peak limits from 25 to 100 MHz.

The test results on the ground line, in the frequency range of 150 kHz–108 MHz, are
shown in Figure A.18.

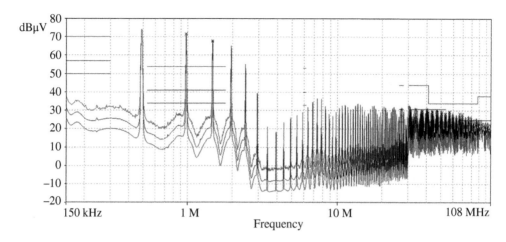

Figure A.17   Conducted emission test results – BAT line – 150 kHz–108 MHz.

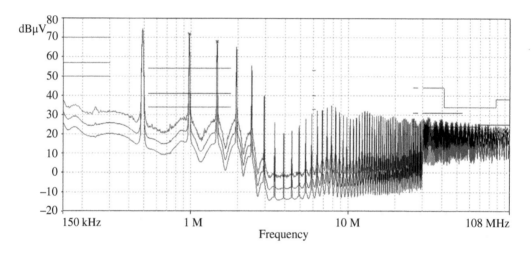

Figure A.18   Conducted emission test results – GND line – 150 kHz–108 MHz.

From this measurement of emissions from the GND line it is seen that there is a similar amount of emissions coming from the battery line and the GND line for this DUT. This measurement was taken with a 9 kHz resolution bandwidth from 150 kHz to 30 MHz, and a 120 kHz resolution bandwidth from 30 to 108 MHz. Like the battery line measurements this shows failures of all three limits at 978 kHz and 1.4685 MHz, and multiple failures of the average and quasi-peak limits from 25 to 100 MHz.

A.2.3 CISPR 25 Conducted Emissions (Current Method) Test Results

Figure A.19 shows the current method conducted emissions test setup.

The test results, at 50 mm, in the frequency range of 150 kHz–245 MHz are shown in Figure A.20. The measurement from 150 kHz to 30 MHz was taken with a 9 kHz resolution bandwidth, and the measurement from 30 to 245 MHz was taken with a 120 kHz resolution bandwidth.

The conducted emissions measurement at 50 mm shows significant broadband emissions from 25 to 100 MHz and at 180 MHz. These emissions are likely due to ringing in the switching waveform.

The test results, at 750 mm, in the frequency range of 150 kHz to 245 MHz are shown in Figure A.21. The measurement from 150 kHz to 30 MHz was taken with a 9 kHz resolution bandwidth, and the measurement from 30 to 245 MHz was taken with a 120 kHz resolution bandwidth.

Like the 50 mm measurements, this measurement, taken at 750 mm, shows significant broadband emissions from 25 to 100 MHz and at 180 MHz. These emissions are likely due to ringing in the switching waveform.

A.3 DC/DC Converter – EMC Countermeasures – Radiated Emissions Results

Section A.2 presented the radiated and conducted emission results from the baseline design which did not contain any EMC countermeasures. The results showed multiple failures in both radiated and conducted emissions.

This section presents a systematic approach to improve these failures by populating the PCB with optional EMC countermeasures on component pads that have already been designed into the PCB layout and showing their impact on the radiated emissions [Adamczyk et al., 2021c]. The countermeasures are presented in an order that we would typically follow in an EMC diagnostic session where, due to time restrictions, not every single permutation of EMC countermeasure will be tested.

The EMC countermeasures are illustrated in Figure A.22, as dashed boxes labeled EMC-A through EMC-F.

The impact of these countermeasures is discussed next.

A.3.1 EMC-A and EMC-E Input and Output Capacitor Impact

Radiated emissions were measured in the frequency range of 150 kHz–30 MHz. The baseline results (Figure A.23a) show high level of emissions in the 25–30 MHz band. To attempt to reduce these emissions two capacitors C9 = 1 nF (EMC-A) and C4 = 1 nF (EMC-E) were populated.

The radiated emissions measurement taken with these countermeasures populated is shown in Figure A.23b.

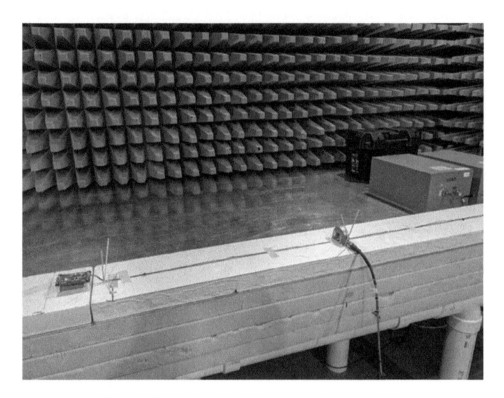

Figure A.19   Conducted emission test setup (current method).

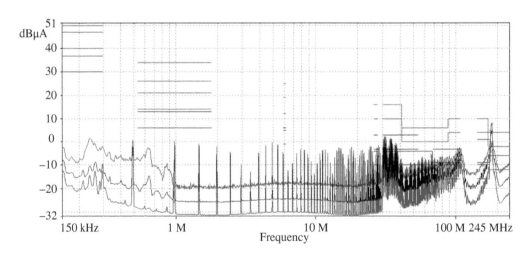

Figure A.20   Conducted emission test results at 50 mm.

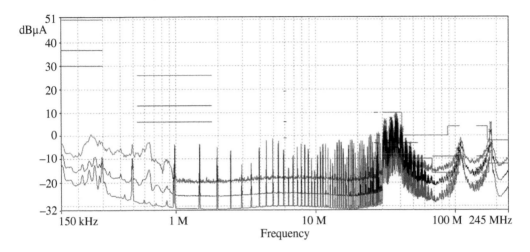

Figure A.21   Conducted emission test results at 750 mm.

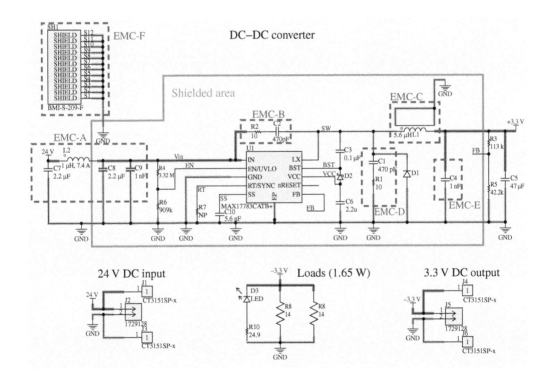

Figure A.22   DC/DC schematic with EMC countermeasures.

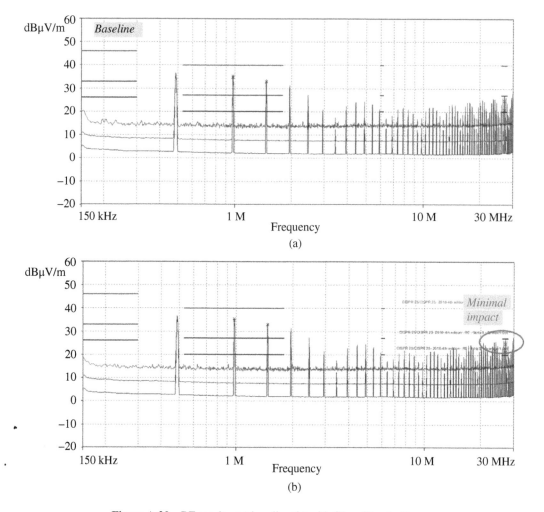

Figure A.23 RE results: (a) baseline (b) with C9 = C4 = 1 nF caps.

Typically, these 1 nF capacitors help filter the noise in higher frequencies however, as the plot in Figure A.23b shows, they had a minimal impact on radiated emissions performance on the upper-frequency range of this band.

A.3.2 EMC-A Input Inductor Impact

Next, we targeted emissions in the lower-frequency range 0.9–2 MHz. A 1 µH input inductor, L2, was placed. The radiated emissions measurement taken with this countermeasure in place is shown in Figure A.24.

Typically, this inductor helps to reduce emissions in this range by filtering the input and attempting to prevent noise from getting onto the harness where the 2 m wire length acts as an effective re-radiator. However, as Figure A.24b shows, it had a minimal impact in the 150 kHz–30 MHz range.

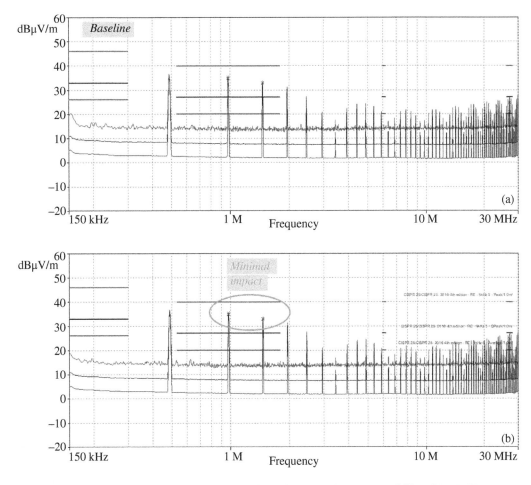

Figure A.24 RE results: (a) baseline (b) with input inductor L2 = 1 μH, and C9 = C4 = 1 nF caps.

The results of this testing suggest that the noise that is causing emissions at these lower frequencies is either radiating directly from the Printed Circuit Board Assembly (PCBA) or may be common-mode emissions (rather than differential-mode noise) conducting out on the wire harness. This leads us to our next countermeasure of shielding the switching inductor L1. Due to the ineffectiveness of both the 1 μH inductor and the 1 nF capacitors, they were removed from the sample before testing the next countermeasure.

A.3.3 EMC-C Switching Inductor Impact

Next, the switching inductor, L1, was changed to a Vishay 3232 IHLE 5.6 μH. These inductors have an integrated *E*-field shield that is tied to ground on both sides of the inductor. They offer a way to capture the emissions that radiate directly from the switch inductor and boast a 20 dB *E*-field reduction at 1 cm [Adamczyk et al., 2021c].

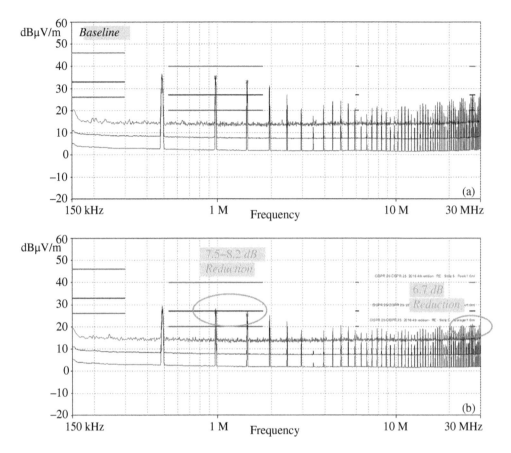

Figure A.25 RE results: (a) baseline (b) with L1 = 5.6 μH IHLE, 150 kHz–30 MHz.

The radiated emission results with the switching inductor changed to a Vishay 3232 IHLE 5.6 μH are shown in Figure A.25.

As the plots show, the inductor had a substantive impact, not only in the 0.9–2 MHz range but also in the 25–30 MHz range. Because the objective of this shielded inductor is to reduce the emissions by capturing the electric field, a significant improvement was observed in the monopole antenna range (150 kHz–30 MHz). This justifies changing the measurement setup to the biconical antenna to evaluate the improvements in the 30–300 MHz range. The results are shown in Figure A.26 (*Note*: From this point forward, measurements below 300 MHz are not captured due to passing results in baseline testing).

As Figure A.26 shows the IHLE 5.6 μH inductor is also successful at reducing emissions in the 30–70 MHz band, but it is not as successful around 180 MHz. As the noise in this range is likely due to ringing in the switching waveform, we often recommend the use of a snubber (series R–C) circuit with the purpose of dampening the ringing waveform and reducing the high-frequency content of the signal. The IHLE 5.6 μH inductor is removed from the PCB during the evaluation of snubbers, in order to clearly see the impact of the snubbers.

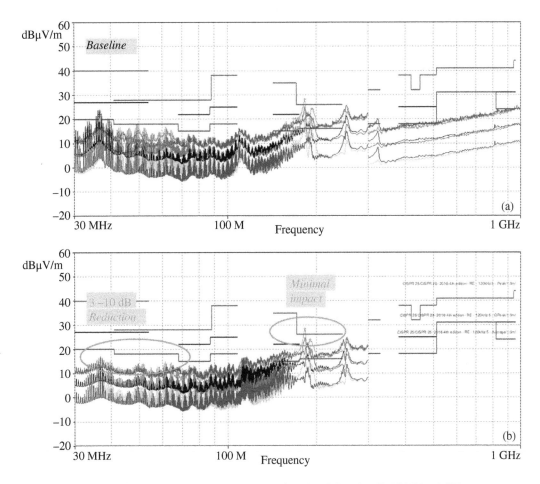

Figure A.26 RE results: (a) baseline (b) with L1 = 5.6 µH IHLE, 30 MHz–1 GHz.

A.3.4 EMC-B and EMC-D Snubber Impact

A snubber was placed across the catch diode D1: R1 = 10 Ω, C1 = 470 pF (EMC-D). These values were chosen based on experience gained from work on other SMPS designs. These values have not been optimized nor calculated using the various methods available, but in a time-restricted scenario they provide a good starting point and allow us to see if a snubber can make a positive impact.

The radiated emissions test results are shown in Figure A.27.

This shows a significant reduction in the emissions measured with the antenna in the vertical polarization around 36 MHz but has minimal impact in the horizontal polarization. This also shows a smaller reduction in emissions around 180 MHz.

Next, the snubber is removed from its location across the catch diode and placed across the FET which is internal to the IC. This snubber is placed on the placeholders R2 and C2 (EMC-B). The radiated emissions results are shown in Figure A.28.

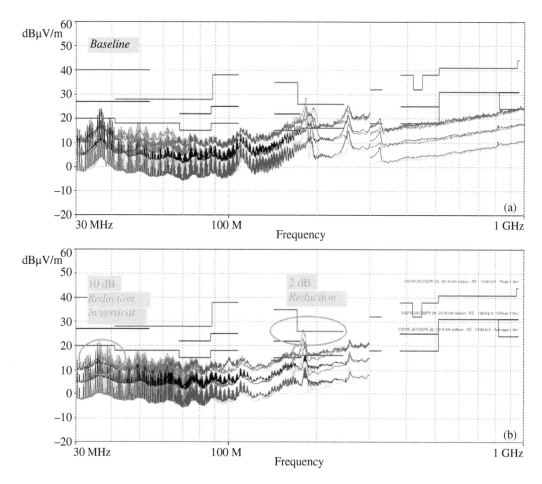

Figure A.27   RE results: (a) baseline (b) with $R1 = 10\,\Omega$, $C1 = 470\,\text{pF}$.

This shows a slight reduction in the emissions measured with the antenna in the vertical polarization around 36 MHz but has minimal impact in the horizontal polarization. This also shows a reduction in emissions around 180 MHz.

The last step in evaluating these snubbers was to populate the placeholders for both the FET snubber (EMC-B) and the catch diode snubber (EMC-D) combined. The radiated emissions results are shown in Figure A.29.

This shows a significant reduction in the emissions in the 30–80 MHz range and around 180 MHz. This technically passes CISPR 25 Class 5, but due to expected lab-to-lab variation greater margin is desired in the average measurement around 180 MHz before finalizing the design.

Note: Due to the RE chamber scheduling constraints, the radiated emission tests described in Section A.2 had to be temporarily put on hold. Since the conducted emissions chambers were available at the time, we began performing the conducted emissions testing and implemented EMC countermeasures to address these failures.

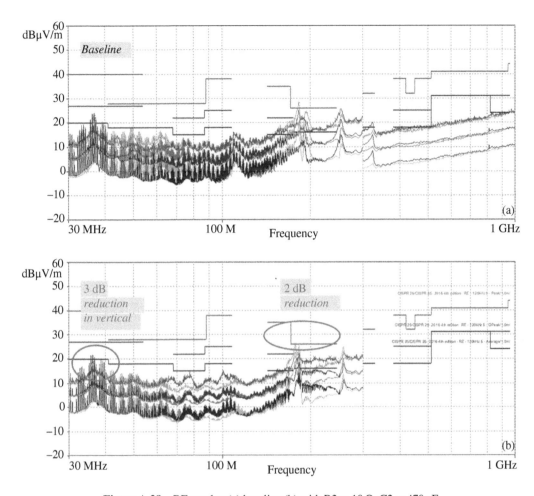

Figure A.28 RE results: (a) baseline (b) with R2 = 10 Ω, C2 = 470 pF.

Upon the completion of the conducted emissions testing we returned to the radiated emissions testing. The radiated emissions testing that was resumed and described next contains the EMC countermeasures implemented during the conducted emission testing. One of these countermeasures was the removal of the RC snubber, as other conducted emissions countermeasures rendered it unnecessary. The other countermeasures are described next.

A.3.5 EMC-A, EMC-E – Conducted Emissions Countermeasures Impact

Conducted emissions countermeasures resulted in the addition of two 2.2 μF capacitors in parallel with C7 and C8. Additionally, the input inductor L2 was changed to 2.2 μH; also, C9 and C4 were populated with 10 nF capacitors. The switching inductor L1 was also populated with a Vishay 3232 IHLE 5.6 μH inductor.

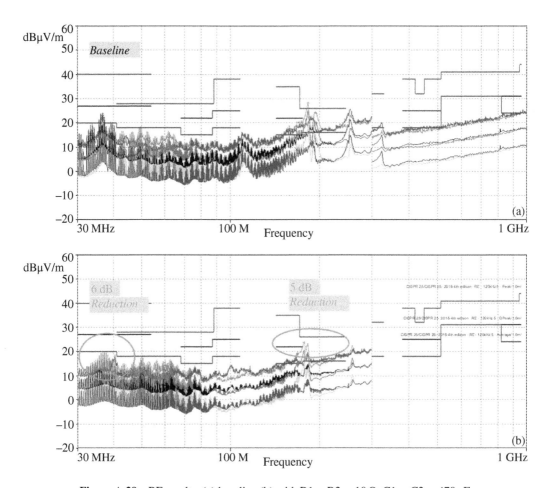

Figure A.29 RE results: (a) baseline (b) with R1 = R2 = 10 Ω, C1 = C2 = 470 pF.

The input filter and inductor changes provided much benefit to the conducted emissions. These results will be shown in detail when discussing the conducted emissions results. For now, we will continue to focus on the radiated emissions results.

Figure A.30 shows the radiated emission measurements in the 150 kHz–30 MHz range, while Figure A.31 shows the results in the 30–1000 MHz range.

The conducted emissions countermeasures were very effective at reducing radiated emissions in both the 150 kHz–30 MHz and 30–300 MHz ranges.

A.3.6 Impact of the Shield Frame

Next, a shield frame SH1 was soldered to the PCB on the perimeter over the shielded area shown in Figure A.22. This is EMC countermeasure EMC-F. This is in addition to the conducted emissions countermeasures previously described. It is important to note that the shield frame was placed without a shield lid in this first evaluation. This is being

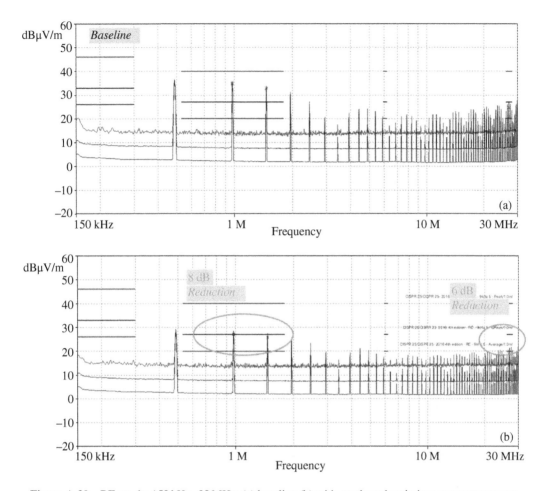

Figure A.30 RE results 150 kHz–30 MHz: (a) baseline (b) with conducted emissions countermeasures.

evaluated since in some cases sufficient emissions reductions can be achieved with only the frame component.

Figure A.32 shows the radiated emissions results in the frequency range 150 kHz–30 MHz, while Figure A.33 shows the results in the range 30–1000 MHz.

If this SMPS was a product that was intended for sale the next step would be to start removing the EMC countermeasures one by one to reduce the Bill of Materials (BOM) cost of each unit produced.

An example of this would be trying to remove the IHLE 5.6 μH inductor in favor of a cheaper inductor. In our previous trials this increased emissions in the 0.9–2 MHz range above the average limit. An evaluation was performed by exchanging the IHLE 5.6 μH E-field shielded inductor (L1) for the original IHLP 5.6 μH magnetically shielded inductor (L1) and a shield lid was placed on the shield frame.

The radiated emissions results with the shield lid and L1 swapped are shown in the frequency range 150 kHz–30 MHz in Figure A.34.

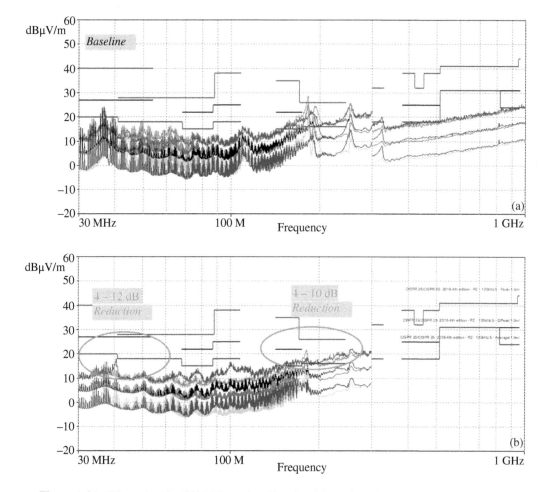

Figure A.31 RE results 30–1000 MHz: (a) baseline (b) with conducted emissions countermeasures.

This shows that the shield with the lid in addition to the conducted emissions input filter greatly reduces emissions in the range of 150 kHz–30 MHz. Placing the shield lid does not negatively affect the emissions in the range of 30–1000 MHz (passing with margin) and therefore are not shown in this last evaluation.

Based on the results presented so far we recommend finalizing the design to meet radiated emissions requirements (CISPR 25 Class 5) with the following countermeasures populated on the baseline design:

EMC A – Front-End Filter
C7 = 2.2 μF (with additional 2.2 μF or change to 4.7 μF)
L2 = 2.2 μH
C7 = 2.2 μF (with additional 2.2 μF or change to 4.7 μF)
C9 = 10 nF
EMC-B – Internal FET Snubber
Not populated

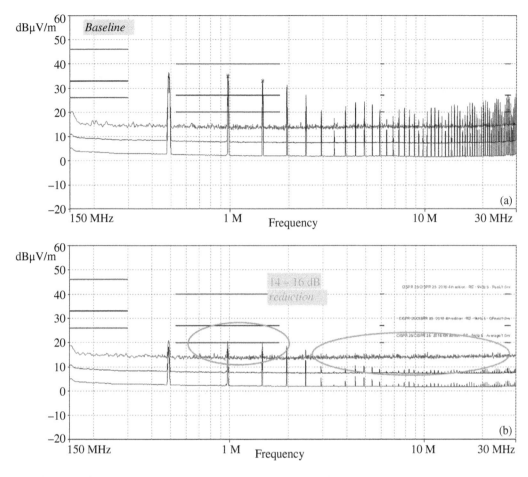

Figure A.32 RE results 150 kHz – 30 MHz: (a) baseline (b) with a shield and CE countermeasures.

EMC-C – Shielded Switch Inductor
Preserve original IHLP 5.6 µH magnetically shielded inductor
EMC-D – Catch Diode Snubber
Not populated
EMC-E – Output High-Frequency Capacitance
C4 = 10 nF
EMC-F – Shield Frame and Lid
Populate both frame and lid

Additional effort can be applied to this configuration of EMC countermeasures to further optimize for cost while meeting EMC requirements. If the target application does not require the stringent levels of CISPR 25 Class 5, it may be possible to remove the EMC shield in favor of cheaper countermeasures. A full analysis has not been performed on this design to pursue these potential objectives.

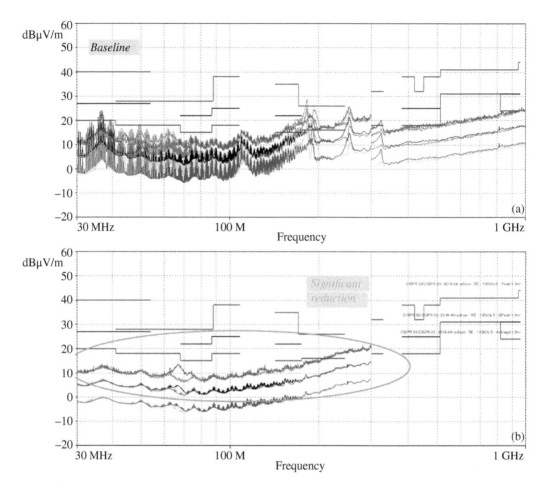

Figure A.33   RE results 30–1000 MHz: (a) baseline (b) with a shield and CE countermeasures.

A.4 DC/DC Converter – EMC Countermeasures – Conducted Emissions Results – Voltage Method

In this section we evaluate the implementation of several EMC countermeasures and present the conducted emissions results, for voltage method, according to CISPR 25 Class 5 limits [Adamczyk et al., 2021d]. The voltage method results are shown only for the supply line, as the ground line results were similar.

A.4.1 EMC-A and EMC-E Input and Output Capacitor Impact

Conducted emissions were measured in the frequency range of 150 kHz–108 MHz. The baseline results (Figure A.35a) show high level of emissions in the 1–2 MHz and 25–100 MHz bands.

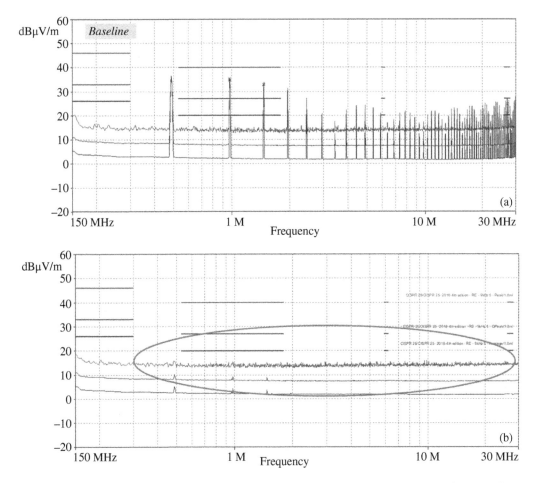

Figure A.34 RE results 150 kHz–30 MHz: (a) baseline (b) with a shield frame and lid, 2.2 μF capacitors added in parallel with C7 and C8, C9 = C4 = 10 nF, and $L2$ = 2.2 μH.

To attempt to reduce these emissions two capacitors C9 = 10 nF (EMC-A) and C4 = 10 nF (EMC-E) were populated. The conducted emissions measurement taken with these countermeasures populated is shown in Figure A.35b.

The 10 nF capacitors are meant to help filter the noise in higher frequencies. As the plot in Figure A.35b shows, the capacitors increase the conducted emissions in the frequency range of 25–40 MHz and decrease the conducted emissions in the frequency range of 40–100 MHz.

The additional capacitors have minimal impact on the conducted emissions in the 1–2 MHz band. The 10 nF capacitors are kept on the board as they had a positive impact in the 40–100 MHz band.

A.4.2 EMC-A Input Inductor Impact

Next, we targeted the conducted emissions using an input filter. A 1 μH input inductor, L2 (EMC-A), was placed on the input to create a PI filter with the input capacitors. The

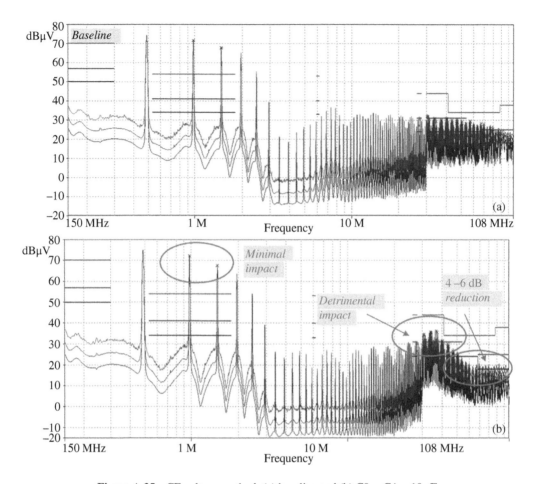

Figure A.35   CE voltage method: (a) baseline and (b) C9 = C4 = 10 nF.

conducted emissions measurement taken with this countermeasure in place is shown in Figure A.36.

As Figure A.36 shows, the inductor had a substantial improvement on the emissions in the 1–2 MHz range; additionally, it reduced the higher-frequency emission by 2–4 dB compared to Figure A.35b.

To further reduce the emissions around 1 MHz, the input inductor L2 was changed to 3.3 μH. The emission results are shown in Figure A.37.

As Figure A.37 shows, an additional 4–12 dB reduction was achieved in the lower-frequency range compared to Figure A.36.

A.4.3 EMC-A Additional Input Capacitors Impact

Next, we added two additional 2.2 μF input capacitors, C7 and C8 in order to increase the impedance of the input PI filter by lowering the frequency of the low pass filter. The emissions results are shown in Figure A.38.

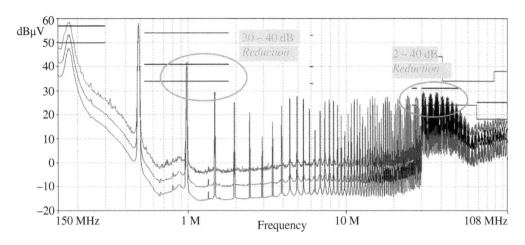

Figure A.36 CE voltage method: L2 = 1 µH, C9 = C4 = 10 nF.

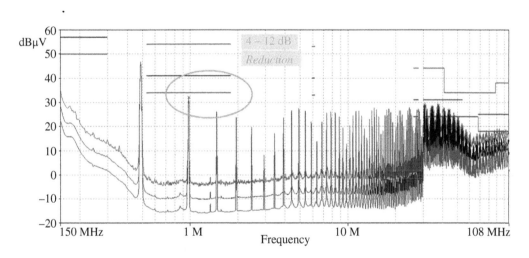

Figure A.37 CE voltage method: L2 = 3.3 µH, C9 = C4 = 10 nF.

As Figure A.38 shows, an additional 6 dB reduction was achieved around 1 MHz compared to Figure A.37.

A.4.4 EMC-A Input Inductor Impact

Next, the input inductor L2 was changed back to 2.2 µH IHLP magnetically shielded inductor. The emission results are shown in Figure A.39.

As Figure A.39 shows, the change from 3.3 to 2.2 µH had virtually no impact, as compared to Figure A.38. This suggests that the increase in capacitance on C7 and C8 offers more improvement and the increase in inductance from 2.2 to 3.3 µH has a negligible impact, and therefore, we retained the 2.2 µH inductor in our design.

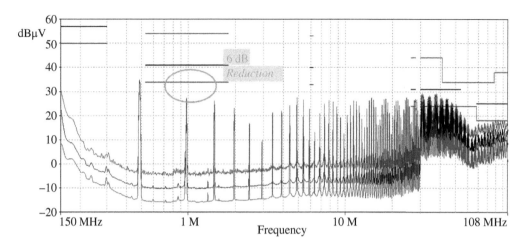

Figure A.38 CE voltage method: L2 = 3.3 μH, C9 = C4 = 10 nF, C7 = C8 = 2 × 2.2 μF.

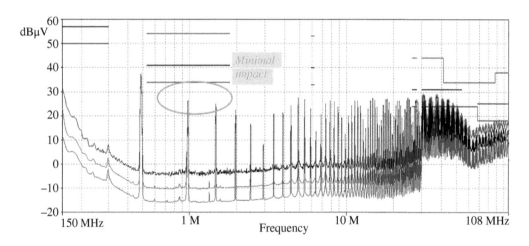

Figure A.39 CE voltage method: L2 = 2.2 μH, C9 = C4 = 10 nF, C7 = C8 = 2 × 2.2 μF.

A.4.5 EMC-C Switching Inductor Impact

Next, the switching inductor, L1 (EMC-C), was changed to a Vishay 3232 IHLE 5.6 μH. These inductors have an integrated *E*-field shield that is tied to ground on two sides of the inductor. The conducted emission results are shown in Figure A.40.

As the plots show the inductor had a substantial impact in the 40–70 MHz range. At this point the DUT is passing CISPR 25 Class 5, and initially this is the solution that we used. However, with the additional time we have had we decided to evaluate some other countermeasures that would be cheaper than the IHLE inductor. This approach is often used in debugging EMC issues for industry as we first prioritize finding a solution then optimize the PCB assembly cost as time allows.

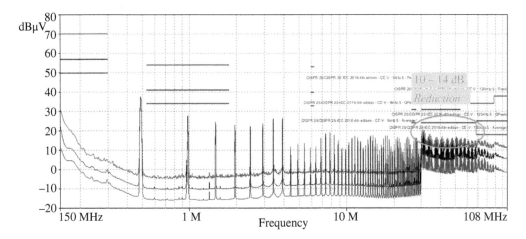

Figure A.40   CE voltage method: L2 = 2.2 μH, C9 = C4 = 10 nF, C7 = C8 = 2 × 2.2 μF caps, L1 = 5.6 μH IHLE.

Next, we investigated the impact of the two series RC snubbers (EMC-B and EMC-D). The IHLE 5.6 μH inductor was removed and replaced by L2 = 2.2 μH IHLP magnetically shielded inductor (in order to reduce the cost). The remaining EMC components were retained.

A.4.6 EMC-B and EMC-D Snubber Impact

One of the snubbers was placed across the catch diode D1: R1 = 10 Ω, C1 = 470 pF (EMC-D), and the other was placed across the FET that controls the switching, R2 = 10 Ω, C2 = 470 pF (EMC-B). This FET is inside of the IC package. The conducted emissions test results are shown in Figure A.41.

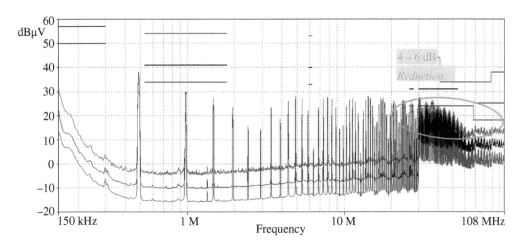

Figure A.41   CE voltage method: R1 = R2 = 10 Ω, C1 = C2 = 470 pF.

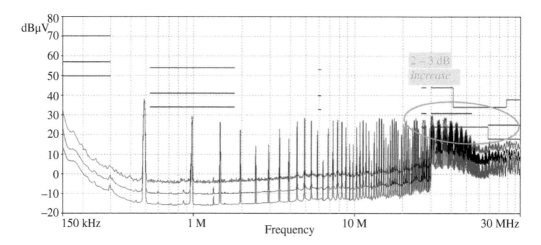

Figure A.42   CE voltage method: R2 = 10 Ω, C2 = 470 pF.

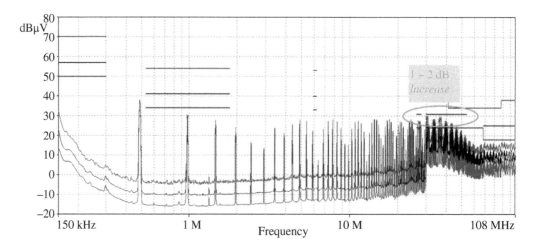

Figure A.43   CE results voltage method: R1 = 10 Ω, C1 = 470 pF.

With both snubbers in place we observe a 4–6 dB decrease in emissions in the 30–100 MHz band. Next, the snubber across the catch diode (R1 and C1) was removed while the R2 and C2 snubber was retained. The conducted emissions results are shown in Figure A.42.

As Figure A.42 shows the removal the catch diode snubber (R1 and C1) results in a 2–3 dB increase in emissions in the 30–100 MHz band. Next, the FET snubber (R2 and C2) was removed, while the catch diode snubber (R1 and C1) was repopulated. The conducted emissions results are shown in Figure A.43.

Figure A.43 shows this change had a 1–2 dB increase in 30–40 MHz band, as compared to Figure A.42. We therefore retained both snubbers.

A.5 DC/DC Converter – EMC Countermeasures – Conducted Emissions Results – Current Method

In this section we evaluate the implementation of several EMC countermeasures and present the conducted emissions results, for current method, according to CISPR 25 Class 5 limits [Adamczyk et al., 2021d]. The current method results are shown only for the 50 mm probe location, as the 750 mm results were similar.

A.5.1 EMC-A, EMC-C, and EMC-E Input and Output Capacitor and Inductor Impact

Conducted emission were measured in the frequency range of 150 kHz–245 MHz using the current method. The configuration tested was as follows: C9 = C4 = 10 nF, L2 = 2.2 μH, C7 = C8 = 2.2 μF, L1 = 5.6 μH IHLE. This is the same configuration that was used in the radiated emissions testing discussed earlier [Adamczyk et al., 2021c]. The conducted emissions, current method test results are shown in Figure A.44.

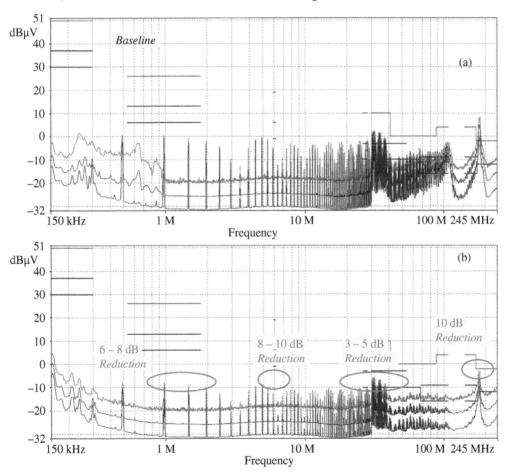

Figure A.44 CE current method: (a) baseline and (b) C9 = C4 = 10 nF, L2 = 2.2 μH, C7 = C8 = 2.2 μF, L1 = 5.6 μH IHLE.

We observe the emissions reduction across several frequency bands. This is not surprising as this configuration performed very well in the conducted emissions voltage method test.

A.5.2 EMC-B and EMC-D Snubber Impact

In order to address some of the higher-frequency emissions we again investigated the impact of snubbers. The IHLE 5.6 µH inductor was removed and replaced by L2 = 2.2 µH IHLP magnetically shielded inductor. The remaining EMC components were retained. As it was done in the voltage method, one of the snubbers was placed across the catch diode D1: R1 = 10 Ω, C1 = 470 pF (EMC-D), and the other was placed across the switching node, R2 = 10 Ω, C2 = 470 pF (EMC-B).

The conducted emissions test results are shown in Figure A.45.

The addition of the two snubbers had a minimal impact on the emissions at the lower-frequency ranges, but resulted in about 2–3 dB reduction in the 20–30 MHz range, and at the 180 MHz spike as compared to Figure A.44b. This configuration does technically pass CISPR 25 Class 5, but in practice we like to see more margin to prevent failures that can be caused by lab-to-lab variation in measurements.

Next, the FET snubber (R2 and C2) was removed, while the catch diode snubber (R1 and C1) was retained. It resulted in the increased emissions in the frequency range of 30–40 MHz and around 180 MHz. Subsequently, we repopulated the FET snubber (R2 and C2) and removed the catch diode snubber (R1 and C1). The result was similar.

The final snubber configuration tested was with R1 = R2 = 5.6 Ω, C1 = C2 = 470 pF. The test results are shown in Figure A.46.

As Figure A.46 shows, this change had a positive impact in the 30–40 MHz range causing a 1–2 dB reduction in emissions, as compared to Figure A.45. This configuration passes CISPR 25 Class 5 with 3 dB of margin.

The analysis of the conducted emissions testing and addition of EMC countermeasures show that in large part the failures identified in baseline testing can be mitigated through

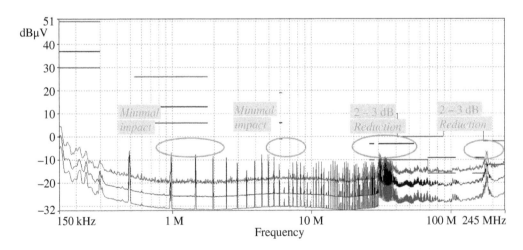

Figure A.45   CE current method: R1 = R2 = 10 Ω, C1 = C2 = 470 pF.

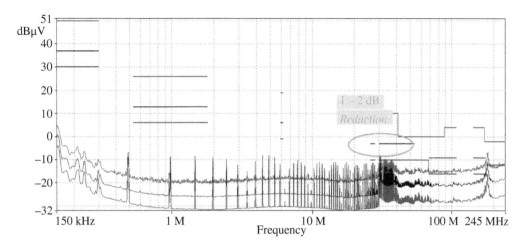

Figure A.46   CE current method: R1 = R2 = 5.6 Ω, C1 = C2 = 470 pF.

the use of front-end filtering and snubbers. However, a larger margin of passing results can be achieved by investing more in the BOM by using an *E*-field shielded IHLE inductor and/or a PCB shield. Depending on the class performance desired, a lower cost may be required to gain compliance with a comfortable margin.

A.6 PCB Layout Considerations

In this section we discuss the PCB layout considerations and the design of the reference return paths for the one- and two-layer boards [Adamczyk et al., 2021e].

A.6.1 Introduction

The PCB layout and the design of reference return connections (may also be referred to as grounding) play a critical role in the EMC performance of any circuit. This is especially critical for power converters. In circuit design it can be easy to focus on power and signal trace connections while overlooking or not focusing enough attention on how circuit current returns. Proper reference return design can especially be challenging in single- and two-layer designs where best practices can't always be applied. It is important to understand and visualize the path of the return current so its entire loop area can be controlled by design. The complete loop area of each circuit tends to be the dominant factor when compared with other parasitic inductances associated with the components or vias. This inductance has a detrimental effect on the EMC performance.

Visualizing the loop areas allows the designer to identify ways to reduce the size and cross-sectional area of the loops, thus reducing the inductance and high-frequency impedance. In single-layer PCB designs, there is fierce competition for copper routing real estate as all routes need to be completed on a single layer. In this setting, we don't have the luxury of a reference return on the secondary side of the PCB. We rely on prioritizing the reference return connections between critical points, and loop areas

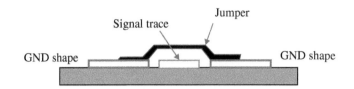

Figure A.47   Example of GND stitching with jumper.

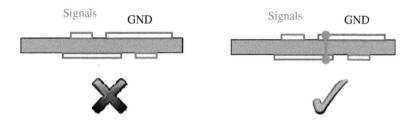

Figure A.48   Two-layer example of ground connection.

are often much larger than we would like to see. This can drive the need for additional decoupling capacitors, filter components, and "jumper" components to "stitch" reference returns back together across other trace routes.

Figure A.47 shows an example of a jumper used to "stitch" the ground areas back together across a signal trace route:

In two-layer PCBs, there is more opportunity for proper reference return design as the additional layer of copper combined with return vias allows us to make a more consistent return path with smaller loop areas. Stitching ground areas together in single- and two-layer PCBs is important not only for better emissions performance but it also aids in reducing immunity issues.

Figure A.48 shows an example of implementing vias connecting different "ground floods" on the top and bottom layers to create a "ground mesh" and improve the flow of return currents. Efforts should be made to reduce the number of signals on the secondary side to create a more solid reference return plane.

Device application notes can sometimes recommend introducing splits into reference returns for returns such as analog and digital circuits. There may be legitimate reasons for splitting the reference returns, but in our experience this almost always causes an increase in EMC emissions or immunity performance issues.

When splits are introduced, efforts are required to "reconnect" these separate shapes either with jumpers or with capacitors. Often times the efforts to reconnect the separate shapes are not as effective as making the original solid connections in the PCB layout.

A.6.2 Visualizing Complete Forward and Return Paths

In the design process, we recommend drawing the forward and return currents of all power and signal paths as a three-step process.

Step 1: Draw these complete paths (loops) on the electrical schematic itself.
Step 2: Draw these complete paths (loops) on the PCB board layout.
Step 3: Minimize the loop areas (and discontinuities) in the PCB layout.

During the design of the DC–DC Buck converter that has been discussed so far, the reference path design has been the main concern and will continue to be a concern of this study.

The schematic and layout of this DC–DC Buck converter is shown in Figure A.49. Figure A.49a shows the power and signal traces as well as the reference return conductors. Figure A.49b shows the corresponding layout traces and the ground pads.

All of the power and signal traces were routed on the top layer, and the reference return paths reside on the bottom layer which is a full solid plane. Implementing this circuit (power and signal traces) on the top layer serves two purposes.

First, in later parts of this study this circuit will need to be constructed on a 1-layer PCB, and routing it on just the top layer now will help to keep the layouts similar between this PCB and the future one.

Second, the full reference plane on layer two provides an ideal return path as it is not constrained. Such an unconstrained path is highly desirable, especially for the high-frequency currents. Without the solid reference plane, the effectiveness of the filtering on the board would be diminished and the loop areas involving the reference return would increase.

Let's demonstrate this by looking at one of the output filtering capacitors shown in Figure A.50. The forward path of the current is $A-B-C-D$, and the current return path is $E-F-G$.

Currents return to the source following the path of least impedance; at DC and low frequencies (below 100 kHz or so) this is predominantly the path of least resistance. At higher frequencies the return predominantly is the path of least inductance. This inductance is the inductance of the loop formed by the currents' forward and return paths.

At the higher frequencies that we are concerned with in EMC, the loop inductance will be the dominating factor in the impedance, meaning the current will likely follow the path of least inductance to return to the source. This loop inductance is kept at a minimum when the return path is directly under the forward path. This means in a completely unbroken ground plane the current will likely flow directly under the forward path as depicted in Figure A.50.

With a completely solid GND plane the currents will have no issues returning directly under the forward path, but in many practical designs this is just not possible. Generally, at least a couple of traces may need to be routed in the GND plane, especially on a two-layer board.

The output section of this DC–DC power supply was modified to allow us to analyze how the current might return to the source when the ideal return path is broken by trace in the GND plane. This modified output section is shown in Figure A.51.

In this modified case the feedback was routed on layer two as opposed to layer one. In this case it would be expected that the current will initially follow the same path as it did in our original layout, until it reaches the feedback trace that is routed in layer two.

At this point the current will have to go around the break in the ground plane and continue to return under the forward path back to the source. The loop area added to the currents path introduced by the cut out in the return plane increases the inductance of the

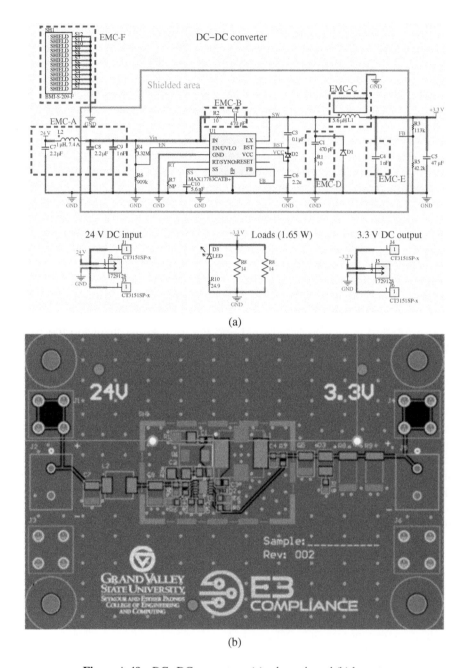

Figure A.49   DC–DC converter – (a) schematic and (b) layout.

loop. This causes an increase in the radiation from the current loop. Having a cut-out in the reference plane is therefore not desirable, in more complicated designs where space is more a premium this might be unavoidable.

Next, let's look at the input filtering section shown in Figure A.52.

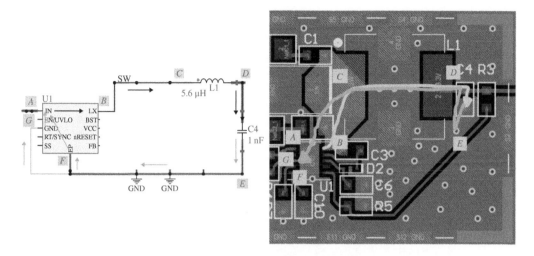

Figure A.50 Output filtering section of PCB.

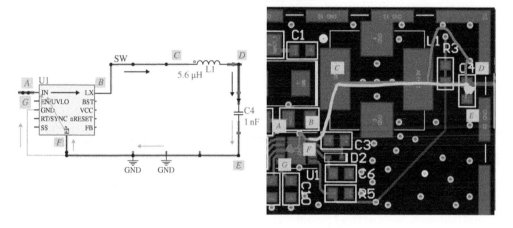

Figure A.51 Modified output DC–DC converter.

As described earlier, the high-frequency current paths were traced. Because we are concerned with the high-frequency noise that is generated by the switching in U1, we assume the current path starts at the Vin in U1.

However, in this case there are multiple possible return paths for the high-frequency currents. The obvious and most likely three paths are through C7, C8, or C9.

To complicate things further the loop that noise chooses to satisfy may also change with frequency. For example, noise at 100 MHz may choose to go through the smaller capacitor, C9, and noise at a lower frequency such as 500 kHz may choose to go through one of the larger capacitors, C7 or C8.

Interrupting any of these return paths has the potential to generate a common-mode noise due to increasing the size of the possible current loops if that current path is being

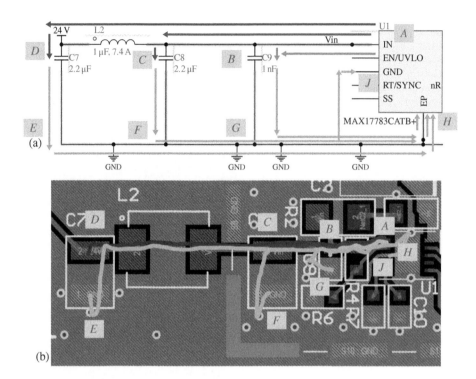

Figure A.52 Input filtering section of buck regulator PCB: (a) schematic and (b) PCB layout.

used. It is not guaranteed that if one of these paths is interrupted there will be an emissions failure, but it becomes more likely.

As the circuit complexity increases it becomes challenging to visualize all possible high-frequency current paths. What can be done then?

This is where the application of some good EMC rules of thumb can help. Here are several that we have identified over the years by working on power converters and solving EMC emissions issues:

1. Wherever possible keep a solid reference plane on the secondary (or adjacent) layer.
2. Place decoupling and by-pass capacitors as close to the IC pins as possible.
3. Ensure short connections and provide adequate reference via connections adjacent to component reference (GND) pins to ensure a low impedance path (smallest loop area).
4. Place all high di/dt components on the same layer of the PCB and in close proximity.
5. Place optional snubber components (series R–C) across internal switch and free-wheeling diode. Locate components in close proximity with short connections.
6. Fill with reference area copper beneath switching components (ICs, inductors, etc.). All of these approaches reduce the current loop area and serve to reduce radiated and conducted emissions from the switched-mode power supply.

A.6.3 Return-Plane Split in AC–DC Converter

Next, we focus on an AC/DC power supply that utilizes an Off-line Flyback circuit. For safety purposes, the primary-side and secondary-side circuits must be isolated.

Figure A.53 shows a part of the schematic for the Maxim MAX5022 Evaluation Kit, MAX5022, Rev 3, Maxim Integrated, 2021, with the current path on the primary side of the transformer shown on the left-hand side of Figure A.53, whereas the path of the current flow on the secondary side of the transformer is shown on the right side of Figure A.53.

Forward currents in both loops are drawn in dark gray color and return currents are drawn in light gray color. Note the dashed line in parallel with C7 that denotes the current flowing back to the source from the secondary to the primary through a stitching capacitor that is safety rated. This capacitor provides a pre-determined path for the noise currents to return back to their source along the PCB surface rather than through the air thus reducing radiated and conducted emissions.

The capacitor C7 stitches the two grounds together at high frequencies. Next, the return current flows through the current sense resistor (R7) back to the switching transistor completing the loop. The placement of the stitching capacitor impacts the size of the current loop, and therefore it should be placed close to the transformer as possible. In some cases a second stitching capacitor is needed so that a stitching capacitor can be provided above and below the body of the transformer. Total values of stitching capacitors must meet the required limitations imposed by isolation requirements.

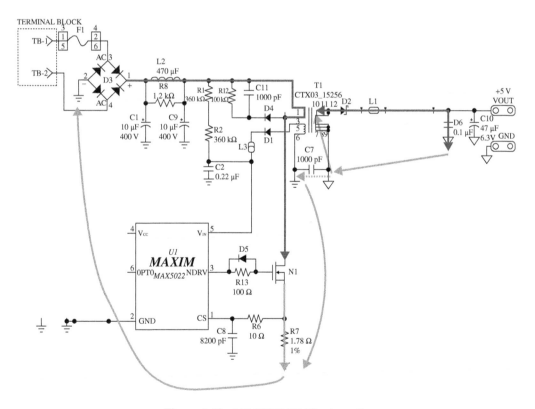

Figure A.53 MAX5022 EV kit schematic.

A.7 AC/DC Converter Design with EMC Considerations

This section is devoted to the design of the AC/DC Off-Line Flyback Converter [Adamczyk et al., 2021f]. This converter straddles Partitions *A* and *B*, as shown in Figure A.54.

In this section we present a schematic and PCB layout along with the EMC considerations and supporting design documentation.

A.7.1 AC/DC Converter Schematics and Design Requirements

Figure A.55 shows the block diagram AC/DC converter schematic.

The detailed schematic of the converter is shown in Figure A.56.

The AC input filter consists of a common-mode choke (L3), line-to-ground Y-caps (C13, C1 and C14, C17 and C18), and line-to-line X-caps (C15 and C16).

The filtered signal is fed to the full-wave bridge rectifier which produces a pulsating positive AC waveform. This waveform is smoothed by the LC filter consisting of L1, C2, C3, and R2 (R2 damps the antiresonant behavior of the filter).

The output of the filter is a DC signal of the value $115 \times \sqrt{2} \approx 162\,\text{V}$. This signal is fed to the primary winding of the flyback transformer.

R1 and C1 on the primary side constitute a protective circuitry for the primary winding when the MOSFET switches off. R14 and D5 constitute a protective circuitry for the MOSFET. R17 and C27 constitute a snubber across the MOSFET.

PWM Transistor Q1, controlled by U3, opens and closes the connection to the transformer primary. This switching of the line voltage generates a current on the secondary side and on the auxiliary winding. The winding ratios of the transformer were chosen to

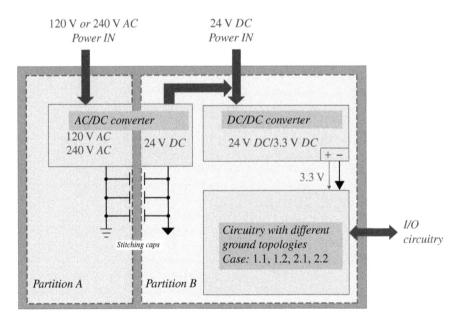

Figure A.54   Top-level schematic – functional blocks and AC/DC converter.

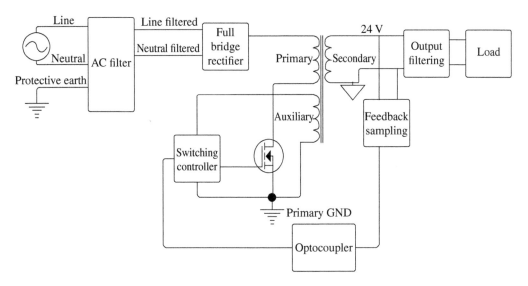

Figure A.55   AC–DC converter block diagram schematic.

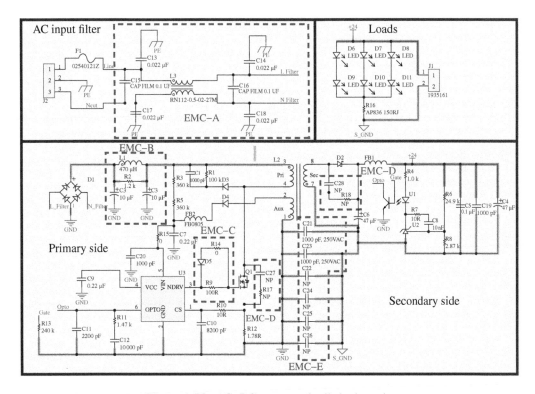

Figure A.56   AC–DC converter detailed schematic.

generate 24 V on the secondary side and 19 V on the auxiliary winding. Auxiliary voltage powers U3. Inductor FB2 and C7 on the auxiliary side provide filtering.

Series combination of R3 and R5 constitutes a start-up resistor that charges a reservoir capacitor C7. VCC is internally regulated down from VIN. It is decoupled to ground with a capacitor C9. VIN is provided by the auxiliary winding of the transformer.

Initially, both VIN and VCC are 0 V. After the line voltage is applied, the current flowing through the start-up resistor charges capacitor C7. Subsequently, the internal regulator charges capacitor C9 (at this point the switching transistor is off and thus the auxiliary winding voltage is zero). Charging of C9 stops when VCC reaches approximately 9.5 V, while the voltage across C7 continues rising until it reaches the wakeup level of 24 V.

Once VIN exceeds the Undervoltage Lockout (UVLO) wakeup level of 24 V, NDVR begins switching MOSFET providing energy to the secondary and auxiliary windings. If the voltage on the auxiliary winding builds up to higher than 10 V (UVLO lower threshold), then the start up has been accomplished and sustained operation commences.

To sustain the operation of the IC, VIN voltage must be in the range of 11–28 V. In our design we chose 19 V. This voltage is provided by the auxiliary side of the transformer and is determined by the winding ratio between the primary and auxiliary side. Decoupling capacitor C20 is connected to VIN. The primary- and secondary-side grounds are connected through the Y-rated stitching capacitors C21 and C23.

The voltage on the secondary side is sampled for a feedback to U3 through the optocoupler U1. The output voltage set point is determined by the shunt regulator U2 and resistor divider, R6 and R8. Output voltage is given by the following equation

$$V_{OUT} = V_{ref} \left(1 + \frac{R_6}{R_8} \right) \tag{A.1}$$

where $V_{ref} = 1.24$ V. R7 and C8 constitute a snubber across the regulator.

Output voltage filtering is provided by C5, C19, and C4. Lastly there is a bank of 6 LEDs and a 150 Ω resistor to load the output, and there is a 2-pin screw terminal so any other load can be added. This amounts to 3 W of power dissipated in the load.

Top layer of the PCB used to create the AC/DC converter is shown in Figure A.57, while the bottom layer is shown in Figure A.58.

Figure A.59 shows the AC/DC PCB converter populated with the components.

A.7.2 EMC Considerations

Similar to the DC–DC converter discussed earlier, provisions were added to this design that allow us to add/change components with the goal of improving the EMC performance of the design. These EMC considerations are shown in dashed boxes, labeled A through E in Figure A.56. These considerations are addressed below.

EMC-A: Provisions for an AC input filter were added to the input of this device. This filter consists of a common-mode choke (L3), 0.1 μF X-capacitors (C15 and C16), and 0.022 μF Y-Capacitors (C13, C14, C17, and C18). The goal of this filter is to filter out the noise generated by the switching circuit and propagating out onto the power cord of the device.

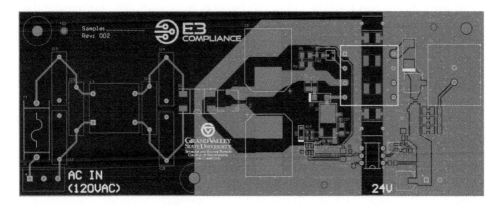

Figure A.57 Top layer of the PCB.

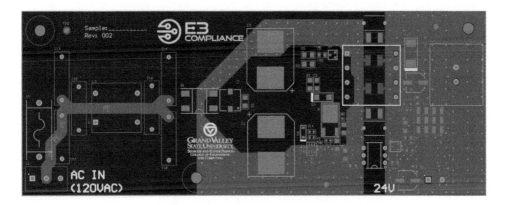

Figure A.58 Bottom layer of the PCB.

Figure A.59 AC–DC converter PCB with components.

EMC-B: A PI-filter was added right after the full bridge rectifier; this filter serves the purpose of smoothing out the rectified AC voltage, but this also provides some additional
input filtering. This filter comprised a 470 µF inductor (L1), a 1.2 kΩ resistor (R2), and
two 10 µF capacitors (C2 and C3).

EMC-C: A pair of gate drive resistors (R9 and R14), and a diode (D5), were added in
series with the connection between the pin 3 of the switching controller (U3) and the
N-Channel MOSFET (Q1). These components allow us to separately control the rise
and fall times of this gate drive signal in order to allow us to slow down the rise and fall
times of the signal, thus reducing the high-frequency content of the signal. The diode
allows us to control the rise and fall times separately in order to ensure that the rise and
fall times are equal.

EMC-D: Snubber circuits were added to both the N-Channel MOSFET (Q1) and the catch
diode (D2). The snubber circuit on the N-Channel MOSFET, comprising C27 and R17,
is used to control the ringing from the MOSFET that results from the step response to
the RLC network. The snubber circuit on the catch diode, comprising C28 and R18, is
added across the catch diode to reduce ringing across the diode junction.

EMC-E: Place holders for stitching capacitors (between primary and secondary ground)
are added (C21 through C26). These allow us to evaluate the number, placement of
capacitors as well as different types of capacitors.

A.8 AC/DC Converter – Baseline EMC Emissions Evaluation

In this section we evaluate the performance of the baseline AC/DC converter [Adamczyk
et al., 2022b]. The baseline AC/DC converter has only the components needed for functionality and does not have any specific EMC components populated. This configuration
will give us a view into what the conducted and radiated emissions issues will be prior
to adding components and cost to specifically address EMC issues. In this section, we
present the test results from the baseline radiated and conducted emissions tests performed according to the CFR Title 47, Part 15, Subpart B, Class B.

The baseline schematic for the AC/DC converter, with EMC components removed, is
shown in Figure A.60.

Figure A.61 shows the baseline AC/DC PCB converter populated with the baseline
components.

A.8.1 Radiated Emissions Test Results

The AC–DC converter was tested according to CFR Title 47, Part 15, Subpart B, Class
B. A legend for the radiated emissions plot is shown in Figure A.62.

Radiated emissions measurements were made using a biconical antenna from 30 to
300 MHz, and a log-periodic antenna from 300 MHz to 1 GHz. The measurements were
taken with the DUT at four different positions (angles) with each side of the PCB facing
the antenna. We only present the results for the zero-degree angle (AC inlet facing the
antenna) as this angle resulted in the highest emissions.

Figure A.63 shows the results from 30–MHz to 1–GHz.

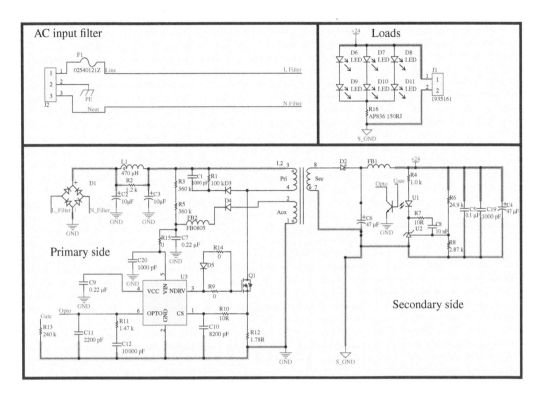

Figure A.60 AC–DC converter baseline schematic (EMC components removed).

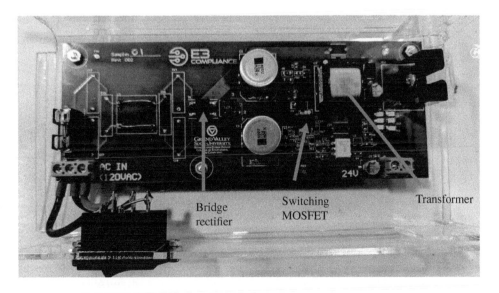

Figure A.61 Baseline AC–DC converter PCB with components.

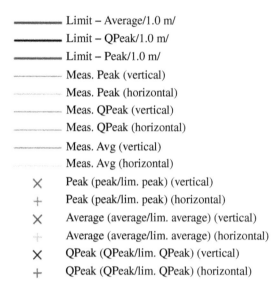

Figure A.62   Radiated emissions legend.

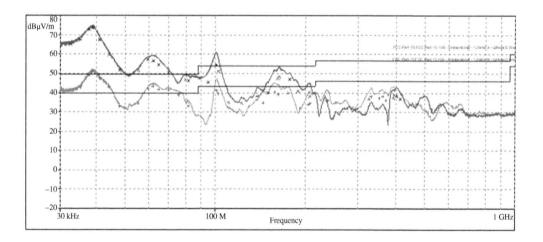

Figure A.63   Radiated emission results in the frequency range 30 MHz–1 GHz.

As shown in Figure A.63, there are numerous failures in the biconical range (30–300 MHz) – these will be investigated later.

The failing emissions are considered broadband noise and come primarily from the switching circuitry and magnetics. At these frequencies, the harness length is the most likely "antenna" where common-mode emissions conduct and re-radiate effectively. Reducing these emissions will most likely involve filtering, using snubber circuits and tuning stitching capacitance between the SGND and GND.

A.8.2 Conducted Emissions Test Results

A legend for the conducted emissions plots is shown in Figure A.64.

The test results on both the line and neutral, in the frequency range of 150 kHz–30 MHz, are shown in Figure A.65.

The conducted emissions results show multiple failures up to the frequency of 20 MHz. The failures comprised the fundamental switching frequency (≈270 kHz) and

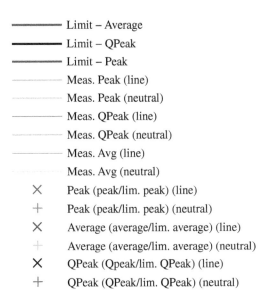

Figure A.64   Conducted emission results legend.

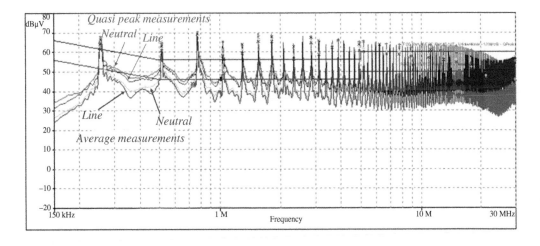

Figure A.65   Conducted emission test results 150 kHz–30 MHz.

the subsequent harmonics. Reducing these emissions will most likely involve front-end filtering components such as a common-mode choke, Y-capacitors, and X-capacitors. These will be investigated next.

A.9 AC/DC Converter – EMC Countermeasures – Conducted and Radiated Emissions Results

So far, we evaluated the performance of the baseline AC/DC converter. The baseline AC/DC converter had only the components needed for functionality and did not have any specific EMC components populated. The results showed multiple failures in both radiated and conducted emissions.

Now, we present a systematic approach to improve these failures by populating the PCB with optional EMC countermeasures on component pads that have already been designed into the PCB layout, and showing their impact on the radiated and conducted emissions [Adamczyk et al., 2022c]. The EMC countermeasures are illustrated in Figure A.66, as dashed boxes labeled *EMC-A* through *EMC-F*.

The conducted emissions results are discussed first, followed by the radiated emissions results.

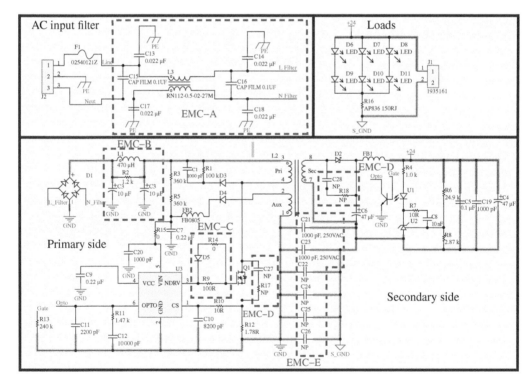

Figure A.66 AC/DC schematic with EMC countermeasures.

A.9.1 Conducted Emissions Test Results

Conducted emissions were measured in the frequency range of 150 kHz–30 MHz. The conducted emissions results show multiple failures up to the frequency of 20 MHz, as shown in Figure A.67a. The failures comprised the fundamental switching frequency (≈270 kHz) and the subsequent harmonics.

In attempt to reduce these emissions we began with the front-end filtering components (EMC-A) such as Y-capacitors (C13, C17) of the value 0.022 μF, between line (L-Filter) and protective earth (PE), and neutral (N-Filter) and PE and an X-capacitor (C15) of the value 0.1 μF, between the line and neutral.

The conducted emissions measurement taken with these countermeasures populated is shown in Figure A.67b. As the plot shows, the capacitors decrease the emissions by 4–15 dB, over the entire frequency range. There was only one quasi-peak failure at 255 kHz (still reduced by about 5 dB from baseline). There are

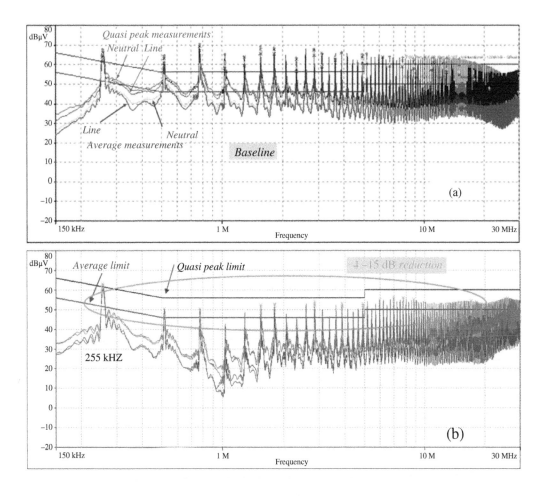

Figure A.67 CE results: (a) baseline (b) with X cap C15 = 0.1 μF and Y caps C13 = C17 = 22 nF.

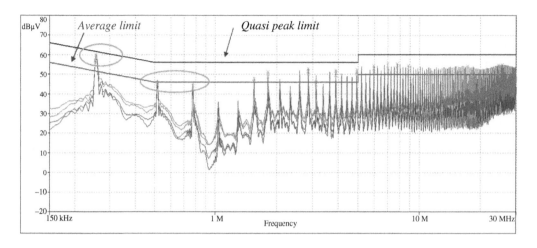

Figure A.68   CE results: X cap C15 = 0.1 µF, Y caps C13 = C17 = 22 nF, and CMC L3 = Schaffner RN112-0.5-02-27M.

still several failures over a broad range of frequency when measured with the average detector.

Next, a Schaffner RN112-0.5-02-27M common-mode choke (L3) was added. The results are shown in Figure A.68.

The addition of the common-mode choke eliminated the quasi-peak failure and lowered the emissions mainly below 1 MHz. It may be possible with further study to reduce conducted emissions by evaluating different common-mode chokes. However, we chose to focus on other EMC design controls to make further reductions.

Next, a 100 Ω gate resistor (R9) for Q1 (switching MOSFET) was added (EMC-C). The results are shown in Figure A.69.

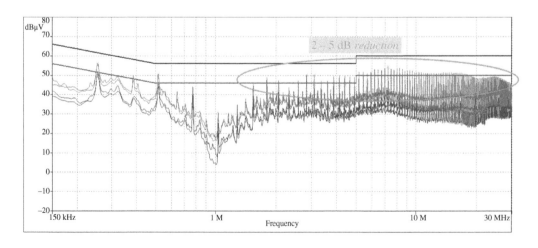

Figure A.69   CE results: X cap C15 = 0.1 µF, Y caps C13 = C17 = 22 nF, CMC L3 = Schaffner RN112-0.5-02-27M, gate drive resistor R9 = 100 Ω.

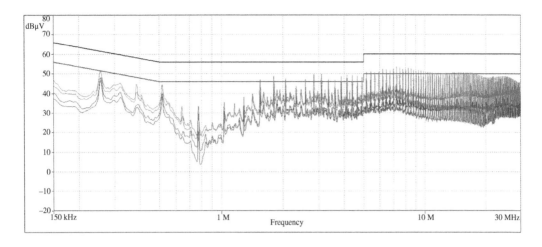

Figure A.70 CE results: X cap C15 = 0.1 μF, Y caps C13 = C17 = 33 nF, CMC L3 = Schaffner RN112-0.5-02-27M, gate drive resistor R9 = 100 Ω.

The main impact of the gate resistor was above 1 MHz, resulting on a 2–5 dB reduction in emissions. Further adjustments to the gate resistor R9 or R14 were not evaluated in conducted emissions, pending the measurements of radiated emissions.

Next, the Y capacitor values (C13 and C17) were increased from 0.022 μF to 0.033 μF (EMC-A). The results are shown in Figure A.70.

This change eliminated the remaining conducted emissions failures.

A.9.2 Radiated Emissions Test Results

Radiated emissions were measured in the frequency range of 30–300 MHz. As we have already identified the countermeasures that need to be added to resolve conducted emissions failures, we start radiated emissions diagnostics with all of the required CE modifications populated as per Figure A.70.

Figure A.71a shows the "baseline" results (DUT with the conducted emissions modifications).

In order to reduce the failing emissions further, two stitching caps (EMC-E) C21 and C23 of the value 1000 pF were added. The stitching capacitors tie the secondary back to the primary at high frequency allowing noise currents to return to their source more directly rather than through the air, thus reducing the radiated emissions. The result was the reduction in emission as shown in Figure A.71b.

At this point, failures can still be observed in the vertical emissions between 30 and 35 MHz, and the margin of the emissions around 35–40 MHz and 80 MHz isn't sufficient. We typically want to see at least 6 dB margin to the required emissions limit to account for lab-to-lab variation and component/build tolerances.

Therefore, a snubber was added across the Drain to Source pin of the switching MOSFET Q1 (EMC-D) consisting of R17 = 10 Ω, C27 = 330 pF to further reduce the emissions. The results are shown in Figure A.72.

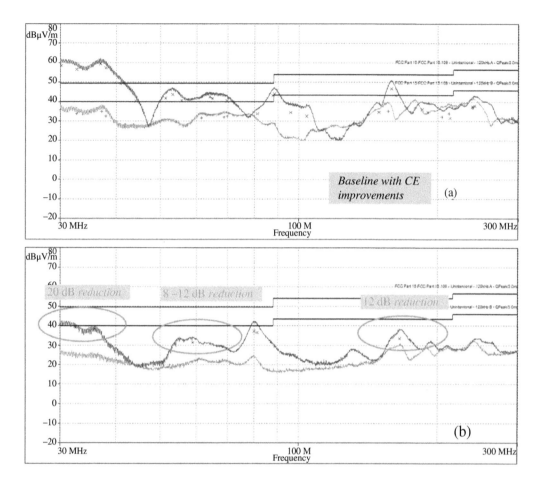

Figure A.71 Radiated emissions results: (a) baseline with CE modifications (b) with added C21 and C23 caps = 1000 pF.

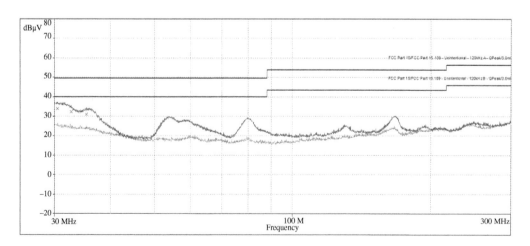

Figure A.72 Radiated emissions results with a snubber R17 = 10 Ω, C27 = 330 pF added across switching MOSFET Q1.

The addition of the snubber resulted in the device passing the radiated emission test with sufficient margins. The cumulative changes resulted in passing both conducted and radiated emissions results. Front-end filtering through the use of X and Y capacitors along with a common-mode choke provided a significant improvement in conducted emissions.

Adjusting the Q1 gate resistor, adding a snubber across Q1 Drain–Source, and stitching capacitance all made significant improvements in radiated emissions. If further radiated or conducted emissions reductions were needed, some experimentation with R14 (turn off Q1 gate resistance) could be evaluated. All design changes are recommended to be evaluated for potential design tradeoffs with other requirements such as thermal power dissipation, functionality over temperature, and input voltage variations as well as other EMC requirements (including immunity).

A.10 Complete System – Conducted and Radiated Emissions Results

This is the final section in a series devoted to the design, test, and EMC emissions evaluation of 1- and 2-layer PCBs that contain AC/DC and/or DC/DC converters, and employ different ground techniques [Adamczyk et al., 2022a]. The goal of this study was to evaluate the impact of different grounding strategies and the tradeoff with other design constraints that designers often face. In this article we present the complete system conducted and radiated emissions results performed according to the CFR Title 47, Part 15, Subpart B, Class B.

A.10.1 Complete System and Board Topologies

Figure A.73 shows the top-side complete system PCB assembly and its top-level schematic with the functional blocks.

The board is capable of accepting either an AC or DC input. The AC to DC conversion takes part in Partition A of the board (not drawn to scale). The DC to DC converter in Partition B accepts 24 V DC input either from the AC/DC converter in Partition A or from an external source.

The AC to DC converter is controlled by a Maxim MAX5022 IC, the DC to DC converter is controlled by MAX17783CATB+. These two circuits provide power to the embedded system section of the board which consists of a ST Microelectronics STM32G030F6P6 microcontroller, a Microchip MCP7940NT-I/MS real-time clock, and a Maxim Integrated MAX31855JASA+ cold-junction compensated thermocouple to digital converter.

These circuits contained several EMC countermeasures, previously described. The AC/DC converter contains an AC input filter, a DC input filter, a slew resistor on the gate of the primary-side switching MOSFET, and a snubber on the primary-side switching MOSFET. The DC/DC converter contains an input pi filter, a Vishay IHLE electric field shielded switching inductor, and a high-frequency output filtering capacitor.

The STM32G030F6P6 microcontroller and MCP7940NT-I/MS real-time clock had high-frequency decoupling capacitors placed close to the power input pins of the device. The Maxim Integrated MAX31855JASA+ had high-frequency decoupling capacitors placed near the power input pins, and a high-frequency filtering capacitor across the two-pin screw terminal connection for the J-type thermocouple.

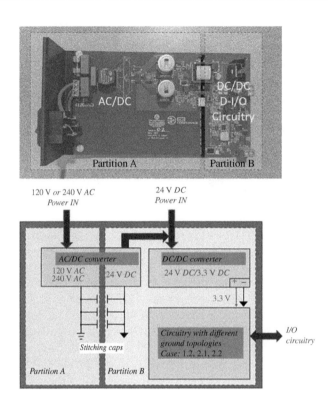

Figure A.73 Complete PCB assembly and its block diagram.

Three different PCB topologies were evaluated. First, we evaluated a one-layer board-layer board with ground traces and ground floods located only on the top side. The back side of this board is shown in Figure A.74.

Next, we evaluated a two-layer board where the bottom layer is a mostly solid ground reference plane with some slots accounting for the need to route signals on the bottom layer. The bottom layer of this board is shown in Figure A.75.

And finally, a two-layer board where the bottom layer is a complete ground flood with via stitching to the top-layer ground reference return areas. This is shown in Figure A.76.

A.10.2 Conducted Emissions Results

Conducted emissions were measured in the frequency range of 150 kHz–30 MHz. Figure A.77 shows a reference legend for these measurements.

Figure A.78 shows the conducted emissions results for a one-layer board.

The conducted emissions results show multiple failures over a wide frequency range.

Figure A.79 shows the conducted emissions results for the two-layer board with slots in the ground reference return.

Figure A.79 shows at least 8 dB of improvement across most of the 150 kHz–30 MHz frequency range. We can conclude that introducing a ground reference return (even with

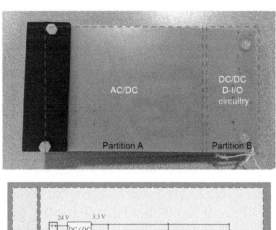

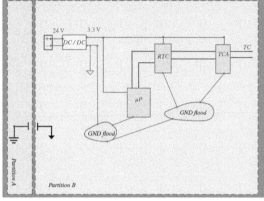

Figure A.74   One-layer board – bottom-side view (no ground reference return on bottom side).

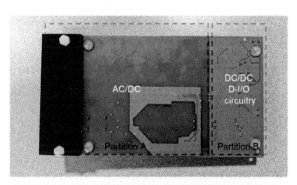

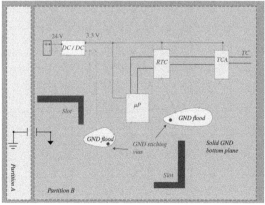

Figure A.75   Two-layer board – bottom-side view (slots in ground reference return).

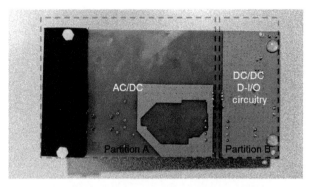

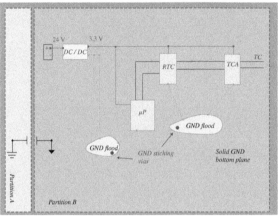

Figure A.76   Two-layer board – bottom-side view (full ground reference return).

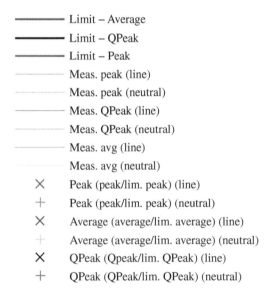

Figure A.77   Reference legend for conducted emission results.

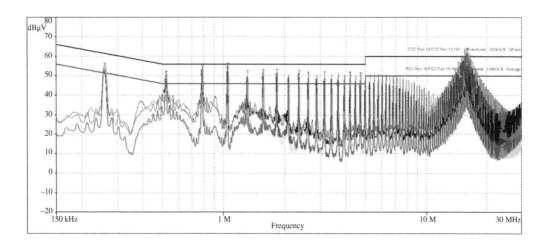

Figure A.78   Conducted emissions results: one-layer board.

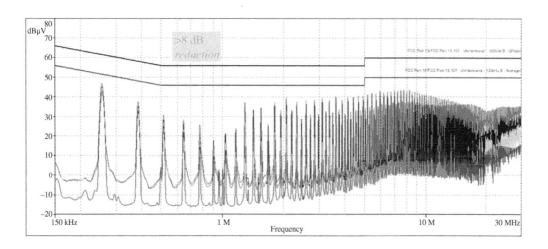

Figure A.79   Conducted emissions results: two-layer board with slots in ground reference return.

slots in it) on the bottom layer improves the conducted emissions performance significantly and results in passing the test.

Figure A.80 shows the conducted emissions results for a two-layer board with a solid ground reference return on the bottom layer.

Comparing the ground reference return with slots on the bottom layer to the case with a solid ground reference return on the bottom layer shows little change in conducted emissions. Approximately 1–3 dB of variation in emissions levels occurred between 500 kHz and 2 MHz. However, comparing the solid ground reference return to the case with a one-layer PCB (no ground reference return on bottom layer) there are substantial improvements across the entire frequency range for the conducted emissions test.

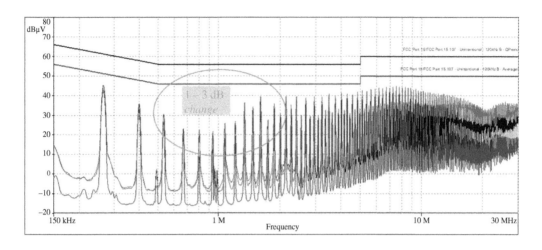

Figure A.80 Conducted emissions results: two-layer board with solid ground.

A.10.3 Radiated Emissions Results

Radiated emissions were measured in the frequency range of 30–300 MHz. Figure A.81 shows a reference legend for these measurements.

Figure A.82 shows the radiated emissions results for a one-layer board.

The single-layer radiated emissions results show multiple broadband failures over the wide frequency range.

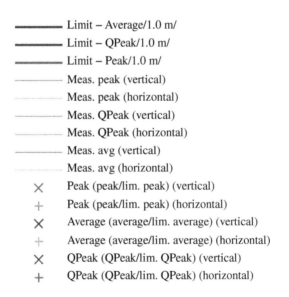

Figure A.81 Reference legend for radiated emission results.

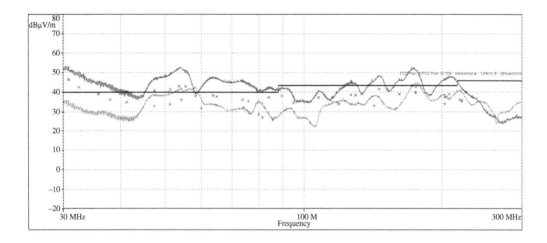

Figure A.82   RE results: one-layer board.

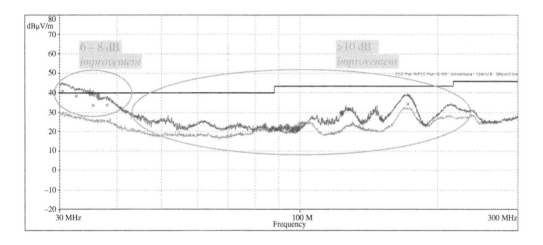

Figure A.83   RE results: two-layer board with slots in ground reference return.

Figure A.83 shows the radiated emissions results for the two-layer board with slots in the ground reference return.

Figure A.83 shows that the introduction of a ground reference return, even with slots in it, provides us with a 6–8 dB improvement from 30 to 45 MHz, and an improvement greater than 10 dB from 45–240 MHz.

Figure A.84 shows the radiated emissions results for a two-layer board with solid ground reference return.

As shown in Figure A.84, using the solid ground reference return causes a 1–2 dB increase from 30 to 33 MHz, and a 5 dB increase from 40 to 48 MHz compared to the two-layer board with slots in the ground reference return.

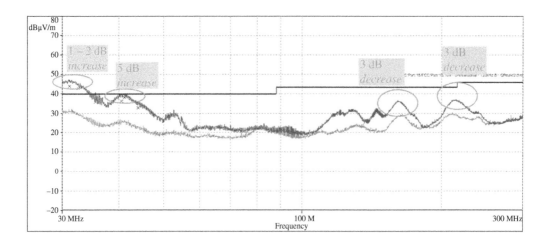

Figure A.84   RE results: two-layer board with solid ground reference return.

Additionally, there is general improvement in emissions between 100 and 300 MHz. Specifically, a 3 dB decrease in emissions around 160–170 MHz and a 3 dB decrease in emissions at 214 MHz can be observed. The differences in the lower frequencies 30–45 MHz are most likely attributed to differences in operating conditions between the two AC/DC converter circuits. Differences in operating conditions may be related to different fundamental switching frequencies, transformer tolerances, and other component tolerances. The improvement in emissions between 120 and 300 MHz is likely to be from the improvement in the ground reference return.

A.10.4 Conclusions

In conclusion, the single-layer board had failures in both conducted and radiated emissions. As one would expect, adding a ground reference return layer either with or without slots caused significant improvements in both conducted and radiated emissions.

When comparing the two-layer designs with and without slots in the ground reference return plane, there were subtle differences in conducted emissions and a net improvement in radiated emissions.

This study has identified and confirmed a number of best practices in EMC design such as:

1. PCB layout can have a dramatic effect on RF emissions performance.
2. The ground reference return in the power converter designs used should be:
   a. Filled on adjacent layers to the extent possible to provide a good reference return.
   b. Connected as direct and low impedance as possible between input filtering reference to output filter reference.
   c. Stitched between layers to improve the reference return path.
3. It is recommended to fill with ground reference beneath switching magnetics when possible (e.g. Isolation issues, efficiency issues allow it).

4. Lower-frequency emissions failures in conducted and radiated emissions can be addressed by ensuring good input and output filtering.

5. Mid-to-high-frequency emissions failures in conducted and radiated emissions can be improved through the use of snubber circuits, gate drive slewing, good decoupling and PCB or inductor shielding.

6. In isolated switching supplies, stitching capacitance is important to evaluate and tune to optimize emissions performance without violating any isolation requirements.

References

Bogdan Adamczyk, Scott Mee, and Nick Koeller. Evaluation of EMC Emissions and Ground Techniques on 1- and 2-layer PCBs with Power Converters - Part 1: Top-Level Description of the Design Problem. *In Compliance Magazine*, May 2021a. URL https://incompliancemag .com/article/evaluation-of-emc-emissions-and-ground-techniques-on-1-and-2-layer-pcbs-with-power-converters/.

Bogdan Adamczyk, Scott Mee, and Nick Koeller. Evaluation of EMC Emissions and Ground Techniques on 1- and 2-layer PCBs with Power Converters - Part 3: DC/DC Converter – Baseline EMC Emissions Evaluation. *In Compliance Magazine*, July 2021b. URL https://incompliancemag .com/article/evaluation-of-emc-emissions-and-ground-techniques-on-1-and-2-layer-pcbs-with-power-converters-2/.

Bogdan Adamczyk, Scott Mee, and Nick Koeller. Evaluation of EMC Emissions and Ground Techniques on 1- and 2-layer PCBs with Power Converters - Part 4: DC/DC Converter – EMC Countermeasures – Radiated Emissions Results. *In Compliance Magazine*, August 2021c. URL https://incompliancemag.com/article/evaluation-of-emc-emissions-and-ground-techniques-on-1-and-2-layer-pcbs-with-power-converters-part4/.

Bogdan Adamczyk, Scott Mee, and Nick Koeller. Evaluation of EMC Emissions and Ground Techniques on 1- and 2-layer PCBs with Power Converters - Part 5: DC/DC Converter – EMC Countermeasures – Conducted Emissions Results. *In Compliance Magazine*, October 2021d. URL https://incompliancemag.com/article/evaluation-of-emc-emissions-and-ground-techniques-on-1-and-2-layer-pcbs-with-power-converters-part5/.

Bogdan Adamczyk, Scott Mee, and Nick Koeller. Evaluation of EMC Emissions and Ground Techniques on 1- and 2-layer PCBs with Power Converters - Part 6: PCB Layout Considerations. *In Compliance Magazine*, November 2021e. URL https://incompliancemag.com/article/evaluation-of-emc-emissions-and-ground-techniques-on-1-and-2-layer-pcbs-with-power-converters-part6/.

Bogdan Adamczyk, Scott Mee, and Nick Koeller. Evaluation of EMC Emissions and Ground Techniques on 1- and 2-layer PCBs with Power Converters - Part 7: AC/DC Converter Design with EMC Considerations. *In Compliance Magazine*, December 2021f. URL https://incompliancemag .com/article/evaluation-of-emc-emissions-and-ground-techniques-on-1-and-2-layer-pcbs-with-power-converters-part7/.

Bogdan Adamczyk, Scott Mee, and Nick Koeller. Evaluation of EMC Emissions and Ground Techniques on 1- and 2-layer PCBs with Power Converters -Part 10: Complete System – Conducted and Radiated Emissions Results. *In Compliance Magazine*, March 2022a. URL https:// incompliancemag.com/article/evaluation-of-emc-emissions-and-ground-techniques-on-1-and-2-layer-pcbs-with-power-converters-part10/.

Bogdan Adamczyk, Scott Mee, and Nick Koeller. Evaluation of EMC Emissions and Ground Techniques on 1- and 2-layer PCBs with Power Converters - Part 8: AC/DC Converter – Baseline EMC Emissions Evaluation. *In Compliance Magazine*, January 2022b. URL https://incompliancemag .com/article/evaluation-of-emc-emissions-and-ground-techniques-on-1-and-2-layer-pcbs-with-power-converters-part8/.

Bogdan Adamczyk, Scott Mee, and Nick Koeller. Evaluation of EMC Emissions and Ground Techniques on 1- and 2-layer PCBs with Power Converters - Part 9: AC/DC Converter – EMC Countermeasures – Conducted and Radiated Emissions Results. *In Compliance Magazine*, February 2022c. URL https://incompliancemag.com/article/evaluation-of-emc-emissions-and-ground-techniques-on-1-and-2-layer-pcbs-with-power-converters-part9/.

Index

Principles of Electromagnetic Compatibility: Laboratory Exercises and Lectures, First Edition. Bogdan Adamczyk.
© 2024 John Wiley & Sons Ltd. Published 2024 by John Wiley & Sons Ltd.
Companion website: www.wiley.com/go/principlesofelectromagneticcompatibility